WORKED EXAMPLES FOR THE DESIGN OF CONCRETE STRUCTURES TO EUROCODE 2

WORKED EXAMPLES FOR THE DESIGN OF CONCRETE STRUCTURES TO EUROCODE 2

TONY THRELFALL

CRC Press
Taylor & Francis Group
Boca Raton London New York

CRC Press is an imprint of the
Taylor & Francis Group, an **informa** business

A SPON BOOK

CRC Press
Taylor & Francis Group
6000 Broken Sound Parkway NW, Suite 300
Boca Raton, FL 33487-2742

© 2013 by Taylor & Francis Group, LLC
CRC Press is an imprint of Taylor & Francis Group, an Informa business

No claim to original U.S. Government works

Printed on acid-free paper
Version Date: 20130214

International Standard Book Number-13: 978-0-415-46819-0 (Paperback)

Library of Congress Cataloging-in-Publication Data

Threlfall, A. J.
 Worked examples for the design of concrete structures to Eurocode 2 / Tony Threlfall.
 pages cm
 Includes bibliographical references and index.
 ISBN 978-0-415-46819-0 (pbk.)
 1. Concrete construction--Europe--Problems, exercises, etc. 2. EN1992 Eurocode 2 (Standard)--Problems, exercises, etc. I. Title.

TA681.T497 2013
624.1'83402184--dc23   2012050477

Visit the Taylor & Francis Web site at
http://www.taylorandfrancis.com

and the CRC Press Web site at
http://www.crcpress.com

Contents

Preface

The purpose of this book is to demonstrate how to apply the recommendations of Eurocode 2, and other related standards, for a number of reinforced concrete structures. The examples have been chosen to include different structural elements and design procedures. The calculations cover the analysis of the structure and the design of the members.

Each step of the calculations, which are presented in a form suitable for design office purposes, is explained. References to specific clauses in the codes and standards that affect the design are included at each stage. For each structural element, a complete reinforcement detail is provided together with a commentary explaining the bar arrangement.

Chapter 1 is an introduction to the structural Eurocodes and explains how partial safety factors and action combination factors are incorporated in the design. The significance of the action combination to be used, when considering the cracking limitations for watertightness in tanks, is also examined.

Chapter 2 summarises the design of members with regard to durability, fire resistance, axial force, bending, shear, torsion, deflection, cracking and other considerations that affect the design details. It refers particularly to the design information given in Appendix A and in *Reynolds's Reinforced Concrete Designer's Handbook*.

The first two examples deal with the design of a multi-storey framed building. For each example, three alternative forms of construction are considered. In Example 1, which covers the design of the superstructure, the floor takes alternative forms of beam and slab, flat slab and integral beam and ribbed slab, respectively. In Example 2, which deals with the design of the substructure including the basement, the foundations take alternative forms of a continuous raft, isolated pad bases and pile foundations, respectively.

Example 3 is for a freestanding cantilever earth-retaining wall with two designs, for bases bearing on non-cohesive and cohesive soils, respectively.

The last three examples are for liquid-retaining structures in which the protection against leakage depends entirely on the integrity of the structure. Example 4 is for an underground service reservoir in which the wall and floor are formed of elements separated by movement joints. Example 5 is for a continuous rectangular tank bearing on an elastic soil with the interaction of the walls and the floor taken into account in the analysis. Example 6 is for a continuous cylindrical tank bearing on an elastic soil with both hydraulic and thermal actions considered in the design.

An important feature of this book is the collection of full-page tables and charts contained in three appendices. Appendix A has nine tables of general information relating to the design of members. Appendix B has 11 tables dealing with the analysis of beams on elastic foundations. Appendix C has 14 tables for the analysis of rectangular and cylindrical tanks.

The examples in this book inevitably reflect the knowledge and experience of the author. Writing the book has also given me the opportunity to investigate problems that I had found difficult to solve during my career. This applies particularly to the analysis of complex structures on elastic foundations for which text book solutions are not readily available. I hope that the information provided in Appendices B and C and the analyses that are included in the examples will be helpful to present-day design engineers faced with similar problems.

I owe a considerable debt of gratitude to many people from whose intellect and expertise I have benefited over the years.

Finally, my sincere thanks go to my dear wife, Joan, for her constant support and encouragement throughout the writing of this book.

Tony Threlfall

Acknowledgements

Permission to reproduce extracts from BS EN 1990, BS EN 1991-4, BS EN 1992-1-1, BS EN 1992-1-2, BS EN 1992-3 and BS EN 1997-1 is granted by BSI (British Standards Institution).

British Standards can be obtained in PDF or hard copy formats from the BSI online shop: www.bsigroup.com/ Shop or by contacting BSI Customer Services for hard-copies only: Tel: +44 (0)20 8996 9001, Email: cservices@ bsigroup.com.

Information in Tables C2 to C13 is reproduced with permission from the Portland Cement Association, Skokie, Illinois, USA.

Author

Tony Threlfall was educated at Liverpool Institute High School for Boys, after which he studied civil engineering at Liverpool University. After eight years working for BRC, Pierhead Ltd and IDC Ltd, he took a diploma course in concrete structures and technology at Imperial College. For the next four years he worked for CEGB and Camus GB Ltd, before joining the Cement and Concrete Association (C&CA) in 1970, being engaged primarily in education and training activities until 1993. After leaving the C&CA, he continued in private practice to provide training in reinforced and prestressed concrete design and detailing. He is the author of several publications concerned with concrete design, including the 11th edition of *Reynolds's Reinforced Concrete Designer's Handbook*.

Symbols and Notes

The symbols adopted in this book comply, where appropriate, with those in the relevant code of practice. Only the principal symbols are listed here: all other symbols are defined in the text and tables concerned.

A_c	Area of concrete section
A_s	Area of tension reinforcement
A'_s	Area of compression reinforcement
A_{sc}	Area of longitudinal reinforcement in a column
C	Torsional constant
E_c	Static modulus of elasticity of concrete
E_s	Modulus of elasticity of reinforcing steel
F	Action, force or load (with appropriate subscripts)
G	Shear modulus of concrete
G_k	Characteristic permanent action or dead load
I	Second moment of area of cross-section
K	A constant (with appropriate subscripts)
L	Length; span
M	Bending moment
N	Axial force
Q_k	Characteristic variable action or imposed load
R	Reaction at support
S	First moment of area of cross-section
T	Torsional moment; temperature
V	Shear force
W_k	Characteristic wind load
a	Dimension; deflection
b	Overall width of cross-section, or width of flange
d	Effective depth-to-tension reinforcement
d'	Depth-to-compression reinforcement
f	Stress (with appropriate subscripts)
f_{ck}	Characteristic (cylinder) strength of concrete
f_{cu}	Characteristic (cube) strength of concrete
f_{yk}	Characteristic yield strength of reinforcement

g_k	Characteristic dead load per unit area
h	Overall depth of cross-section
i	Radius of gyration of concrete section
k	A coefficient (with appropriate subscripts)
l	Length; span (with appropriate subscripts)
m	Mass
q_k	Characteristic imposed load per unit area
r	Radius
$1/r$	Curvature
t	Thickness; time
u	Perimeter (with appropriate subscripts)
v	Shear stress (with appropriate subscripts)
x	Neutral axis depth
z	Lever arm of internal forces
α, β	Angle; ratio
α_e	Modular ratio E_s/E_c
γ	Partial safety factor (with appropriate subscripts)
ε_c	Compressive strain in concrete
ε_s	Strain in tension reinforcement
ε'_s	Strain in compression reinforcement
λ	Slenderness ratio
v	Poisson's ratio
ϕ	Diameter of reinforcing bar
φ	Creep coefficient (with appropriate subscripts)
ρ	Proportion of tension reinforcement A_s/bd
ρ'	Proportion of compression reinforcement A'_s/bd
σ	Stress (with appropriate subscripts)
ψ	Factor defining representative value of action

Note 1: In this book, the decimal point is denoted by a full stop rather than a comma as shown in the Eurocodes.

Note 2: In the calculation sheets, the references are to clauses in BS EN 1992-1-1 unless stated otherwise.

1 Eurocodes and Design Actions

Structural Eurocodes are an international set of unified codes of practice. They comprise the following standards generally consisting of a number of parts:

EN 1990 Basis of structural design
EN 1991 Actions on structures
EN 1992 Design of concrete structures
EN 1993 Design of steel structures
EN 1994 Design of composite steel and concrete structures
EN 1995 Design of timber structures
EN 1996 Design of masonry structures
EN 1997 Geotechnical design
EN 1998 Design of structures for earthquake resistance
EN 1999 Design of aluminium structures

National standards implementing the Eurocodes are issued in conjunction with a National Annex that contains information on those parameters that are left open in the Eurocode for national choice. In addition, when guidance is needed on an aspect not covered by the Eurocode, a country can choose to publish documents containing non-contradictory information.

EN 1992 Eurocode 2: *Design of concrete structures* contains four parts, each with its own National Annex, and additional documents as follows:

EN 1992-1-1 General rules and rules for buildings
EN 1992-1-2 General rules – Structural fire design
EN 1992-2 Reinforced and prestressed concrete bridges
EN 1992-3 Liquid retaining and containment structures
PD 6687-1 Background paper to the UK National Annexes to BS EN 1992-1
PD 6687-2 Recommendations for the design of structures to BS EN 1992-2

In the Eurocodes, design requirements are set out in relation to specified limit state conditions. Calculations to determine the ability of members to satisfy a particular limit state are undertaken by using design actions (loads or deformations) and design strengths. The design values are determined from representative values of actions and characteristic strengths of materials by the application of partial safety factors.

1.1 ACTIONS

EN 1991 Eurocode 1: *Actions on structures* contains ten parts, each with its own National Annex, as follows:

1991-1-1 General actions – Densities, self-weight, imposed loads for buildings

1991-1-2 Actions on structures exposed to fire
1991-1-3 Snow loads
1991-1-4 General actions – Wind actions
1991-1-5 Thermal actions
1991-1-6 Actions during execution
1991-1-7 Accidental actions due to impact and explosions
1991-2 Traffic loads on bridges
1991-3 Actions induced by cranes and machinery
1991-4 Actions on silos and tanks

A variable action (e.g., imposed load, snow load, wind load, thermal action) can have the following representative values:

Characteristic value	Q_k
Combination value	$\psi_0 Q_k$
Frequent value	$\psi_1 Q_k$
Quasi-permanent value	$\psi_2 Q_k$

The characteristic and combination values are used for the verification of the ultimate and irreversible serviceability limit states. The frequent and quasi-permanent values are used for the verification of ultimate limit states involving accidental actions, and reversible serviceability limit states. The quasi-permanent values are also used for the calculation of long-term effects.

Design actions (loads) are given by

$$\text{Design action (load)} = \gamma_F \times \psi F_k$$

where F_k is the specified characteristic value of the action, γ_F is the value of the partial safety factor for the action (γ_A for accidental actions, γ_G for permanent actions, γ_Q for variable actions) and the limit state being considered, and ψ is 1.0, ψ_0, ψ_1 or ψ_2. Recommended values of γ_F and ψ are given in EN 1990 Eurocode: *Basis of structural design*.

1.2 MATERIAL PROPERTIES

The characteristic strength of a material f_k means the value of either the cylinder strength f_{ck} or the cube strength $f_{ck,cube}$ of concrete, or the yield strength f_{yk} of steel reinforcement, below which not more than 5% of all possible test results are expected to fall. The concrete strength is selected from a set of strength classes, which in Eurocode 2 are based on the cylinder strength. The deformation properties of concrete are summarised in *Reynolds*, Tables 4.2 and 4.3. The application rules in Eurocode 2 are valid for reinforcement in accordance with EN 10080, whose specified yield strength is in the range 400–600 MPa.

Design strengths are given by

$$\text{Design strength} = f_k/\gamma_M$$

where f_k is either f_{ck} or f_{yk} as appropriate and γ_M is the value of the partial safety factor for the material (γ_C for concrete, γ_S for steel reinforcement) and the limit state being considered.

1.3   BUILDINGS

Details of the design requirements and partial safety factors for buildings are summarised in *Reynolds*, Table 4.1.

The design action combinations to be considered and values of the factor ψ to be used are shown in Table 1.1.

TABLE 1.1
Design Considerations, Action Combinations and Values of ψ for Variable Actions on Buildings

Limit State and Design Consideration[a]	Combination of Design Actions (see EN 1990)		
Ultimate (persistent and transient actions)	$\Sigma\gamma_{G,j}\,G_{k,j} + \gamma_{Q,1}\,Q_{k,1} + \Sigma\gamma_{Q,i}\,\psi_{0,i}\,Q_{k,i}$ $(j \geq 1, i > 1)$		
Ultimate (accidental action)	$A_d + \Sigma G_{k,j} + (\psi_{1,1} \text{ or } \psi_{2,1})\,Q_{k,1} + \Sigma\psi_{2,i}\,Q_{k,i}$ $(j \geq 1, i > 1)$		
Serviceability (function, including damage to structural and non-structural elements, e.g., partition walls)	$\Sigma G_{k,j} + Q_{k,1} + \Sigma\psi_{0,i}\,Q_{k,i}$ $(j \geq 1, i > 1)$		
Serviceability (comfort to user, use of machinery, avoiding ponding of water, etc.)	$\Sigma G_{k,j} + \psi_{1,1}\,Q_{k,1} + \Sigma\psi_{2,i}\,Q_{k,i}$ $(j \geq 1, i > 1)$		
Serviceability (appearance)	$\Sigma G_{k,j} + \Sigma\psi_{2,i}\,Q_{k,i}$ $(j \geq 1, i \geq 1)$		

Imposed Loads (Category and Type, See EN 1991-1-1)	ψ_0	ψ_1	ψ_2
A: domestic, residential area, B: office area	0.7	0.5	0.3
C: congregation area, D: shopping area	0.7	0.7	0.6
E: storage area	1.0	0.9	0.8
F: traffic area (vehicle weight ≤ 30 kN)	0.7	0.7	0.6
G: traffic area (30 kN < vehicle weight ≤ 160 kN)	0.7	0.5	0.3
H: roof	0.7	0	0

Snow Loads (See EN 1991-1-3)			
Sites located at altitude >1000 m above sea level	0.7	0.5	0.2
Sites located at altitude ≤1000 m above sea level	0.5	0.2	0
Wind loads (see EN 1991-1-4)	0.5[b]	0.2	0
Thermal actions (see EN 1991-1-5)	0.6	0.5	0

Note: In the combination of design actions shown above, $Q_{k,1}$ is the leading variable action and $Q_{k,i}$ are any accompanying variable actions. Where necessary, each action in turn should be considered as the leading variable action.

[a] Serviceability design consideration and associated combination of design actions as specified in the UK National Annex.

[b] As specified in the UK National Annex.

1.3.1   ULTIMATE LIMIT STATE

The design ultimate actions to be taken for structural design are shown in Table 1.2. Either option 1 or the less favourable of options 2a and 2b may be used. For option 2b, the value of the unfavourable multiplier for permanent actions is given by $\xi\gamma_G = 0.925 \times 1.35 = 1.25$. For all permanent actions from one source, for example, the self-weight of the structure, either the unfavourable or the favourable value should be used for all parts. When variable actions are favourable, $Q_k = 0$ should be used. Where necessary, each variable action in turn should be considered as the leading action.

If $Q_{k,1}$ relates to a storage area, for which $\psi_0 = 1.0$, options 1 and 2 are identical. In other cases, it is advantageous to use option 2, where option 2b governs for values of $G_k \leq 4.5Q_k$ when $\psi_0 = 0.7$, and for values of $G_k \leq 7.5Q_k$ when $\psi_0 = 0.5$.

In this book, option 2b has been used in Examples 1 and 2.

1.3.2   SERVICEABILITY LIMIT STATES

In EN 1992-1-1, a check under quasi-permanent loading is normally allowed when considering cracking and deflection. This appears to comply with the recommendation in EN 1990 with regard to appearance. With regard to function including possible damage to elements of the structure, a check under characteristic loading is indicated. In this book, to avoid possible damage to partitions, characteristic loading has been used for the deflection check in Example 1.

1.4   CONTAINMENT STRUCTURES

Silos and tanks are different from many other structures in that they can be subjected to the full loads from particulate solids or liquids for most of their life. The actions to be considered are detailed in Eurocode 1: Part 4: *Silos and tanks*, where the contents of informative annexes A and B are replaced by the recommendations given in the UK National Annex. Values of the combination factor appropriate to each design action are shown in Table 1.3.

1.4.1   ULTIMATE LIMIT STATE

In tanks, $\gamma_Q = 1.2$ may be used for the loads induced by the stored liquid, at the maximum design liquid level. During testing, at the maximum test liquid level, and for accidental design situations, $\gamma_Q = 1.0$ may be used. In silos, $\gamma_Q = 1.5$ should be used for loads induced by stored particulate solids.

1.4.2   SERVICEABILITY LIMIT STATES

For the serviceability limit state of cracking, a classification of liquid-retaining structures in relation to the required degree of protection against leakage and the corresponding design requirements as given in Eurocode 2: Part 3 are summarised in Table 1.4. Silos containing dry materials may generally be designed as Class 0.

TABLE 1.2

Design Ultimate Actions for Buildings

| Option | EN 1990 | Permanent Actions | | Variable Actions | |
		Unfavourable	Favourable	Leading	Others (i > 1)
1	Equation 6.10	$1.35G_k$	$1.0G_k$	$1.5Q_{k,1}$	$1.5\sum\psi_{0,i}Q_{k,i}$
2a	Equation 6.10a	$1.35G_k$	$1.0G_k$	$1.5\psi_{0,1}Q_{k,1}$	$1.5\sum\psi_{0,i}Q_{k,i}$
2b	Equation 6.10b	$1.25G_k$	$1.0G_k$	$1.5Q_{k,1}$	$1.5\sum\psi_{0,i}Q_{k,i}$

TABLE 1.3

Values of ψ for Variable Actions on Silos and Tanks (as Specified in the UK National Annex)

Action	ψ_0	ψ_1	ψ_2	Action	ψ_0	ψ_1	ψ_2
Liquid loads	1.0	0.9	0.3	Foundation settlement	1.0	1.0	1.0
Solids filling	1.0	0.9	0.3	Imposed loads or deformation	0.7	0.5	0.3
Solids discharge	1.0	0.3	0.3	Snow loads	0.5	0.2	0
Thermal actions	0.6	0.5	0	Wind action	0.5	0.2	0

TABLE 1.4

Classification of Water-Tightness and Cracking Limitations in EN 1992-3

Class	Leakage Requirements	Design Provisions
0	Leakage acceptable or irrelevant.	The provisions in EN 1992-1-1 may be adopted.
1	Leakage limited to small amount. Some surface staining or damp patches acceptable.	The width of any cracks that can be expected to pass through the full thickness of the section should be limited to w_{k1} given by $0.05 \leq w_{k1} = 0.225(1 - h_w/45h) \leq 0.2$ mm where h_w/h is the hydraulic gradient (i.e., head of liquid divided by thickness of section) at the depth under consideration. Where the full thickness of the section is not cracked, the provisions in EN 1992-1-1 apply.
2	Leakage minimal. Appearance not to be impaired by staining.	Cracks that might be expected to pass through the full thickness of the section should be avoided, unless measures such as liners or water bars are included.
3	No leakage permitted.	Special measures (e.g., liners or prestress) are required to ensure water-tightness.

It is implied but not clearly stated in Eurocode 2: Part 3 that the cracking check may be carried out under quasi-permanent loading. In this case, since $\psi_2 = 0.3$ for hydrostatic load, the cracking check is less onerous than the design ultimate requirement. This is a significant departure from previous United Kingdom practice, in which characteristic loading was used for the cracking check, and this check was nearly always critical.

It also appears that thermal actions have no effect on the cracking check, since $\psi_2 = 0$ in this case. Since thermal actions can usually be ignored at the ultimate limit state, on the basis that 'elastic' stresses reduce with increasing strain, it would appear that the effect of thermal actions can be discounted altogether in the design.

The author of this book considers that the check for cracking should be carried out under the frequent loading, and that the recommended values of ψ_2 need to be reviewed. In Examples 4 and 5, a conservative approach has been adopted and the characteristic value has been taken for the hydrostatic load. In Example 6, the frequent loading combination has been taken and $\psi_2 = 0.9$ has been applied to the hydrostatic load.

1.5 GEOTECHNICAL DESIGN

Eurocode 7: *Geotechnical design* provides in outline all the requirements for the design of geotechnical structures. It classifies structures into three categories according to their complexity and associated risk, but concentrates on the design of conventional structures with no exceptional risk. These include spread, raft and pile foundations, retaining structures, bridge piers and abutments, embankments and tunnels. Limit states of stability, strength and serviceability need to be considered. The requirements of the ultimate and serviceability limit states may be met by several methods, alone or in combination. The calculation method adopted in the United Kingdom for the ultimate limit state requires the consideration of two combinations of partial safety factors for actions and soil parameters, as shown in Table 1.5.

Generally, combination 2 determines the overall size of the structure and combination 1 governs the structural design of the members. Characteristic soil parameters are defined as cautious estimates of the values affecting the occurrence of a limit state. Thus, for combination 2, design values for the soil strength at the ultimate limit state are given by

TABLE 1.5

Partial Safety Factors for the Ultimate Limit State for Geotechnical Design

Combination	Safety Factor on Actions[a], γ_F		Safety Factor on Soil Parameters, γ_M		
	γ_G	γ_Q	$\gamma_{\phi'}$	$\gamma_{c'}$	γ_{cu}
1	1.35	1.5	1.0	1.0	1.0
2	1.0	1.3	1.25	1.25	1.4

[a] If the action is favourable, values of $\gamma_G = 1.0$ and $\gamma_Q = 0$ should be used.

$$\tan \phi'_d = (\tan \phi')/1.25 \quad \text{and} \quad c'_d = c'/1.25$$

where c' and ϕ' are characteristic values for the cohesion intercept and the angle of shearing resistance (in terms of effective stress), respectively.

Design values for shear resistance at the interface of the base and the sub-soil, for the drained (base friction) and undrained (base adhesion) conditions, respectively, are given by

$$\tan \delta_d = \tan \phi'_d \text{ (for cast } in \text{ } situ \text{ concrete) and } c_{ud} = c_u/1.4$$

where c_u is the undrained shear strength.

Free-standing earth-retaining walls need to be checked for the ultimate limit state regarding overall stability, ground bearing resistance and sliding. For bases on clay soils, the bearing and sliding resistances should be checked for both long-term (drained) and short-term (undrained) conditions. In Example 3, designs for bases on both sand and clay are shown.

The traditional practice in which characteristic actions and allowable bearing pressures are considered, to limit ground deformation and check bearing resistance, may be adopted by mutual agreement. In this case, a linear variation of ground bearing pressure is assumed for eccentric loading.

2 Design of Members

2.1 PRINCIPLES AND REQUIREMENTS

In the European structural codes, a limit state design concept is used. Ultimate limit states (ULS) and serviceability limit states (SLS) are considered, as well as durability and, in the case of buildings, fire resistance. Partial safety factors are included in both design loads and material strengths, to ensure that the probability of failure (i.e., not satisfying a design requirement) is acceptably low. Members are first designed to satisfy the most critical limit state, and then checked to ensure that the other limit states are not reached.

In buildings, for most members, the critical consideration is the ULS, on which the required resistances of the members in bending, shear and torsion are based. The requirements of the various SLS, such as deflection and cracking, are considered later.

Since the selection of a suitable span/effective depth ratio to prevent excessive deflection, and the choice of a suitable bar spacing to avoid excessive cracking, is affected by the stress level in the reinforcement, limit state design is an interactive process. Nevertheless, it is normal to begin with the ULS requirements.

In the following section, the concrete cover to the first layer of bars, as shown in the drawings, is described as the nominal cover. It is defined as a minimum cover plus an allowance in the design for deviation. A minimum cover is required to ensure the safe transmission of bond forces, the protection of steel against corrosion and an adequate fire resistance. To transmit the bond forces safely and to ensure adequate concrete compaction, the minimum cover should be not less than the bar diameter or, for bundled bars, should be not less than the equivalent diameter of a notional bar having the same cross-sectional area as the bundle.

2.2 DURABILITY

Concrete durability is dependent mainly on its constituents, and limitations on the maximum free water/cement ratio and the minimum cement content are specified according to the conditions of exposure. These limitations result in minimum concrete strength classes for particular types of cement. For reinforced concrete, protection of the reinforcement against corrosion depends on the concrete cover.

2.2.1 EXPOSURE CLASSES

Details of the classification system used in BS EN 206-1 and BS 8500-1, with informative examples applicable in the United Kingdom, are shown in *Reynolds*, Table 4.5. When the concrete can be exposed to more than one of the actions described in the table, a combination of the exposure classes will apply.

2.2.2 CONCRETE STRENGTH CLASSES AND COVERS

The required thickness of the cover is related to the exposure class, the concrete quality and the intended working life of the structure. Information taken from the recommendations in BS 8500 is shown in *Reynolds*, Table 4.6. The values for the minimum cover apply for ordinary carbon steel in concrete without special protection, and for structures with an intended working life of at least 50 years.

The values given for the nominal cover include an allowance for tolerance of 10 mm, which is recommended for buildings and is also normally sufficient for other types of structures. The cover should be increased by at least 5 mm for uneven concrete surfaces (e.g., ribbed finish or exposed aggregate).

If *in situ* concrete is placed against another concrete element (precast or *in situ*), the minimum cover to the reinforcement at the interface needs to be not more than that recommended for an adequate bond, provided the following conditions are met: the value of $f_{ck} \geq 25$ MPa, the exposure time of the concrete surface to an outdoor environment is not more than 28 days, and the interface has been roughened.

The nominal cover should be at least 50 mm for concrete cast against prepared ground (including blinding), and 75 mm for concrete cast directly against the earth.

2.3 FIRE RESISTANCE

2.3.1 BUILDING REGULATIONS

The minimum periods of fire resistance required for the elements of the structure, according to the purpose group of a building and its height or, for a basement, the depth relative to the ground are shown in *Reynolds*, Table 3.12. Insurers require longer fire periods for buildings containing storage facilities.

2.3.2 DESIGN PROCEDURES

BS EN 1992-1-2 contains prescriptive rules, in the form of both tabulated data and calculation models, for the standard fire exposure. A procedure for a performance-based method using fire-development models is also provided.

The tabulated data tables give minimum dimensions for the size of a member and the axis distance of the reinforcement. The axis distance is the nominal distance from the centre of the main reinforcing bars to the surface of the concrete as shown in Figure 2.1.

Tabulated data are given for beams, slabs and braced columns, for which provision is made for the load level to be taken into account. In many cases, for fire periods up to about 2 h, the cover required for other purposes will be the controlling factor.

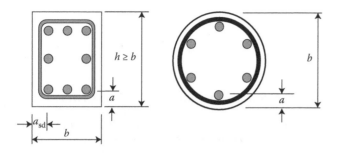

FIGURE 2.1   Cross section showing the nominal axis distances.

2.4 BENDING AND AXIAL FORCE

Typically, beams and slabs are members subjected mainly to bending while columns are subjected to a combination of bending and axial force. In this context, a beam is defined as a member whose span is not less than 3 times its overall depth. Otherwise, the member is treated as a deep beam for which different design methods are appropriate. A column is defined as a member whose greater overall cross-sectional dimension does not exceed 4 times the smaller dimension. Otherwise, the member is considered as a wall. In this case, bending in the plane of the wall is treated in a different way.

2.4.1 Basic Assumptions

For the analysis of the section at the ULS, the tensile strength of concrete is neglected, and strains are based on the assumption that plane sections before bending remain plane after bending. The strain distribution to be assumed is shown in Figure 2.2.

For sections subjected to pure axial compression, the strain is limited to ε_{c2}. For sections partly in tension, the compressive strain is limited to ε_{cu}. For intermediate conditions, the strain diagram is obtained by taking the compressive strain as ε_{c2} at a level equal to 3/7 of the section depth from the more highly compressed face. For values of $f_{ck} \leq 50$ MPa, the limiting strains are $\varepsilon_{c2} = 0.002$ and $\varepsilon_{cu} = 0.0035$.

Reinforcement stresses are determined from bilinear design stress–strain curves. Two alternatives are prescribed in which the top branch of the curve is taken as either horizontal with no limit to the strain (curve A), or rising to a specified maximum strain (curve B).

For concrete in compression, alternative design stress–strain curves give stress distributions forming either a parabola and a rectangle, or a triangle and a rectangle. Another option is to assume a uniform stress distribution. Whichever alternative is used, the proportions of the stress block and the maximum strain are constant for values of $f_{ck} \leq 50$ MPa. In reality, the alternative assumptions lead to only minor differences in the values obtained for the resistance of the section.

For a rectangular concrete area of width b and depth x, the total compressive force can be written as $k_1 f_{ck} bx$ and the distance of the force from the compression face can be written as $k_2 x$. If a uniform stress distribution is assumed, then, for $f_{ck} \leq 50$ MPa, values of $k_1 = 0.453$ and $k_2 = 0.4$ are obtained.

2.4.2 Beams and Slabs

Beams and slabs are generally subjected to only bending, but can also be required to resist an axial force, for example, in a portal frame, or in a floor acting as a prop between basement walls. Axial thrusts not greater than $0.12f_{ck}$ times the area of the cross section may generally be ignored, since the effect of the axial force is to increase the moment of resistance.

If, as a result of moment redistribution allowed in the analysis of a member, the design moment is less than the maximum elastic moment at any section; the necessary ductility may be assumed without explicit verification if, for $f_{ck} \leq 50$ MPa, the neutral axis satisfies the condition $x/d \leq (\delta - 0.4)$.

d is the effective depth, x the neutral axis depth, δ the ratio of the design moment to the maximum elastic moment for values of $1.0 > \delta \geq 0.7$ for ductility class B or C reinforcement and values of $1.0 > \delta \geq 0.8$ for ductility class A reinforcement.

Where plastic analysis is used, the necessary ductility may be assumed without explicit verification if, for $f_{ck} \leq 50$ MPa, the neutral axis at any section satisfies the condition $x/d \leq 0.25$.

2.4.2.1 Singly Reinforced Rectangular Sections

The lever arm between the forces indicated in Figure 2.3 is given by $z = (d - k_2 x)$, from which $x = (d - z)/k_2$.

Taking moments for the compressive force about the line of action of the tensile force gives

$$M = k_1 f_{ck} bxz = k_1 f_{ck} bz(d - z)/k_2$$

The solution of the resulting quadratic equation in z gives

$$z/d = 0.5 + \sqrt{0.25 - (k_2/k_1)\mu} \quad \text{where } \mu = M/bd^2 f_{ck}$$

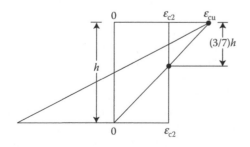

FIGURE 2.2   Strain diagram at the ultimate limit state.

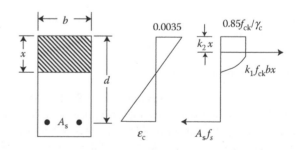

FIGURE 2.3   Strain diagram and forces on a singly reinforced section.

Taking moments for the tensile force about the line of action of the compressive force gives

$$M = A_s f_s z, \text{ from which } A_s = M/f_s z$$

The strain in the reinforcement $\varepsilon_s = 0.0035(1 - x/d)/(x/d)$ and from the design stress–strain curves, the stress is given by

$$f_s = \varepsilon_s E_s = 700(1 - x/d)/(x/d) \leq k_s f_{yk}/1.15$$

If the top branch of the design stress–strain curve is taken as horizontal (curve B), $k_s = 1.0$ and $f_s = f_{yk}/1.15$ for values of

$$x/d \leq 805/(805 + f_{yk}) = 0.617 \quad \text{for } f_{yk} = 500 \text{ MPa}$$

2.4.2.2 Doubly Reinforced Rectangular Sections

The forces provided by the concrete and the reinforcement are indicated in Figure 2.4. Taking moments about the line of action of the tensile force gives

$$M = k_1 f_{ck} bx(d - k_2 x) + A'_s f'_s(d - d')$$

The strain in the reinforcement $\varepsilon'_s = 0.0035(1 - d'/x)$ and from the design stress–strain curve B, the stress is given by

$$f'_s = \varepsilon'_s E_s = 700(1 - d'/x) \leq f_{yk}/1.15$$

Thus, $f'_s = f_{yk}/1.15$ for values of

$$x/d \geq [805/(805 - f_{yk})](d'/d) = 2.64(d'/d) \quad \text{for } f_{yk} = 500 \text{ MPa}$$

Equating the tensile and the compressive forces gives

$$A_s f_s = k_1 f_{ck} bx + A'_s f'_s$$

where the stress in the tension reinforcement is given by the expression derived for singly reinforced sections.

2.4.2.3 Design Formulae for Rectangular Sections

No design formulae are given in the code but the following are valid for values of $f_{ck} \leq 50$ MPa and $f_{yk} \leq 500$ MPa. The formulae are based on the rectangular stress block for the

concrete and stresses of $0.87 f_{yk}$ in tension and compression reinforcement. The compression reinforcement requirement depends on the value of $K = M/bd^2 f_{ck}$ compared to K' where

$$K' = 0.210 \qquad\qquad\qquad \text{for } \delta \geq 1.0$$
$$K' = 0.453(\delta - 0.4) - 0.181(\delta - 0.4)^2 \quad \text{for } \delta < 1.0$$

δ is the ratio of the design moment to the maximum elastic moment, where $\delta \geq 0.7$ for class B and class C reinforcement, and $\delta \geq 0.8$ for class A reinforcement.

For $K \leq K'$, compression reinforcement is not required and

$$A_s = M/0.87 f_{yk} z$$

where

$$z = d\{0.5 + \sqrt{0.25 - 0.882K}\} \quad \text{and} \quad x = (d - z)/0.4$$

For $K > K'$, compression reinforcement is required and

$$A'_s = (K - K')bd^2 f_{ck}/0.87 f_{yk}(d - d')$$
$$A_s = A'_s + K'bd^2 f_{ck}/0.87 f_{yk} z$$

where

$$z = d\{0.5 + \sqrt{0.25 - 0.882K'}\} \quad \text{and} \quad x = (d - z)/0.4$$

For $d'/x > 0.375$ (for $f_y = 500$ MPa), A'_s should be replaced by $1.6(1 - d'/x)A'_s$ in the equations for A'_s and A_s.

A design table, based on the formulae, is given in Table A1. In the table, the lever arm factor z/d is limited to a maximum value of 0.95. Although not a requirement of Eurocode 2, this restriction is common in UK practice.

2.4.2.4 Flanged Sections

In monolithic beam and slab construction, where the web of the beam projects below the slab, the beam is considered as a flanged section for sagging moments. The effective width of flange, over which uniform stress conditions can be assumed, may be taken as $b_{eff} = b_w + b'$, where

$$b' = 0.1(a_w + l_0) \leq 0.2 l_0 \leq 0.5 a_w \quad \text{for L beams}$$
$$b' = 0.2(a_w + l_0) \leq 0.4 l_0 \leq 1.0 a_w \quad \text{for T beams}$$

In the above expressions, b_w is the web width, a_w is the clear distance between the webs of adjacent beams and l_0 is the distance between successive points of zero-bending moment for the beam. If l_{eff} is the effective span, l_0 may be taken as $0.85 l_{eff}$ when there is continuity at one end of the span, and $0.7 l_{eff}$ when there is continuity at both ends. For up-stand beams, when considering hogging moments, l_0 may be taken as $0.3 l_{eff}$ at internal supports and $0.15 l_{eff}$ at end supports.

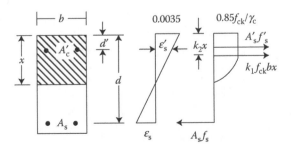

FIGURE 2.4 Strain diagram and forces on a doubly reinforced section.

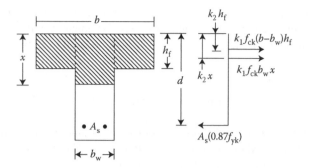

FIGURE 2.5 Forces on flanged section with $x > h_f$.

In sections where the flange is in compression, the depth of the neutral axis will generally be not greater than the thickness of the flange. In this case, the section can be considered to be rectangular with b taken as the flange width. The condition regarding the neutral axis depth can be confirmed initially by showing that $M \leq k_1 f_{ck} b h_f (d - k_2 h_f)$, where h_f is the thickness of the flange. Alternatively, the section can be considered to be rectangular initially, and the neutral axis depth can be checked subsequently.

Figure 2.5 shows a flanged section in which the neutral axis depth exceeds the flange thickness, and the concrete force is divided into two components.

The required area of the tension reinforcement is given by

$$A_s = A_{s1} + k_1 f_{ck} (b - b_w) h_f / 0.87 f_{yk}$$

where A_{s1} is the area of reinforcement required to resist a moment M_1 applied to a rectangular section of width b_w, where

$$M_1 = M - k_1 f_{ck} (b - b_w) h_f (d - k_2 h_f) \leq \mu' b d^2 f_{ck}$$

Using the rectangular concrete stress block in the forgoing equations gives $k_1 = 0.45$ and $k_2 = 0.4$. This approach gives solutions that are 'correct' when $x = h_f$, but becomes slightly more conservative as $(x - h_f)$ increases.

2.4.2.5 Analysis of a Given Section

The analysis of a section of any shape, with any arrangement of reinforcement, involves a trial-and-error process. An initial value is assumed for the neutral axis depth, from which the concrete strains at the positions of the reinforcement can be calculated. The corresponding stresses in the reinforcement are determined, and the resulting forces in the reinforcement and the concrete are obtained. If the forces are out of balance, the value of the neutral axis depth is changed and the process is repeated until equilibrium is achieved. Once the balanced condition has been found, the resultant moment of the forces about the neutral axis, or any convenient point, is calculated.

2.4.3 Columns

Columns are compression members that can bend about any axis. In design, an effective length and a slenderness ratio are determined in relation to major and minor axes of bending. The effective length of the column is a function of the clear

height and depends upon the restraint conditions at the ends. A slenderness ratio is defined as the effective length divided by the radius of gyration of the uncracked concrete section.

Columns should generally be designed for both first-order and second-order effects, but second-order effects may be ignored provided the slenderness ratio does not exceed a particular limiting value. This can vary considerably and has to be determined from an equation involving several factors. These can be calculated but default values are also given.

Columns are subjected to combinations of bending moment and axial force, and the cross section may need to be checked for more than one combination of values. Several methods of analysis, of varying complexity, are available for determining second-order effects. Many columns can be treated as isolated members, and a simplified method of design using equations based on an estimation of curvature is commonly used. The equations contain a modification factor K_r, the use of which results in an iterative process with K_r taken as 1.0 initially. The procedures are shown in *Reynolds*, Tables 4.15 and 4.16.

In the code, for sections subjected to pure axial load, the concrete strain is limited to 0.002 for values of $f_{ck} \leq 50$ MPa. In this case, the design stress in the reinforcement should be limited to 400 MPa. However, in other parts of the code, the design stress in this condition is shown as $f_{yd} = f_{yk}/\gamma_s = 0.87 f_{yk}$. In the derivation of the charts in this chapter, which apply for all values of $f_{ck} \leq 50$ MPa and $f_{yk} \leq 500$ MPa, the maximum compressive stress in the reinforcement was taken as $0.87 f_{yk}$. The charts contain sets of K_r lines to aid the design process.

2.4.3.1 Rectangular Columns

Figure 2.6 shows a rectangular column section in which the reinforcement is disposed equally on two opposite sides of a horizontal axis through the mid-depth. By resolving forces and taking moments about the mid-depth of the section, the following equations are obtained for $0 < x/h \leq 1.0$:

$$N/bh f_{ck} = k_1(x/h) + 0.5(A_s f_{yk}/bh f_{ck})(k_{s1} - k_{s2})$$

$$M/bh^2 f_{ck} = k_1(x/h)\{0.5 - k_2(x/h)\} + 0.5(A_s f_{yk}/bh f_{ck})(k_{s1} + k_{s2}) \times (d/h - 0.5)$$

The stress factors, k_{s1} and k_{s2}, are given by

$$k_{s1} = 1.4(x/h + d/h - 1)/(x/h) \leq 0.87$$

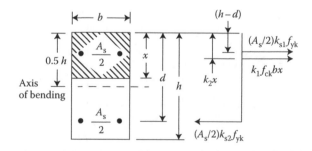

FIGURE 2.6 Forces acting on a rectangular column section.

$$k_{s2} = 1.4(d/h - x/h)/(x/h) \le 0.87$$

The maximum axial force N_u is given by the equation

$$N_u/bhf_{ck} = 0.567 + 0.87(A_s f_{yk}/bhf_{ck})$$

Design charts, based on the rectangular stress block for the concrete, and for the values of $d/h = 0.8$ and 0.85, are given in Tables A2 and A3, respectively. On each curve, a straight line has been taken between the point where $x/h = 1.0$ and the point where $N = N_u$. The charts, which were determined for $f_{yk} = 500$ MPa, may be safely used for $f_{yk} \le 500$ MPa. In determining the forces in the concrete, no reduction has been allowed for the area of concrete displaced by the compression reinforcement. In the design of slender columns, the K_r factor is used to modify the deflection corresponding to a load N_{bal} at which the moment is at maximum. A line corresponding to N_{bal} passes through a cusp on each curve. For $N \le N_{bal}$, the K value is taken as 1.0. For $N > N_{bal}$, K can be determined from the lines on the chart.

2.4.3.2 Circular Columns

Figure 2.7 shows a circular column section in which six bars are equally spaced around the circumference. Solutions based on six bars will be slightly conservative if more bars are used. The bar arrangement relative to the axis of bending affects the resistance of the section, and some combinations of bending moment and axial force can result in a slightly more critical condition, if the arrangement shown is rotated through 30°. These small variations can reasonably be ignored.

The following analysis is based on a uniform stress block for the concrete, of depth λx and width $h \sin \alpha$ at the base (as shown in Figure 2.7). Negative axial forces are included to cater for members such as tensile piles. By resolving forces and taking moments about the mid-depth of the section, the following equations are obtained, where $\alpha = \cos^{-1}(1 - 2\lambda x/h)$ for $0 < x \le 1.0$, and h_s is the diameter of a circle through the centres of the bars:

$$N/h^2 f_{ck} = k_c(2\alpha - \sin 2\alpha)/8 + (\pi/12)(A_s f_{yk}/A_c f_{ck}) \\ \times (k_{s1} - k_{s2} - k_{s3})$$

$$M/h^3 f_{ck} = k_c(3\sin \alpha - \sin 3\alpha)/72 + (\pi/27.7)(A_s f_{yk}/A_c f_{ck})(h_s/h) \\ \times (k_{s1} + k_{s3})$$

Since the width of the compression zone decreases in the direction of the extreme compression fibre, the design stress in the concrete has to be reduced by 10%. Thus, in the above equations: $k_c = 0.9 \times 0.567 = 0.51$ and $\lambda = 0.8$.

The stress factors, k_{s1}, k_{s2} and k_{s3}, are given by

$$-0.87 \le k_{s1} + 1.4(0.433h_s/h - 0.5 + x/h)/(x/h) \le 0.87$$

$$-0.87 \le k_{s2} + 1.4(0.5 - x/h)/(x/h) \le 0.87$$

$$-0.87 \le k_{s3} = 1.4(0.5 + 0.433h_s/h - x/h)/(x/h) \le 0.87$$

To avoid irregularities in the charts, the reduced design stress in the concrete is used to determine the maximum axial force N_u, which is given by the equation:

$$N_u/h^2 f_{ck} = (\pi/4)\{0.51 + 0.87(A_s f_{yk}/A_c f_{ck})\}$$

The minimum axial force N_{min} is given by the equation:

$$N_{min}/h^2 f_{ck} = -0.87(\pi/4)(A_s f_{yk}/A_c f_{ck})$$

Design charts for the values of $h_s/h = 0.6$ and 0.7, are given in Tables A4 and A5, respectively. The previous statements on the derivation and use of the charts for rectangular sections also apply to those for circular sections.

2.4.3.3 Analysis of a Given Section

Any given cross-section can be analysed by a trial-and-error process. For a section bent about one axis, an initial value is assumed for the neutral axis depth, from which the concrete strains at the positions of the reinforcement can be calculated. The resulting stresses in the reinforcement are determined, and the forces in the reinforcement and concrete are evaluated. If the resultant force is not equal to the design axial force N, the value of the neutral axis depth is changed and the process is repeated until equality is achieved. The resultant moment of all the forces about the mid-depth of the section is then the moment of resistance appropriate to N.

2.4.3.4 Example

The column section shown in Figure 2.8 is reinforced with 8H32 arranged as shown. The moment of resistance about the major axis is to be obtained for the following requirements:

$$N = 2300 \text{ kN}, \quad f_{ck} = 32 \text{ MPa}, \quad f_{yk} = 500 \text{ MPa}$$

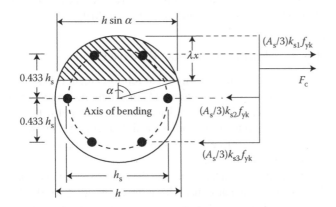

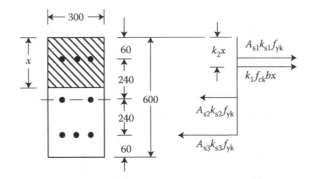

FIGURE 2.7 Forces acting on a circular column section.

FIGURE 2.8 Forces acting on a given column section.

Consider the bars in each half of the section to be replaced by an equivalent pair of bars. The depth to the centroid of the bars in one-half of the section = $60 + 240/4 = 120$ mm. The section is now considered to be reinforced with four equivalent bars, where $d = 600 - 120 = 480$ mm.

$$A_s f_{yk}/bhf_{ck} = 6434 \times 500/(300 \times 600 \times 32) = 0.56$$

$$N/bhf_{cu} = 2300 \times 10^3/(300 \times 600 \times 32) = 0.40$$

From the design chart for $d/h = 480/600 = 0.8$,

$$M_u/bh^2 f_{ck} = 0.18 \text{ (Table A2)}$$

$$M_u = 0.18 \times 300 \times 600^2 \times 32 \times 10^{-6} = 622 \text{ kN m}$$

The solution can be checked using a trial-and-error process to analyse the original section, as follows:
The axial load on the section is given by

$$N = k_1 f_{ck} bx + (A_{s1} k_{s1} - A_{s2} k_{s2} - A_{s3} k_{s3}) f_{yk}$$

where
$d/h = 540/600 = 0.9$, and k_{s1}, k_{s2} and k_{s3} are given by
$k_{s1} = 1.4(x/h + d/h - 1)/(x/h) \leq 0.87$
$k_{s2} = 1.4(0.5 - x/h)/(x/h) \leq 0.87$
$k_{s3} = 1.4(d/h - x/h)/(x/h) \leq 0.87$

With $x = 300$ mm, $x/h = 0.5$, $k_{s1} = 0.87$, $k_{s2} = 0$ and $k_{s3} = 0.87$

$N = 0.45 \times 32 \times 300 \times 300 \times 10^{-3} = 1296$ kN (< 2300)

With $x = 360$ mm, $x/h = 0.6$, $k_{s2} = -0.233$ and $k_{s3} = 0.7$

$N = 0.45 \times 32 \times 300 \times 360 \times 10^{-3} + (2413 \times 0.87 + 1608$
$\quad \times 0.233 - 2413 \times 0.7) \times 500 \times 10^{-3}$
$\quad = 1555 + 392 = 1947$ kN (< 2300)

With $x = 390$ mm, $x/h = 0.65$, $k_{s2} = -0.323$ and $k_{s3} = 0.538$

$N = 0.45 \times 32 \times 300 \times 390 \times 10^{-3}$
$\quad + (2413 \times 0.87 + 1608 \times 0.323 - 2413 \times 0.538)$
$\quad \times 500 \times 10^{-3}$
$\quad = 1685 + 660 = 2345$ kN (> 2300)

With $x = 387$ mm, $x/h = 0.645$, $k_{s2} = -0.315$ and $k_{s3} = 0.553$

$N = 0.45 \times 32 \times 300 \times 387 \times 10^{-3}$
$\quad + (2413 \times 0.87 + 1608 \times 0.315 - 2413 \times 0.553)$
$\quad \times 500 \times 10^{-3}$
$\quad = 1672 + 636 = 2308$ kN

Since the internal and external forces are now sensibly equal, taking moments about the mid-depth of the section gives

$M_u = k_1 f_{ck} bx(0.5h - k_2 x) + (A_{s1} k_{s1} + A_{s3} k_{s3})(d - 0.5h) f_{yk}$
$\quad = 0.45 \times 32 \times 300 \times 387 \times (300 - 0.4 \times 387) \times 10^{-6}$
$\quad + (2413 \times 0.87 + 2413 \times 0.553)(540 - 300) \times 500 \times 10^{-6}$
$\quad = 243 + 412 = 655$ kN m (> 622 obtained earlier)

The method in which the reinforcement was replaced by four equivalent bars can be seen to give a conservative estimate.

2.5 SHEAR

In an uncracked section, shear results in a system of mutually orthogonal diagonal tension and compression stresses. When the diagonal tension stress reaches the tensile strength of the concrete, a diagonal crack occurs. This simple concept rarely applies to reinforced concrete, since members such as beams are already cracked in flexure, and sudden failure can occur in members without shear reinforcement. Resistance to shear can be increased by adding shear reinforcement but, at some stage, the resistance is limited by the capacity of the inclined struts that form within the web.

2.5.1 MEMBERS WITHOUT SHEAR REINFORCEMENT

The design resistance at any cross-section of a member not requiring shear reinforcement can be calculated as

$$V_{Rd,c} = v_{Rd,c} b_w d$$

where
b_w is the minimum width of the section in the tension zone
d is the effective depth to the tension reinforcement and
$v_{Rd,c}$ is the design concrete shear stress.

The design concrete shear stress is a function of the concrete strength, the effective depth and the reinforcement percentage at the section considered. To be effective, this reinforcement should extend for a minimum distance of $(l_{bd} + d)$ beyond the section, where l_{bd} is the design anchorage length.

At a simple support, for a member carrying predominantly uniform load, the length l_{bd} may be taken from the face of the support. The design shear resistance of members with and without axial load can be determined from the information provided in *Reynolds*, Table 4.17.

In the UK National Annex, it is recommended that for values of $f_{ck} > 50$ MPa, the shear strength of the concrete should be determined by tests, unless there is evidence of satisfactory past performance of the particular concrete mix including the aggregates used. Alternatively, the shear strength should be limited to that given for $f_{ck} = 50$ MPa.

2.5.2 MEMBERS WITH SHEAR REINFORCEMENT

The design of members with shear reinforcement is based on a truss model, shown in Figure 2.9, in which the compression and tension chords are spaced apart by a system consisting of inclined concrete struts and vertical or inclined reinforcing bars. Angle α between the reinforcement and the axis of the member should be $\geq 45°$.

Angle θ between the struts and the axis of the member may be selected by the designer within the limits $1.0 \leq \cot \theta \leq 2.5$ generally. However, for elements in which shear co-exists with externally applied tension, $\cot \theta$ should be taken as 1.0.

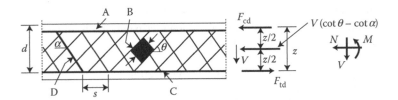

FIGURE 2.9 Truss model and notation for members with shear reinforcement. A–compression chord, B–concrete strut, C–tension chord and D–shear reinforcement.

The web forces are $V \sec \theta$ in the struts and $V \sec \alpha$ in the shear reinforcement over a panel length $l = z(\cot \alpha + \cot \theta)$, where z may normally be taken as $0.9d$. The width of each strut is $z(\cot \alpha + \cot \theta) \sin \theta$, and the design value of the maximum shear force $V_{Rd,max}$ is limited by the compressive resistance provided by the struts, which includes a strength reduction factor for concrete cracked in shear. The least shear reinforcement is required when $\cot \theta$ is such that $V = V_{Rd,max}$.

The truss model results in a force ΔF_{td} in the tension chord that is additional to the force M/z due to bending, but the sum $\Delta F_{td} + M/z$ need not be taken greater than M_{max}/z, where M_{max} is the maximum moment in the relevant hogging or sagging region. The additional force ΔF_{td} can be taken into account by shifting the bending moment curve on each side of any point of maximum moment by an amount $a_1 = 0.5 \, z(\cot \theta - \cot \alpha)$.

For members without shear reinforcement, $a_1 = d$ should be used. The curtailment of the longitudinal reinforcement can then be based on the modified bending moment diagram. A design procedure to determine the required area of shear reinforcement, and details of the particular requirements for beams and slabs, are shown in *Reynolds*, Table 4.18.

For most beams, a minimum amount of shear reinforcement in the form of links is required, irrespective of the magnitude of the shear force. Thus, there is no need to determine $V_{Rd,c}$.

In members with inclined chords, the shear components of the design forces in the chords may be added to the design shear resistance provided by the reinforcement. In checking that the design shear force does not exceed $V_{Rd,max}$, the same shear components may be deducted from the shear force resulting from the design loads.

2.5.3 Shear under Concentrated Loads

In slabs and column bases, the maximum shear stress at the perimeter of a concentrated load should not exceed $v_{Rd,max}$. Shear in solid slabs under concentrated loads can result in punching failures on the inclined faces of truncated cones or pyramids. For design purposes, a control perimeter forming the shortest boundary that nowhere comes closer to the perimeter of the loaded area than a specified distance should be considered. The basic control perimeter may generally be taken at a distance $2d$ from the perimeter of the loaded area.

If the maximum shear stress here is not greater than $v_{Rd,c}$, then no shear reinforcement is required. Otherwise, the position of the control perimeter at which the maximum shear stress is equal to $v_{Rd,c}$ should be determined, and shear reinforcement should be provided in the zone between this control perimeter and the perimeter of the loaded area.

For flat slabs with enlarged column heads (or drop panels), where d_H is the effective depth at the face of the column and the column head (or drop) extends a distance $l_H > 2d_H$ beyond the face of the column, a basic control perimeter at a distance $2d_H$ from the column face should be considered. In addition, a basic control perimeter at a distance $2d$ from the column head (or drop) should be considered.

Control perimeters (in part or as a whole) at distances less than $2d$ should also be considered where a concentrated load is applied close to a supported edge, or is opposed by a high pressure (e.g., soil pressure on bases). In such cases, values of $v_{Rd,c}$ may be multiplied by $2d/a$, where a is the distance from the edge of the load to the control perimeter. For bases, the favourable action of the soil pressure may be included when determining the shear force acting at the control perimeter.

Where a load or reaction is eccentric in relation to a shear perimeter (e.g., at the edge of a slab, and in cases of moment transfer between a slab and a column), a magnification factor is included in the calculation of the maximum shear stress. The details of the design procedures for shear under concentrated loads are shown in *Reynolds*, Table 4.19.

2.5.4 Bottom-Loaded Beams

Where load is applied near the bottom of a section, sufficient vertical reinforcement to transmit the load to the top of the section should be provided in addition to any reinforcement required to resist shear.

2.6 TORSION

In normal beam-and-slab or framed construction, calculations for torsion are not usually necessary, since adequate control of any torsional cracking in beams will be provided by the required minimum shear reinforcement. When it is judged as necessary to include torsional stiffness in the analysis of a structure, or torsional resistance is vital for static equilibrium, members should be designed for the resulting torsional moment.

The torsional resistance may be calculated on the basis of a thin-walled closed section, in which equilibrium is satisfied by a plastic shear flow. A solid section may be modelled as an equivalent thin-walled section. Complex shapes may be divided into a series of sub-sections, each of which is modelled as an equivalent thin-walled section, and the total torsional resistance is taken as the sum of the resistances of the individual elements. When torsion reinforcement is required, this should consist of rectangular closed links together with longitudinal reinforcement. Such reinforcement is additional to the requirements for shear and bending. The details of a design procedure for torsion are shown in *Reynolds*, Table 4.20.

2.7 DEFLECTION

The behaviour of a reinforced concrete beam under service loading can be divided into two basic phases: before and after cracking. During the uncracked phase, the member behaves elastically as a homogeneous material. This phase ends when the load reaches a value at which the first flexural crack forms. The cracks result in a gradual reduction in stiffness with increasing load during the cracked phase. The concrete between the cracks continues to provide some tensile resistance though less, on average, than the tensile strength of the concrete. Thus, the member is stiffer than the value calculated on the assumption that concrete carries no tension. These concepts are illustrated in Figure 2.10.

The deflections of members under the service loading should not impair the function or the appearance of a structure. In buildings, the final deflection of members below the support level, after an allowance for any pre-camber, is limited to span/250. To minimise possible damage to non-structural elements such as finishes, cladding and partitions, deflection that occurs after the construction stage should also be limited to span/500.

Generally, explicit calculation of the deflections is unnecessary to satisfy the code requirements, and simple rules in the form of limiting span/effective depth ratios are provided. These are considered adequate for avoiding deflection problems in most circumstances and, subject to particular assumptions

made in their derivation, give a useful basis for estimating long-term deflections of members in buildings, as follows:

$$\text{Deflection} = \frac{\text{actual span/effective depth ratio}}{\text{limiting span/effective depth ratio}} \times \text{span/250}$$

Although a check under quasi-permanent loading is normally allowed, the author of this book believes that a check under characteristic loading is advisable when the need to minimise possible damage to the elements of a building is a consideration, as explained in Chapter 1.

In special circumstances, when the calculation of deflection is considered necessary, an adequate prediction can be made by calculating the curvature at positions of maximum bending moment, and then assuming that the curvature variation along the member is proportional to the bending moment diagram. Some useful deflection coefficients are given in *Reynolds*, Table 3.42.

The deformation of a section, which could be a curvature or, in the case of pure tension, an extension, or a combination of these, is evaluated first for a homogeneous uncracked section, δ_1, and second for a cracked section ignoring tension in the concrete, δ_2. The actual deformation of the section under the design loading is then calculated as

$$\delta = \zeta \, \delta_2 + (1 - \zeta) \, \delta_1$$

where ζ is a distribution coefficient that takes into account the degree of cracking according to the nature and duration of the loading, and the stress in the tension reinforcement under the load causing first cracking in relation to the stress under the design service load.

When assessing long-term deflections, allowances need to be made for the effect of concrete creep and shrinkage. Creep can be taken into account by using an effective modulus of elasticity $E_{c,eff} = E_c/(1 + \varphi)$, where E_c is the short-term value and φ is a creep coefficient. Shrinkage deformations can be calculated separately and added to those due to loading.

Careful consideration is needed in the case of cantilevers, where the usual formulae assume that the cantilever is rigidly fixed and remains horizontal at the root. Where the cantilever forms the end of a continuous beam, the deflection at the end of the cantilever is likely to be either increased or decreased by an amount $l\theta$, where l is the cantilever length measured to the centre of the support, and θ is the rotation at the support. If a cantilever is connected to a substantially rigid structure, the effective length should be taken as the length to the face of the support plus half the effective depth.

The details of span/effective depth ratios and explicit calculation procedures are shown in *Reynolds*, Tables 4.21 and 4.22.

2.8 CRACKING

Cracks in members under service loading should not impair the appearance, durability or water tightness of a structure. In buildings, the calculated crack width under quasi-permanent

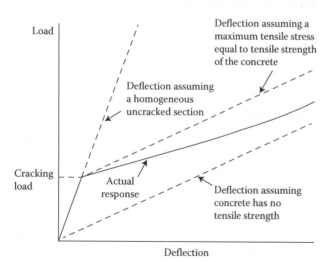

FIGURE 2.10 Load–deflection behaviour.

loading, or as a result of restrained deformations, is generally limited to 0.3 mm.

To control cracking, it is necessary to ensure that the tensile capacity of the reinforcement at yielding is not less than the tensile force in the concrete just before cracking. As a result, a minimum amount of reinforcement is required, according to the strength of the steel, and the tensile strength of the concrete at the time when cracks are likely to form. Cracking due to restrained early thermal effects can occur in continuous walls and slabs within a few days of the concrete being placed. In other cases, it can be several weeks before the applied load reaches a level at which cracking occurs.

Where minimum reinforcement is provided, the crack width requirements may be met by direct calculation, or by limiting either the bar size or the bar spacing. The details of the design procedures are shown in *Reynolds*, Tables 4.23 and 4.24.

For the calculation of crack widths due to restrained imposed deformation, information is provided in PD 6687. The mean strain may be taken as $0.8R\varepsilon_{imp}$, where R is a restraint factor and ε_{imp} is the imposed strain due to early thermal shortening or drying shrinkage. Values of the restraint factor R are given for various pour configurations.

For structures containing liquids, the design requirements are related to leakage considerations. Where a small amount of leakage and the associated surface staining or damp patches is acceptable, the calculated crack width, for cracks that can be expected to pass through the full thickness of the section, is limited to a value that depends on the hydraulic gradient (i.e., head of the liquid divided by thickness of the section). The limits are 0.2 mm for hydraulic gradients ≤ 5, reducing uniformly to 0.05 mm for hydraulic gradients ≥ 35.

Although a cracking check under quasi-permanent loading is implied in the UK National Annex, the author of this book considers that either the frequent or the characteristic load combination should be taken, as explained in Section 1.4.2. For members in axial tension, where at least the minimum reinforcement is provided, the limiting values for either the bar size or the bar spacing may be obtained from that are shown in *Reynolds*, Table 4.25.

In sections subjected to bending, with or without axial force, where the full thickness of the section is not cracked, and at least 0.2 times the section thickness ≤ 50 mm remains in compression, the crack width limit may be taken as 0.3 mm.

For cracking due to the restraint of imposed deformations such as shrinkage and early thermal movements, an estimate needs to be made of the effective tensile strength of the concrete when the first cracks are likely to occur. For walls and slabs less than 1 m in thickness, it is often assumed that such cracking will occur within 3 days of the concrete being placed.

The nature of the cracking depends on the type of restraint. For an element restrained at the ends (e.g., an infill bay with construction joints between the new section of concrete and the pre-existing sections), the crack formation is similar to that caused by external loading. For effective crack control, reinforcement can be determined from *Reynolds*, Table 4.26.

For a panel restrained along one edge (e.g., a wall cast onto a pre-existing stiff base), the formation of the crack only influences the distribution of stresses locally, and the crack width becomes a function of the restrained strain rather than the tensile strain capacity of the concrete.

In EN 1992-3, the mean strain contributing to the crack width is taken as $R_{ax}\varepsilon_{free}$. For early thermal movements, $\varepsilon_{free} = \alpha\Delta T$, where α is the coefficient of thermal expansion for concrete and ΔT is the temperature fall between the hydration peak and ambient at the time of construction. Typical values of ΔT can be estimated from the data in *Reynolds*, Table 2.18. The restraint factor R_{ax} may be taken as 0.5 generally, or may be obtained from *Reynolds*, Table 3.45, where the values are shown for particular zones of panels restrained along one, two or three edges, respectively. For effective crack control, reinforcement can be determined from *Reynolds*, Table 4.27.

It will be found that the calculated strain contributing to the crack width for a panel restrained at its ends is normally more than $R_{ax}\varepsilon_{free}$. Thus, the reinforcement required to limit a crack width to the required value is greater for a panel restrained at its ends than for a panel restrained along one or two adjacent edges.

2.9 CONSIDERATIONS AFFECTING DESIGN DETAILS

Bars may be set out individually, or grouped in bundles of two or three in contact. Bundles of four bars may also be used for vertical bars in compression, and for bars in a lapped joint. For the safe transmission of bond forces, the cover provided to the bars should be not less than the bar diameter or, for a bundle of bars, the equivalent diameter (≤ 55 mm) of a notional bar with a cross-sectional area equal to the total area of the bars in the bundle.

Gaps between bars (or bundles of bars) generally should be not less than the greatest of $(d_g + 5$ mm) where d_g is the maximum aggregate size, the bar diameter (or equivalent diameter for a bundle) or 20 mm. The minimum and maximum amounts for the reinforcement content of different members are shown in *Reynolds*, Table 4.28.

Additional rules for large diameter bars (> 40 mm in the UK National Annex), and for bars grouped in bundles, are given in *Reynolds*, Table 4.32.

At intermediate supports of continuous flanged beams, the total area of tension reinforcement should be spread over the effective width of the flange, but a greater concentration may be provided over the web width.

2.9.1 TIES IN STRUCTURES

Building structures not specifically designed to withstand accidental actions should be provided with a suitable tying system, to prevent progressive collapse by providing alternative load paths after local damage. Where the structure is divided into structurally independent sections, each section should have an appropriate tying system. The reinforcement providing the ties may be assumed to act at its characteristic strength, and only the specified tying forces need to be taken into account. Reinforcement required for other purposes may be considered to form part of, or the whole of the ties. The details of the tying requirements specified in the UK National Annex are shown in *Reynolds*, Table 4.29.

2.9.2 ANCHORAGE LENGTHS

At both sides of any cross section, bars should be provided with an appropriate embedment length or other form of end anchorage. The basic required anchorage length, assuming a constant bond stress f_{bd}, is given by

$$l_{b,rqd} = (\phi/4) \times (\sigma_{sd}/f_{bd})$$

where σ_{sd} is the design stress in the bar at the particular section, and f_{bd} is the design the ultimate bond stress, which depends on the bond condition. This is considered as either 'good' or 'poor', according to the position of the bar during concreting.

The design anchorage length, measured along the centreline of the bar from the section in question to the end of the bar, is given by

$$l_{bd} = \alpha_1 \, \alpha_2 \, \alpha_3 \, \alpha_4 \, \alpha_5 \, l_{b,rqd} \geq l_{b,min}$$

where α_1, α_2, α_3, α_4 and α_5 are coefficients depending on numerous factors. Conservatively, $l_{bd} = l_{b,rqd}$ can be taken.

As a simplified alternative, a tension anchorage for a standard bend, hook or loop may be provided as an equivalent length $l_{b,eq} = \alpha_1 \, l_{b,rqd}$ (see Figure 2.11), where α_1 is taken as 0.7 for covers perpendicular to the bend $\geq 3\phi$. Otherwise, $\alpha_1 = 1.0$.

Bends or hooks do not contribute to compression anchorages. The anchorage requirements are shown in *Reynolds*, Table 4.30.

2.9.3 LAPS IN BARS

Forces can be transferred between reinforcement by lapping, welding or joining bars with mechanical devices (couplers). Laps should be located, if possible, away from positions of maximum moment and should generally be staggered. The design lap length is given by

$$l_0 = \alpha_1 \, \alpha_2 \, \alpha_3 \, \alpha_4 \, \alpha_5 \, \alpha_6 \, l_{b,rqd} \geq l_{0,min}$$

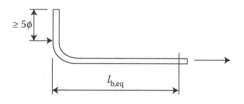

FIGURE 2.11   Equivalent anchorage length for a standard bend.

where α_6 is a coefficient that varies between 1.0 and 1.5, depending on the percentage of lapped bars relative to the total area of bars at the section. Conservatively, $l_0 = \alpha_6 \, l_{b,rqd}$ can be taken.

Transverse reinforcement is required at each end of the lap zone to resist transverse tension forces. In some minor cases, transverse reinforcement or links required for other purposes may be assumed as sufficient. The details of lap lengths are shown in *Reynolds*, Table 4.31.

2.9.4 BENDS IN BARS

The radius of any bend in a reinforcing bar should conform to the minimum requirements of BS 8666, and should ensure that failure of the concrete inside the bend is prevented. For bars bent to the minimum radius according to BS 8666, it is not necessary to check for concrete failure if the anchorage of the bar does not require a length more than 5ϕ beyond the end of the bend. A check for concrete failure is also unnecessary where the plane of the bend is not close to a concrete face, and there is a transverse bar of at least the same size inside the bend. A shear link may be considered as fully anchored, if it passes around another bar not less than its own size, through an angle of $90°$, and continues beyond the end of the bend for a length not less than $10\phi \geq 70$ mm. The details of the minimum bends in bars are given in *Reynolds*, Table 2.19.

In other cases when a bend occurs at a position where the bar is highly stressed, the bearing stress inside the bend needs to be checked, and the radius of the bend will need to be more than the minimum value given in BS 8666. This situation occurs typically at monolithic connections between members; for example, the junction of a beam and an end column, and in short members such as corbels and pile caps.

The design bearing stress depends on the concrete strength, and the containment provided by the concrete perpendicular to the plane of the bend. The details of designed bends in bars are given in *Reynolds*, Table 4.31.

2.9.5 CURTAILMENT OF REINFORCEMENT

In flexural members, it is generally advisable to stagger the curtailment points of the tension reinforcement as allowed by the bending moment envelope. Bars to be curtailed need to extend beyond the points where in theory they are no longer needed for flexural resistance. The extension a_1 is related to the shear force at the section. For members with upright shear links, $a_1 = 0.5 \, z \cot \theta$ where z is the lever arm, and θ is the slope of the concrete struts assumed in the design for shear. For members with no shear reinforcement, $a_1 = d$ is used.

No reinforcement should be curtailed at a point less than a full anchorage length l_{bd} from a section where it is required to be fully stressed. Curtailment rules are shown in *Reynolds*, Table 4.32, and illustrated in Figure 2.12.

At a simple end support, bottom bars should be provided with a tension anchorage beyond the face of the support, where the tensile force to be anchored is given by $F = 0.5V \cot \theta$.

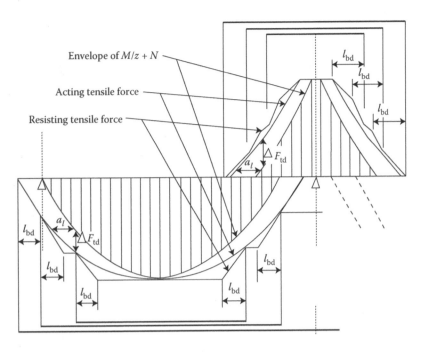

Envelope of $M/z + N$

Acting tensile force

Resisting tensile force

FIGURE 2.12 Curtailment of longitudinal reinforcement taking into account the resistance within the anchorage lengths.

2.10 REINFORCEMENT

Reinforcement for concrete generally consists of steel bars, or welded steel mesh fabric, which depend upon the provision of a durable concrete cover for protection against corrosion. The essential properties of bars to BS 4449 and wires to BS 4482 are summarised in *Reynolds*, Table 2.19.

2.10.1 Bars

BS 4449 provides for bars with a characteristic yield strength of 500 MPa in ductility classes A, B and C. Class A ductility is not suitable where more than 20% moment redistribution is assumed in the design. The bars are round in cross section, with sets of parallel transverse ribs separated by longitudinal ribs. The nominal size is the diameter of a circle with an area equal to the effective cross-sectional area of the bar. Values

of the total cross-sectional area of the reinforcement in a concrete section, according to the number or spacing of the bars, for different bar sizes, are given in Table A9.

In BS 8666, a reference letter is used to identify bar types and grades. Reference H allows the reinforcement supplier to use ductility class A, B or C for bars ≤12 mm diameter, and ductility class B or C for larger bars. Reference B is used if it is imperative, or considered desirable, to use ductility class B or C in all sizes. The details of standard bar shapes, designated by shape codes, are shown in *Reynolds*, Tables 2.21 and 2.22.

2.10.2 Fabric

BS 4483 provides for fabric produced from bars to BS 4449 or, for wrapping fabric, wire produced to BS 4482. The details of the standard fabric types are given in Table A9.

3 Example 1: Multi-Storey Building

Description

The general arrangement of a five-storey building plus basement is shown in drawing 1. The suspended floors are of beam and solid slab construction, but two other forms are also considered. A flat slab, and an integral beam and ribbed slab, respectively, are shown in drawing 2. The upper four floors are designed for general office loading plus an allowance for partitions, the ground floor for use as a retail premises and the basement for general storage purposes.

The building is provided with a central core containing two lifts, a general staircase, and an access well for services to all floors. The walls enclosing these areas provide lateral stability to the structure as a whole. An additional fire escape staircase (not shown) will be attached to the outside of the building. The intended working life is at least 50 years.

The minimum fire periods required by the Building Regulations (*Reynolds,* Table 3.12) are as follows:

Ground and upper storeys: offices and shops (not sprinklered), building height above ground ≤18 m, 1 h
Basement storey: storage (not sprinklered), depth of basement ≤10 m, 1.5 h

Note 1. In the reference column of the calculation sheets, clause references are to BS EN 1992-1-1 unless shown otherwise.
Note 2. The designation H is used generally to identify the bar type and grade. This allows the reinforcement supplier to use ductility class A, B or C for bars up to and including 12 mm diameter, and ductility class B or C for larger bars. An exception occurs at sections where the moment redistribution exceeds 20%, and B12 is used instead of H12.

Schedule of Drawings and Calculations

Drawing	Components	Type of Construction	Calc. Sheets
1	General arrangement (plan and cross section)		
2	Alternative floor plans		
	Beam and Solid Slab Floor Construction		
3	Floor slab	Continuous one-way spanning slab	1– 4
4	Beam on line B	Continuous beam supporting uniform load	5–11
5	Beam on line 1	Continuous beam supporting combined uniform and triangular load	12–13
6	Columns B1 and B2	Columns subjected to axial load and bending in one direction	14–17
	Column A1	Column subjected to axial load and bending in two directions	18
	Columns B1 and B2	Braced columns (fire-resistance)	19–20
	Flat slab floor construction		
7– 9	Floor slab	Flat slab (with and without drops)	21–31
10	Columns B1 and B2	Columns subjected to axial load and bending in one direction	32–35
	Column A1	Column subjected to axial load and bending in two directions	36
	Integral Beam and Ribbed Slab Floor Construction		
11	Ribbed slab	Continuous ribbed slab	37–41
12	Beam on line 2	Continuous band beam supporting uniform load	42–47
13	Column A2 and B2	Columns subjected to axial load and bending in one direction	48–52
	Column A1	Column subjected to axial load and bending in two directions	53
	All Forms of Construction		
14	Stairs	Longitudinal flights spanning onto transverse landings	54–56
15–16	Internal walls	Interconnected wall system providing lateral stability to building	57–65

Note:

For design of basement retaining wall and foundation structure, see Example 2.

Example 1: General Arrangement (Plan and Cross-Section)   **Drawing 1**

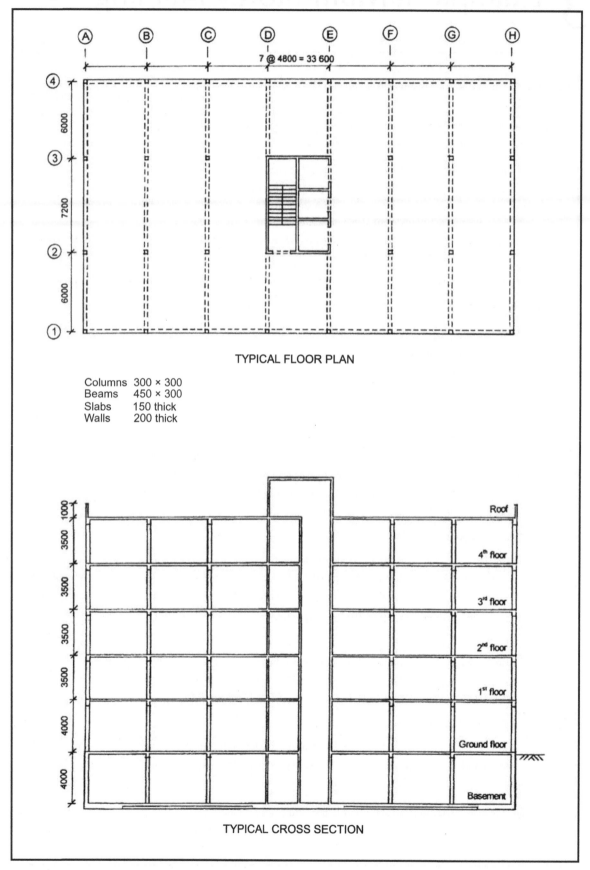

TYPICAL FLOOR PLAN

Columns 300 × 300
Beams   450 × 300
Slabs   150 thick
Walls   200 thick

TYPICAL CROSS SECTION

Example 1: Alternative Floor Plans

Drawing 2

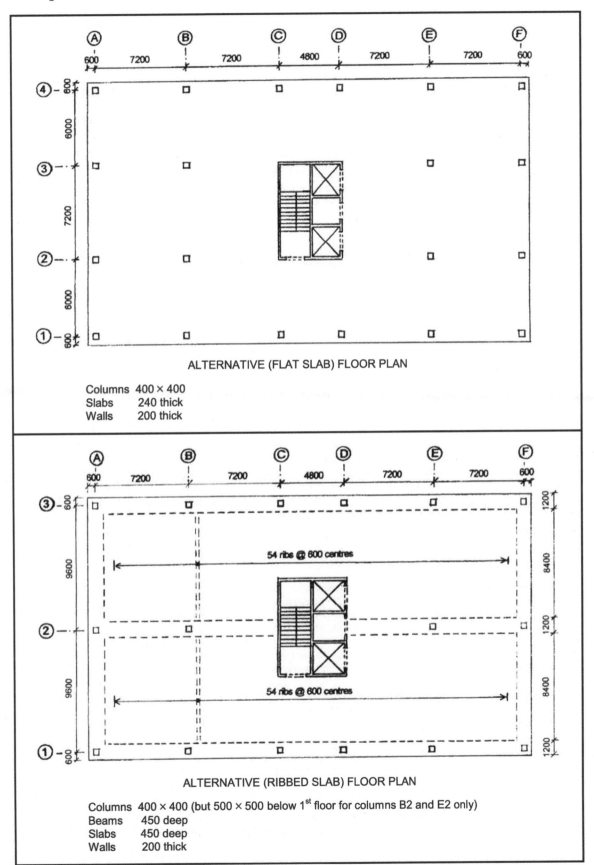

ALTERNATIVE (FLAT SLAB) FLOOR PLAN

Columns 400 × 400
Slabs   240 thick
Walls   200 thick

ALTERNATIVE (RIBBED SLAB) FLOOR PLAN

Columns 400 × 400 (but 500 × 500 below 1st floor for columns B2 and E2 only)
Beams   450 deep
Slabs   450 deep
Walls   200 thick

Example 1   **Calculation Sheet 1**

Reference	CALCULATIONS	OUTPUT

BS 8500	**CONCRETE QUALITY AND COVER FOR DURABILITY**
	
	External surfaces of the perimeter columns and beams are likely to be exposed to cyclic wet and dry conditions, class XC4, and moderate water saturation without de-icing agents, class XF1 (*Reynolds*, Table 4.5). A suitable concrete strength class is C32/40 with 35 mm nominal cover (*Reynolds*, Table 4.6). For internal surfaces, exposure class XC1 applies and 25 mm nominal cover is appropriate.
		Concrete strength class C32/40
		Nominal cover
		  external 35 mm
		  internal 25 mm

	BEAM AND SLAB FLOOR CONSTRUCTION	
	FLOOR SLABS (GROUND AND UPPER FLOORS) For a one-way continuous slab and a characteristic imposed load ≤ 5 kN/m^2, try a span/effective depth ratio of 40. Allowing for 25 mm cover and 12 mm bars gives an estimated thickness = 4800/40 + (25 + 12/2) = 150 mm say	$h = 150$ mm
BS EN 1992-1-2 5.7.3 (2) Table 5.8	**Fire resistance** Allowing for the design to be based on more than 15% redistribution of moment, the slab should be taken as simply supported. For the ground floor slab (minimum fire period 1.5 h), the required minimum dimensions are: Slab thickness: 100 mm Axis distance (to centre of bars): 30 mm Since the cover required for durability is 25 mm, the axis distance is sufficient.	Sufficient for 1.5 h fire period

BS EN 1991-1-1 6.3.1.2 Table NA.3	**Loading**
	
	The National Annex gives the following characteristic values for imposed loads:
	  Offices for general use: 2.5 kN/m^2 Shopping areas: 4.0 kN/m^2
	Allowing 1.5 kN/m^2 for partitions gives the following loads for the office floors:
	
	Permanent load kN/m^2 Variable load kN/m^2
	Self-weight of slab 0.150×25 = 3.75 Imposed = 2.5
	Finishes and services = <u>1.25</u> Partitions = <u>1.5</u>
	             g_k = <u>5.00</u> q_k = <u>4.0</u>
		$g_k = 5.0$ kN/m^2
		$q_k = 4.0$ kN/m^2

BS EN 1990 6.4.3.2 (3) Table A1.2(B)	Two options are given for the action combinations to be considered at the ultimate limit state in relation to internal failure of a structural member. In the Eurocode, either Eq. (6.10) or the less favourable of Eq. (6.10a) and Eq. (6.10b) apply. The National Annex gives values: $\gamma_G = 1.35$, $\gamma_Q = 1.5$, $\xi = 0.925$ and $\psi_0 = 0.7$. In this case, Eq. (6.10b) is the most advantageous (for values of $G_k \leq 4.5Q_k$).
	
	  Design ultimate load, $n = \xi\gamma_G G_k + \gamma_Q Q_k = 1.25 \times 5.0 + 1.5 \times 4.0 = 12.25$ kN/m^2
	
	  Total design ultimate load for 4.8 m span = $12.25 \times 4.8 = 58.8$ kN/m width
		$F = 58.8$ kN/m width

5.1.3 (1)P Table NA.1	**Analysis** The slab spans one way and is continuous over either three or seven equal spans. At an end support the connection will be treated as pinned in the case of a beam, and fixed in the case of a wall. The National Annex allows the design to be based on the single load case of all spans carrying design variable and permanent load, provided the area of each bay exceeds 30 m^2, and $q_k \leq 1.25g_k \leq 5$ kN/m^2 excluding partitions. The resulting support moments are then reduced by 20% and the span moments are increased to suit. The following approximate values can be used.	

Uniformly loaded one-way slab with three or more approximately equal spans						
	End support/slab connection				First interior support	All interior spans	Other interior support
	Pinned		Fixed			
	End support	End span	End support	End span			
Moment	0	0.086*Fl*	−0.063*Fl*	0.063*Fl*	−0.086*Fl*	0.063*Fl*	−0.063*Fl*
Shear	0.4*F*	–	0.48*F*	–	0.6*F*	–	0.5*F*

F is the total design ultimate load and *l* is the effective span. The values shown allow for 20% redistribution of the support moments. For a fixed connection at the end support, the end span moment assumes an end support moment $\geq 0.04Fl$.

Since the redistribution of moment does not exceed 20%, class A ductility reinforcement is acceptable

Example 1

Calculation Sheet 2

Reference	CALCULATIONS	OUTPUT
	The resulting values per m width, with $F = 58.8$ kN/m width and $l = 4.8$ m, are:	

The resulting values per m width, with $F = 58.8$ kN/m width and $l = 4.8$ m, are:

	End support/slab connection				First interior support	All interior spans	Other interior support
	Pinned		Fixed				
	End support	End span	End support	End span			
M kNm	0	24.3	−17.8	17.8	−24.3	17.8	−17.8
V kN	23.5	–	28.2	–	35.3	–	29.4

Flexural Design

Design is based on concrete strength class C32/40 and reinforcement grade 500.

Allowing for 25 mm cover and 12 mm bars, $d = 150 - (25 + 12/2) = 119$ mm

According to the values of M/bd^2f_{ck}, where $b = 1000$ mm, appropriate values of z/d and A_s can be determined (Table A1), and suitable bars selected (Table A9). For the end span (pinned end) and at the first interior support:

$M/bd^2f_{ck} = 24.3 \times 10^6/(1000 \times 119^2 \times 32) = 0.054$ $z/d = 0.95$ (maximum)

$A_s = M/(0.87f_{yk}z) = 24.3 \times 10^6/(0.87 \times 500 \times 0.95 \times 119)$

    $= 494$ mm²/m (H12-200 gives 565 mm²/m)

Location	M/bd^2f_{ck}	z/d	A_s (mm²/m)	Bars[a]
End span (pinned end)	0.054	0.95	494	H12-200
End support (fixed)	0.039	0.95	362	H10-200
End span (fixed end)	0.039	0.95	362	H10-200
First interior support	0.054	0.95	494	H12-200
Interior spans	0.039	0.95	362	H10-200
Other interior supports	0.039	0.95	362	H10-200

[a] A bar spacing of 200 mm has been chosen at all locations so that after 50% curtailment, the resulting spacing will not exceed 400 mm.

Output column:
$f_{ck} = 32$ MPa
$f_{yk} = 500$ MPa
$d = 119$ mm

Reinforcement as shown in table

5.5 (4)
Table NA.1

The National Annex allows redistribution without an explicit check on the rotation capacity, for reinforcement with $f_{yk} \le 500$ MPa, provided $x/d \le (\delta - 0.4)$ where, for reinforcement class A, $\delta \ge 0.8$ and, for reinforcement classes B and C, $\delta \ge 0.7$.

At the first interior support, 20% redistribution (i.e., $\delta = 0.8$) is allowable, provided $x/d \le (\delta - 0.4) = 0.4$. Since $z/d = 0.95$, $x/d = 2.5(1 - z/d) = 0.125$ (< 0.4).

Shear Design

6.2.1 (8)

Shear may be checked at distance d from face of support. At first interior support,

$V = 35.3 - 12.25 \times (0.15 + 0.119) = 32.0$ kN/m
$v = V/b_w d = 32.0 \times 10^3/(1000 \times 119) = 0.27$ MPa

6.2.2 (1)
Table NA.1

The design shear strength of a flexural member without shear reinforcement is given by

$$v_c = \left(\frac{0.18k}{\gamma_c}\right)\left(\frac{100A_{sl}f_{ck}}{b_w d}\right)^{1/3} \ge v_{min} = 0.035k^{3/2}f_{ck}^{1/2}$$

where $k = 1 + \sqrt{\frac{200}{d}} \le 2.0$, $\left(\frac{100A_{sl}}{b_w d}\right) \le 2.0$ and $\gamma_c = 1.5$

With $100A_{sl}/b_w d = 100 \times 565/(1000 \times 119) = 0.47$, and $k = 2$ for $d \le 200$ mm

$v_{min} = 0.035 \times 2^{3/2} \times 32^{1/2} = 0.56$ MPa
$v_c = (0.18 \times 2/1.5)(0.47 \times 32)^{1/3} = 0.59$ MPa ($> v_{min}$)

Note: This value of v_c can also be obtained from *Reynolds*, Table 4.17.

6.2.1 (4)

Since $v < v_c$, no shear reinforcement is required.

No shear reinforcement

Example 1   Calculation Sheet 3

Reference	CALCULATIONS	OUTPUT
	Deflection (see *Reynolds*, Table 4.21).	
7.4.1 (4) BS EN 1990 A1.4.2 NA.2.2.6	Recommended deflection requirements with regard to the appearance and general utility of a structure are given in relation to the quasi-permanent load combination. However, the National Annex to BS EN 1990 indicates that the characteristic load combination should be taken into account when considering damage to structural and non-structural elements, including partitions.	
7.4.1 (6)	Deflection requirements may be met by limiting the span/effective depth ratio to specified values. The actual span/effective depth ratio = 4800/119 = 40.3.	
	Since the end support provides partial fixity, it is reasonable to assume that the value obtained for the span moment (end pinned) is not less than the elastic value for all appropriate load cases (i.e., no redistribution). Thus, the service stress in the reinforcement under the characteristic load is given approximately by $\sigma_s = (f_{yk}/\gamma_s)(A_{s,req}/A_{s,prov})[(g_k + q_k)/n]$ $= (500/1.15)(480/565)(9.0/12.25) = 271$ MPa	
7.4.2 Table NA.5	From Table A2, limiting l/d = basic ratio $\times \alpha_s \times \beta_s$ where: For $100A_s/bd = 100 \times 480/(1000 \times 119) = 0.40 < 0.1f_{ck}^{0.5} = 0.1 \times 32^{0.5} = 0.56$, $\alpha_s = 0.55 + 0.0075f_{ck}/(100A_s/bd) + 0.005f_{ck}^{0.5}[f_{ck}^{0.5}/(100A_s/bd) - 10]^{1.5}$ $= 0.55 + 0.0075 \times 32/0.40 + 0.005 \times 32^{0.5} \times (32^{0.5}/0.40 - 10)^{1.5} = 1.39$ (*Note*: The value of α_s can also be obtained from *Reynolds*, Table 4.21, for the given values of $f_{ck} = 32$ MPa and $100A_s/bd = 0.40$) $\beta_s = 310/\sigma_s = 310/271 = 1.14$ For an end span of a continuous slab, basic ratio = 26, and hence Limiting $l/d = 26 \times \alpha_s \times \beta_s = 26 \times 1.39 \times 1.14 = 41.2$ (>actual $l/d = 40.3$)	Check complies
	Cracking (see *Reynolds*, Table 4.23)	
7.3.2 (2)	Minimum area of reinforcement required in tension zone for crack control: $A_{s,min} = k_c k f_{ct,eff} A_{ct}/\sigma_s$ Taking values of $k_c = 0.4$, $k = 1.0$, $f_{ct,eff} = f_{ctm} = 0.3f_{ck}^{(2/3)} = 3.0$ MPa (for general design purposes), $A_{ct} = bh/2$ (for plain concrete section) and $\sigma_s \leq f_{yk} = 500$ MPa $A_{s,min} = 0.4 \times 1.0 \times 3.0 \times 1000 \times (150/2)/500 = 180$ mm²/m	H10-400
	(*Note*: A value for $100A_{s,min}/A_{ct} = 0.24$ can be obtained from *Reynolds*, Table 4.23, giving $A_{s,min} = 0.0024 A_{ct} = 0.0024 \times 1000 \times 150/2 = 180$ mm²/m)	
7.3.3 (1)	No other specific measures are necessary provided overall depth does not exceed 200 mm, and detailing requirements are observed.	Check complies
	Detailing Requirements (see *Reynolds*, Tables 4.28 and 4.32)	
9.3.1.1 (1)	Minimum area of longitudinal tension reinforcement: $A_{s,min} = 0.26(f_{ctm}/f_{yk})bd = 0.26 \times (3.0/500)bd = 0.00156bd \geq 0.0013bd$ $= 0.00156 \times 1000 \times 119 = 186$ mm²/m	H10-400
9.3.1.1 (2)	Minimum area of secondary reinforcement (20% of principal reinforcement): $A_{s,min} = 0.2 \times 480 = 96$ mm²/m Use H10-400.	H10-400
9.3.1.1 (3)	Maximum spacing of principal reinforcement in area of maximum moment: $2h = 300 \leq 250$ mm. Elsewhere: $3h = 450 \leq 400$ mm	Spacing satisfactory
	Maximum spacing of secondary reinforcement in area of maximum moment: $3h = 450 \leq 400$ mm. Elsewhere: $3.5h = 525 \leq 450$ mm	Spacing satisfactory
9.3.1.2 (1)	At a simply supported end, half the calculated span reinforcement should continue to the support and be anchored. The tensile force is given by $F = (a_l/z)V$, with $a_l = d$ and $z = 0.9d$. With $V = 23.5$ kN/m and $A_s = 283$ mm²/m, $F = 23.5/0.9 = 26.1$ kN/m and $\sigma_s = V/A_s = 26.1 \times 10^3/283 = 92$ MPa	H12-400
8.4.3 (2)	For good bond, $f_{ck} = 32$ MPa and $\sigma_s = 435$ MPa, $l_{b,rqd} = 35\phi$ (*Reynolds*, Table 4.30)	

Example 1   **Calculation Sheet 4**

Reference	CALCULATIONS	OUTPUT
8.4.4	The tabulated value may be multiplied by $\sigma_s/435$, where $\sigma_s = 92$ MPa, giving $l_{b,rqd} = (92/435) \times 35 \times 12 = 89$ mm $\geq l_{b,min} = 10\phi = 10 \times 12 = 120$ mm	
9.3.1.2 (2)	At an end support where partial fixity occurs, top reinforcement to resist at least 15% of the maximum moment in the end span should be provided. Use H10-400.	H10-400
	At the edge beams, transverse partial fixity occurs and top reinforcement to resist at least 25% of the maximum longitudinal moment in the adjacent spans should be provided. The transverse reinforcement should extend at least 0.2 times the length of the adjacent span from the face of the edge beam. Use H10-400.	H10-400
	Curtailment of longitudinal tension reinforcement	
	In the absence of an elastic moment envelope covering all appropriate load cases, the following simplified curtailment rules will be used.	
	For bottom reinforcement, continue 50% onto support for a distance $\geq 10\,\phi$ from the face, and 100% to within a distance from the centre of support as follows:	
	$\leq 0.1 \times$ span at pinned end support	
	$\leq 0.2 \times$ span at interior and fixed end supports	
	For top reinforcement, continue for distance beyond face of support as follows:	
	100% for $\geq 0.2 \times$ span $= 960$ mm $(\geq l_{b,rqd} + d = 35 \times 12 + 119 = 540$ mm$)$	
	50% for $\geq 0.3 \times$ span $= 1440$ mm at interior and fixed end supports	
	Tying Requirements (see *Reynolds*, Table 4.29).	
9.10.2.3 Table NA.1	The principal reinforcement in the bottom of each span can be utilised to provide continuous internal ties. The tensile force to be resisted:	
	$F_{tie,int} = [(g_k + q_k)/7.5](l_r/5)F_t \geq F_t$ kN/m	
	With $l_r = 4.8$ m and $F_t = (20 + 4n_0) \leq 60$ where n_0 is number of storeys,	
	$F_{tie,int} = (9.0/7.5)(4.8/5)(20 + 4 \times 6) = 50.7$ kN/m	
9.10.1 (4)	Minimum area of reinforcement required with $\sigma_s = 500$ MPa (i.e., $\gamma_s = 1.0$)	
	$A_{s,min} = 50.7 \times 1000/500 = 102$ mm^2/m, Use H10-400.	H10-400
8.7.3 (1)	If all bars are lapped at the same position, design lap length (*Reynolds*, Table 4.31):	
	$l_0 = \alpha_6 l_{bd} \geq l_{0,min} = 200$ mm, where $\alpha_6 = 1.5$ for $> 50\%$ bars lapped	
2.4.2.4 Table 2.1N	For accidental design situations, $\gamma_c = 1.2$ and $\gamma_s = 1.0$ (*Reynolds*, Table 4.1), and l_{bd} may be taken as for normal design situations, where $\gamma_c = 1.5$ and $\gamma_s = 1.15$.	
	$l_0 = \alpha_6 \times (35\phi) \times A_{s,req}/A_{s,prov} = 1.5 \times (35 \times 10) \times (102/196) = 300$ mm say	Lap = 300 mm

Bar Marks	Commentary on Bar Arrangement (Drawing 3)
01	Principal bottom bars at a maximum spacing of 400 mm, providing 50% of total span reinforcement. Bars are arranged to lap 300 mm with bars at a similar spacing in adjacent span to provide continuous internal tie. At end support, bars are provided with 250 mm anchorage ($> l_{b,min}$) into beam containing peripheral tie.
02	Principal bottom bars providing rest of span reinforcement, and curtailed 450 mm ($< 0.1 \times$ span) from the centre of end support and 950 mm ($< 0.2 \times$ span) from the centre of interior support.
03	Secondary bars, at a maximum spacing in area of maximum moment of 400 mm, are provided in 6 m lengths for ease of handling and economy of use (12 m stock lengths). Bottom bars extend 250 mm into edge beams and are provided with laps of 550 mm ($> l_0 = 1.5 \times 35\phi$) for optimum crack control. Top bars are given laps of 250 mm ($> l_{0,min} = 200$ mm), and lap with bars 07 at end of run.
04	Secondary bottom bars as 03 but 8.5 m long.
05	Principal top bars at a maximum spacing of 400 mm, with alternate bars staggered and curtailed at distances of say 1100 and 1600 mm from the centre of support (i.e., 950 mm and 1450 mm from face of support).
06	Top bars to resist at least 15% of span moment, anchored into beam and extending 720 mm into slab.
07	Top bars to resist at least 25% of span moment, anchored into beam and extending 960 mm into slab.
08	Secondary top bars lapping with bars 05 and 06.

Example 1: Reinforcement in Floor Slab (Ground and Upper Floors)   **Drawing 3**

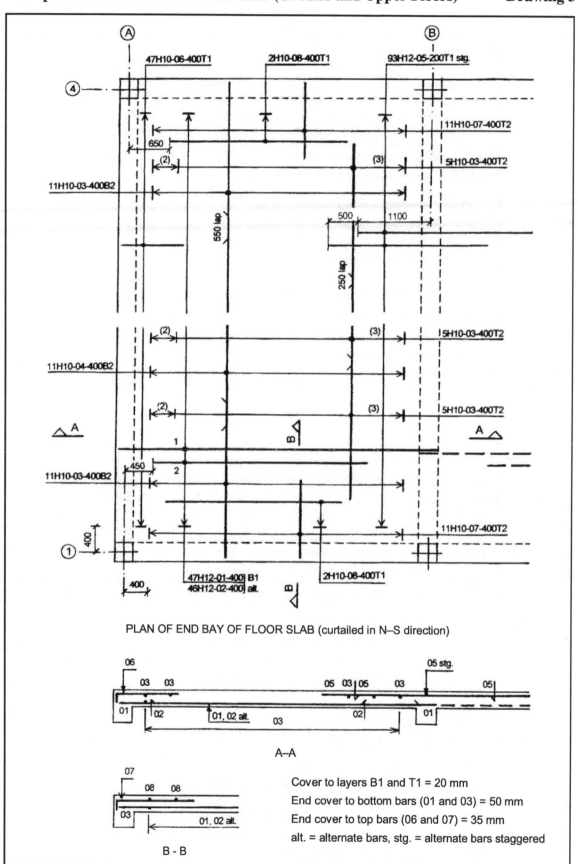

PLAN OF END BAY OF FLOOR SLAB (curtailed in N–S direction)

A–A

B - B

Cover to layers B1 and T1 = 20 mm

End cover to bottom bars (01 and 03) = 50 mm

End cover to top bars (06 and 07) = 35 mm

alt. = alternate bars, stg. = alternate bars staggered

Example 1

Calculation Sheet 5

Reference	CALCULATIONS	OUTPUT
	MAIN BEAMS (GROUND AND UPPER FLOORS)	
	For a continuous beam with a design ultimate load of 70 kN/m say, consider a span/effective depth ratio of 15. Based on the end span, and allowing for 25 mm cover with 8 mm links and 25 mm longitudinal bars, the estimated overall depth:	
	$h = 6000/15 + (25 + 8 + 25/2) = 450$ mm say	$h = 450$ mm
	Fire resistance	
BS EN 1992-1-2 5.6.3 (2) Table 5.5	Allowing for the design to be based on more than 15% moment redistribution, the beam should be taken as simply supported. For the ground floor beams (minimum fire period 1.5 h), the required minimum dimensions are:	
	Beam width: 300 mm axis distance (to centre of bars in one layer): 40 mm	
	Axis distance to side of beam for corner bars: $(40 + 10) = 50$ mm	
	Since the cover required for durability is 25 mm, assuming the use of H8 links and H32 main bars, the axis distances are sufficient.	Sufficient for 1.5 h fire period
	Loading	
	For the slabs, the maximum design load is 12.25 kN/m^2 and the minimum load is $1.25 \times 5.0 = 6.25$ kN/m^2. The loads on the first interior beam taking shear force coefficients for the slab of 0.6 for the end span and 0.5 for the interior span, are:	
	$\quad\quad\quad\quad\quad$ (max) $\quad\quad\quad\quad\quad\quad\quad\quad$ (min) Slab $\quad 1.1 \times 4.8 \times 12.25 = 64.7 \quad\quad 1.1 \times 4.8 \times 6.25 = 33.0$ Beam $\quad 1.25 \times 0.3 \times 0.3 \times 25 = \underline{2.8} \quad\quad\quad\quad\quad\quad\quad = \underline{2.8}$ $\quad\quad\quad\quad\quad\quad\quad\quad \underline{67.5}$ kN/m $\quad\quad\quad\quad\quad\quad \underline{35.8}$ kN/m	67.5 kN/m (max) 35.8 kN/m (min)
	Analysis	
5.1.3 (1)P Table NA.1	The beams, except for those in the area adjacent to the central core of the building, are continuous over three spans. Design moments and shears will be derived from an elastic analysis of a sub-frame consisting of the beam at one level together with the columns above and below. The columns are assumed fixed at the ends remote from the beam. The National Annex allows designs to be based on the following load cases: all spans carrying design variable and permanent load, and alternate spans carrying design variable and permanent load with other spans carrying only design permanent load. Since the sub-frame and the load cases are symmetrical about the centreline, an analysis can be carried out for one half of the sub-frame by taking the stiffness of the central span of the beam as half the actual value. The analysis of a sub-frame where the columns above and below the floor are identical (i.e. 2nd, 3rd and 4th floors) will be shown. The sub-frames at ground and 1st floor levels are slightly different due to the different storey heights.	
	  Dimensions of sub-frame	
5.3.2.1 (4)	Relative stiffness values, given by $K = I/l$, will be based on the properties of the uncracked rectangular section for both beam and column. (Alternatively, for the beam, values based on the effective flange section within the span could be used.)	
	$I_b = 300 \times 450^3/12 = 2.278 \times 10^9$ mm^4 $\quad I_c = 300 \times 300^3/12 = 0.675 \times 10^9$ mm^4	
	$K_{b,end} = I_b/6000 = 0.380 \times 10^6$ mm^3 $\quad\quad K_{b,int} = I_b/7200 = 0.316 \times 10^6$ mm^3	
	$K_{c,upper} = K_{c,lower} = I_c/3500 = 0.193 \times 10^6$ mm^3	

Example 1   Calculation Sheet 6

Reference	CALCULATIONS	OUTPUT
	Distribution factors for unit moment applied at an end joint are:	

$D_b = 0.380/(0.380 + 2 \times 0.193) = 0.496$, $D_c = (1 - 0.496)/2 = 0.252$

Distribution factors for unit moment applied at an interior joint are:

$D_{b,end} = 0.380/(0.380 + 0.5 \times 0.316 + 2 \times 0.193) = 0.380/0.924 = 0.411$

$D_{b,int} = 0.5 \times 0.316/0.924 = 0.171$, $D_c = 0.193/0.924 = 0.209$

Fixed-end moments due to maximum load on beams are:

$M_{end} = 67.5 \times 6^2/12 = 202.5$ kN m $M_{int} = 67.5 \times 7.2^2/12 = 291.6$ kN m

Fixed-end moments due to minimum load on beams are:

$M_{end} = 35.8 \times 6^2/12 = 107.4$ kN m $M_{int} = 35.8 \times 7.2^2/12 = 154.7$ kN m

Although the sub-frame can be conveniently analysed by computer program, the following moment distribution procedure will be used here. In the table below, the basic operations are shown in rows 1 and 2 with a 50% moment carry-over in the end span. There is no moment carry-over in the interior span as a result of taking only 50% of the actual stiffness in calculating the distribution factors. Rows 3 and 4 are obtained by combining rows 1 and 2 in such a way that a moment at one joint can be balanced without disturbing the equilibrium of the moments at the other joint. Rows 5 and 6, which are obtained simply from rows 3 and 4, can now be used to balance the moments at each joint in turn.

Row	Joint and member — Moments in members due to application of moments at joints	End joint Upper column	Beam	Lower column	Interior joint Upper column	End beam	Interior beam	Lower column
1	Unit moment applied at end joints	0.252	0.496	0.252		0.248		
2	Unit moment applied at interior joints		0.205		0.209	0.411	0.171	0.209
3	(Row 1)/0.248 – (Row 2)	1.016	1.795	1.016	–0.209	0.589	–0.171	–0.209
4	(Row 2)/0.205 – (Row 1)	–0.252	0.504	–0.252	1.017	1.752	0.832	1.017
5	(Row 3)/(1.016 + 1.795 + 1.016)	0.2655	0.469	0.2655	–0.055	0.154	–0.044	–0.055
6	(Row 4)/(1.017 + 1.752 + 0.832+ 1.017)	–0.055	0.110	–0.055	0.220	0.380	0.180	0.220

Case	Moments (kN m) in members for load case							
1	Maximum load (67.5 kN/m) on all spans							
	Fixed-end moments		–202.5			202.5	–291.6	
	(Row 5) × 202.5	53.8	94.9	53.8	–11.1	31.2	–9.0	–11.1
	(Row 6) × (291.6 – 202.5)	–4.9	9.8	–4.9	19.6	33.9	16.0	19.6
	Sum to obtain final moments	48.9	–97.8	48.9	8.5	267.6	–284.6	8.5

Case								
2	Maximum load (67.5kN/m) on end spans and minimum load (35.8 kN/m) on interior span							
	Fixed-end moments		–202.5			202.5	–154.7	
	(Row 5) × 202.5	53.8	94.9	53.8	–11.1	31.2	–9.0	–11.1
	(Row 6) × (154.7 – 202.5)	2.6	–5.2	2.6	–10.5	–18.2	–8.6	–10.5
	Sum to obtain final moments	56.4	–112.8	56.4	–21.6	215.5	–172.3	–21.6

Case								
3	Minimum load (35.8 kN/m) on end spans and maximum load (67.5 kN/m) on interior span							
	Fixed-end moments		–107.4			107.4	–291.6	
	(Row 5) × 107.4	28.5	50.4	28.5	–5.9	16.5	–4.7	–5.9
	(Row 6) × (291.6 – 107.4)	–10.1	20.2	–10.1	40.5	70.0	33.2	40.5
	Sum to obtain final moments	18.4	–36.8	18.4	34.6	193.9	–263.1	34.6

Example 1   **Calculation Sheet 7**

Reference	CALCULATIONS	OUTPUT
	The shear forces at the ends of the span and the maximum sagging moment can now be calculated from the following expressions, where the support moments M_L and M_R both take positive values. For the end span,	

$$V_L = nl/2 - (M_R - M_L)/l \text{ and } V_R = nl - V_L$$

Distance from end of span to point of zero shear, $a = V_L/n$

Maximum sagging moment, $M = V_L \times a/2 - M_L$

For the interior span,

$$V_L = V_R = nl/2 \text{ and } M = nl^2/8 - M_L \text{ (or } M_R)$$

Load Case	Location and Member	End Support	End Span	Interior Support	Interior Support	Interior Span
No.	Bending Moment (kN m) in Members for Load Case					
1	Beam	−97.8	127.0	267.6	−284.6	152.8
	Upper column	48.9		8.5		
	Lower column	48.9		8.5		
2	Beam	−112.8	141.8	215.5	−172.3	59.7
	Upper column	56.4		−21.6		
	Lower column	56.4		−21.6		
3	Beam	−36.8	55.3	193.9	−263.1	174.3
	Upper column	18.4		34.6		
	Lower column	18.4		34.6		

No.	Shear Force (kN) in Members for Load Case					
1	Beam	174.2		230.8	243.0	
2	Beam	185.4		219.6	128.9	
3	Beam	81.2		133.6	243.0	

Allowing for some redistribution of moment, the maximum hogging moments in the beam will be taken as 112.8 kN m at the end supports for load cases 1 and 2, and 215.5 kN m at the interior supports for load cases 1 and 3. As a result, the maximum sagging moment in the end span for load case 1 will be the same as that for load case 2. In the interior span, the maximum sagging moment for load cases 1 and 3 will increase in order to maintain equilibrium, as follows:

$$M = 67.5 \times 7.2^2/8 - 215.5 = 221.9 \text{ kN m}$$

Flexural design

At the top of the beam, allowing for 25 mm cover, 12 mm transverse bars in slab and 25 mm longitudinal bars, $d = 450 - (25 + 12 + 25/2) = 400$ mm

OUTPUT: $d = 400$ mm

5.5 (4)
Table NA.1

At the interior supports, the ratio of design moment to maximum elastic moment is $\delta = 215.5/284.6 = 0.76$, and the ductility criterion $x/d \le (\delta - 0.4) = 0.36$ applies.

$$M/bd^2 f_{ck} = 215.5 \times 10^6/(300 \times 400^2 \times 32) = 0.140$$

From Table A1, $A_s f_{yk}/bd f_{ck} = 0.188$ and $x/d = 0.361 \ (\cong 0.36)$.

$$A_s = 0.188 \times 300 \times 400 \times 32/500 = 1444 \text{ mm}^2$$

From Table A9, 2H25 plus 4B12 provides 1434 mm². This is sufficient since the presence of 2H25 in the compression zone will reduce the value of x/d and the required value of $A_s f_{yk}/bd f_{ck}$. (Note: B12 rather than H12 required since $\delta < 0.8$).

OUTPUT: At interior support (200 225 150 225 200; 2B12 2H25 2B12)

5.3.2.1 (3)
9.2.1.2 (2)

The tension flange is considered to extend beyond the side face of the beam say for a distance given by $b_{eff,i} = 0.2 \times 0.15(l_1 + l_2) = 0.03 \times (6000 + 7200) = 400$ mm. The bars will be spread over the effective width of the flange, with 2H25 inside the links and 2B12 either side of the web.

At the end supports, where the bars will be contained inside the links,

$$M/bd^2 f_{ck} = 112.8 \times 10^6/(300 \times 400^2 \times 32) = 0.074, \ A_s f_{yk}/bd f_{ck} = 0.092$$

$$A_s = 0.092 \times 300 \times 400 \times 32/500 = 707 \text{ mm}^2 \ (2\text{H25 gives } 982 \text{ mm}^2)$$

OUTPUT: At end support (2H25)

Example 1   Calculation Sheet 8

Reference	CALCULATIONS	OUTPUT
5.3.2.1 (3)	For sagging moments, the effective flange width is given by: $b_{eff} = b_w + 2 \times 0.2 \times 0.7l = 300 + 0.28\,l$ For the end spans: $b_{eff} = 300 + 0.28 \times 6000 = 1980$ mm, with $d = 400$ mm $M/bd^2 f_{ck} = 141.8 \times 10^6/(1980 \times 400^2 \times 32) = 0.014$, $z/d = 0.95$ (max) $A_s = 141.8 \times 10^6/(0.87 \times 500 \times 0.95 \times 400) = 858$ mm^2 (2H25 gives 982 mm^2) For the interior spans: $b_{eff} = 300 + 0.28 \times 7200 = 2316$ mm, with $d = 400$ mm $M/bd^2 f_{ck} = 221.9 \times 10^6/(2316 \times 400^2 \times 32) = 0.0187$, $z/d = 0.95$ (max) $A_s = 221.9 \times 10^6/(0.87 \times 500 \times 0.95 \times 400) = 1343$ mm^2 (3H25 gives 1473 mm^2)	 2H25 End span 3H25 Interior span
	Shear design	
6.2.1 (8)	Since the load is uniformly distributed, the critical section for shear can be taken at distance d from the face of support, that is, 550 mm from the centre of support. At the end support, the maximum value is $V = 185.4 - 67.5 \times 0.55 = 148.3$ kN The required inclination of the concrete strut (defined by cot θ), to obtain the least amount of shear reinforcement, can be shown to depend on the following factor: $v_w = V/[b_w\,z\,(1 - f_{ck}/250)f_{ck}]$ $\quad = 148.3 \times 10^3/[300 \times 0.9 \times 400 \times (1 - 32/250) \times 32] = 0.049$ From *Reynolds*, Tables 4.18, for vertical links and values of $v_w < 0.138$, $\cot\theta = 2.5$ can be used. The area of links required is then given by	
6.2.3 (3)	$A_{sw}/s = V/f_{ywd}\,z\cot\theta$ $\quad = 148.3 \times 10^3/(0.87 \times 500 \times 0.9 \times 400 \times 2.5) = 0.38$ mm^2/mm From *Reynolds*, Tables 4.20, H8-250 links gives 0.40 mm^2/mm At the interior support, the maximum values are: $V_L = 230.8 - 67.5 \times 0.55 = 193.7$ kN, $A_{sw}/s = 0.495$ mm^2/mm, H8-200 links $V_R = 243.0 - 67.5 \times 0.55 = 205.9$ kN, $A_{sw}/s = 0.526$ mm^2/mm, H8-175 links Minimum requirements for vertical links are given by	H8-250 links H8-200 links H8-175 links
9.2.2 (5) 9.2.2 (6)	$A_{sw}/s = (0.08\sqrt{f_{ck}})\,b_w/f_{yk} = (0.08\sqrt{32}) \times 300/500 = 0.27$ mm^2/mm $s \le 0.75d = 0.75 \times 400 = 300$ mm. Using H8-300 links provides 0.33 mm^2/mm giving a design shear resistance: $V_{Rd,s} = (A_{sw}/s)\,f_{ywd}\,z\cot\theta = 0.33 \times 0.87 \times 500 \times 0.9 \times 400 \times 2.5 \times 10^{-3} = 129.2$ kN The region within which each group of links is required can now be determined, and is shown on the shear force diagram. Shear force diagram showing link requirements	H8-300 links <u>Note</u> The link arrangement shown will need to be modified, to comply with the requirement for transverse bars at the lap zones of the main reinforcement (see commentary on bar arrangement in calculation sheet 11).

Example 1

Calculation Sheet 9

Reference	CALCULATIONS	OUTPUT
	Deflection (see *Reynolds*, Table 4.21)	
7.4.1 (6)	Deflection requirements may be met by limiting the span/effective depth ratio. For the interior span, the actual span/effective depth ratio = 7200/400 = 18	
	The characteristic load is given by	
	$g_k + q_k = 1.1 \times 4.8 \times 9.0 + 2.8/1.25 = 49.8$ kN/m	
	Taking account of the moment redistribution in the analysis, the service stress in the reinforcement under the characteristic load is given approximately by	
	$\sigma_s = (f_{yk}/\gamma_s)(M_{elastic}/M_{design})(A_{s,req}/A_{s,prov})[(g_k + q_k)/n]$	
	$= (500/1.15)(174.3/221.9)(1245/1473)(49.8/67.5) = 213$ MPa	
7.4.2 Table NA.5 PD 6687	From *Reynolds*, Table 4.21, limiting l/d = basic ratio $\times \alpha_s \times \beta_s$ where:	
	With bd taken as $b_{eff}h_f + b_w(d - h_f) = 2316 \times 150 + 300 \times 250 = 422.4 \times 10^3$,	
	$100A_s/bd = 100 \times 1245/(422.4 \times 10^3) = 0.30 < 0.1f_{ck}^{0.5} = 0.1 \times 32^{0.5} = 0.56$	
	$\alpha_s = 0.55 + 0.0075f_{ck}/(100A_s/bd) + 0.005f_{ck}^{0.5}[f_{ck}^{0.5}/(100A_s/bd) - 10]^{1.5}$	
	$= 0.55 + 0.0075 \times 32/0.30 + 0.005 \times 32^{0.5} \times (32^{0.5}/0.30 - 10)^{1.5} = 2.10$	
	$\beta_s = 310/\sigma_s = 310/213 = 1.45$	
	For an interior span of a continuous beam, basic ratio = 30. For flanged sections with $b/b_w = 2316/300 = 7.72 > 3$, the basic ratio should be multiplied by 0.8. For beams with spans >7 m, supporting partitions liable to be damaged by excessive deflections, the basic ratio should be multiplied by 7/span.	
	Limiting $l/d = 30 \times 0.8 \times 7.0/7.2 \times \alpha_s \times \beta_s = 23.3 \times 2.10 \times 1.45 = 71$ (>17.8)	Check complies
	Cracking (see *Reynolds*, Tables 4.23 and 4.24)	
7.3.1 (5) Table NA.4	The National Annex recommends with regard to appearance, for X0 and XC1 exposure classes, that w_k should be limited to 0.3 mm. In the absence of specific requirements for appearance this limit may be relaxed (*Reynolds*, Table 4.1).	
7.3.2 (2)	Minimum area of reinforcement required in tension zone for crack control:	
	$A_{s,min} = k_c k f_{ct,eff} A_{ct}/\sigma_s$	
	For the interior support region, the effective tension flange extends approximately 400 mm beyond the side face of the beam. For the uncracked section, the depth of the tension zone ignoring the effect of the reinforcement is given by	
	$h_{cr} = \dfrac{b_w h^2 + (b_f - b_w)h_f^2}{2[b_w h + (b_f - b_w)h_f]} = \dfrac{300 \times 450^2 + 800 \times 150^2}{2[300 \times 450 + 800 \times 150]} = 154$ mm (>h_f)	
	For the flange, $k_c = 0.5$ (since the neutral axis of the section is only just below the flange), $k = 0.65$ (since $b \geq 800$ mm), $A_{ct} = 1100 \times 150 = 165 \times 10^3$ mm^2. Thus,	
	$A_{s,min} = 0.5 \times 0.65 \times 3.0 \times 165 \times 10^3/500 = 322$ mm^2 (<1434 mm^2 provided)	
BS EN 1990 Table NA.A1.1	The quasi-permanent load, where $\psi_2 = 0.3$ is obtained from the National Annex to the Eurocode (Table 1.1), is given by	
	$g_k + \psi_2 q_k = 1.1 \times 4.8 \times (5.0 + 0.3 \times 4.0) + 2.8/1.25 = 35.0$ kN/m	
	Taking account of the moment redistribution in the analysis, the service stress in the reinforcement under the quasi-permanent load is given approximately by	
	$\sigma_s = (f_{yk}/\gamma_s)(M_{elastic}/M_{design})(A_{s,req}/A_{s,prov})[(g_k + \psi_2 q_k)/n]$	
	$= (500/1.15)(284.6/215.5)(1436/1434)(35.0/67.5) = 298$ MPa	
7.3.3 (2)	The crack width criterion can be satisfied by limiting either the bar size or the bar spacing. For the top of the beam, it is reasonable to ignore any requirement based on appearance, since the surface of the beam will not be visible below the finishes.	
Table 7.2 Table 7.3	Nevertheless, for $w_k = 0.4$ mm and $\sigma_s = 300$ MPa, the recommended maximum values, by interpolation, are bar spacing 175 mm or $\phi^*_s = 14$ mm. The maximum bar size is then given by	
	$\phi_s = \phi^*_s (f_{ct,eff}/2.9)[h_{cr}/4(h - d)] = 14 \times (3.0/2.9) \times [154/(4 \times 50)] = 11$ mm	

Example 1   **Calculation Sheet 10**

Reference	CALCULATIONS	OUTPUT
	The bar arrangement comprises 12 mm bars at 200 mm centres, and 25 mm bars at 150 mm centres, respectively. Thus, although there is no specific requirement to be satisfied, it can be inferred that w_k will be of the order of 0.4 mm.	
	In the end span region, with no redistribution, the stress in the reinforcement under the quasi-permanent loading is given approximately by	
	$\sigma_s = (500/1.15)(811/982)(35.0/67.5) = 186$ MPa	
Table 7.2 Table 7.3	For $w_k = 0.3$ mm, the maximum values (by interpolation) are: $\phi^*_s = 28$ mm, or bar spacing 270 mm. The actual bar size is 25 mm, and the bar spacing is 200 mm. For the interior span, the conditions are less critical since the reinforcement stress is lower, and the actual bar spacing is 100 mm.	
	Detailing requirements	
9.2.1.1 (1)	Minimum area of longitudinal tension reinforcement (*Reynolds*, Table 4.28).	
	$A_{s,min} = 0.26(f_{ctm}/f_{yk})b_t d = 0.26 \times (3.0/500)\, b_t d = 0.00156 bd \geq 0.0013\, b_t d$	
	For the hogging regions,	
	$A_{s,min} = 0.00156 \times 1100 \times 400 = 687$ mm^2 (<982 mm^2 provided)	
	For the sagging regions,	
	$A_{s,min} = 0.00156 \times 300 \times 400 = 188$ mm^2 (<982 mm^2 provided)	
9.2.1.5 (1) 9.2.1.5 (2)	At the bottom of each span, at least 25% of the area provided in the span should continue to the supports and be provided with an anchorage length beyond the face of the support not less than 10ϕ. In the final detail, 2H25 are made effectively continuous for the whole length of the beam. $\qquad l_{b,min} = 10 \times 25 = 250$ mm	$l_{b,min} = 250$ mm
Figure 8.2	At the end supports, even though the top bars will be in the form of U-bars in the vertical plane, poor bond conditions will be assumed. From *Reynolds*, Table 4.30, for poor bond and $f_{ck} = 32$ MPa, $l_{b,rqd} = (A_{s,req}/A_{s,prov}) \times 50\phi \geq l_{b,min}$. Thus,	
8.4.3 (2)	$l_{b,rqd} = (A_{s,req}/A_{s,prov}) \times 50\phi = (707/982) \times 50 \times 25 = 900$ mm	$l_{b,rqd} = 900$ mm
	For U-bars with 50 mm cover top and bottom so as to fit comfortably inside the links, the largest practical radius is obtained by using a semi-circular bend (shape code 13). In this case, $r = 0.5h - (50 + \phi) = 150$ mm (i.e. 6ϕ).	
8.3 (3)	The minimum radius of bend of the bars depends on the value of a_b/ϕ, where a_b is taken as half the centre-to-centre distance between the bars. Allowing for the bars having 70 mm side cover to avoid the column bars, $a_b = 0.5 \times 130 = 65$ mm.	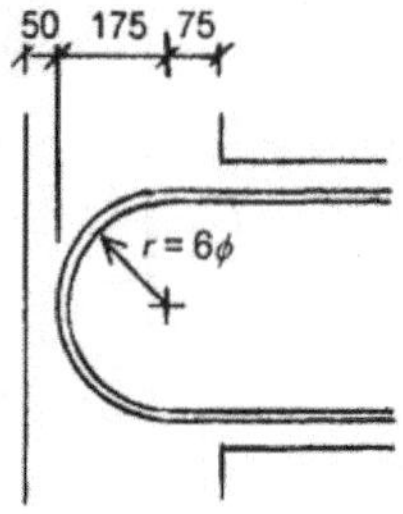
	From *Reynolds*, Table 4.31, with $f_{ck} = 32$ MPa and $a_b/\phi = 65/25 = 2.6$, $r_{min} = 8.4\phi$.	
	This value can be reduced by allowing for $A_{s,req} < A_{s,prov}$, and taking into account the stress reduction in the bar between the edge of the support and the start of the bend. Thus, if $r = 6\phi$, distance to start of bend $= 300 - (50 + 7 \times 25) = 75$ mm.	
	Reduced value of $r_{min} = (707/982)(1 - 75/900) \times 8.4\,\phi = 5.6\phi$ ($< 6\phi$ shown)	
	Curtailment of longitudinal tension reinforcement (see *Reynolds*, Table 4.32)	
	For the interior span, the resistance moment provided by 2H25 at the bottom of the beam can be determined from Table A1 as follows:	
	$A_s f_{yk}/bdf_{ck} = 982 \times 500/(2316 \times 400 \times 32) = 0.016$, $z/d = 0.95$ (max)	
	$M = A_s(0.87 f_{yk})z = 982 \times 0.87 \times 500 \times 0.95 \times 400 \times 10^{-6} = 162.3$ kN m	
	At the interior support, $M_s = 215.5$ kN m and distance x from a support to a point where $M = 162.3$ kN m is given by $Vx - nx^2/2 - 215.5 = 162.3$ kN m	
	Hence, with $V = 243$ kN and $n = 67.5$ kN/m,	
	$0.5x^2 - 3.6x + 5.6 = 0$ solutions of which are $x = 2.27$ and 4.93 m	
	Thus, of the 3H25 required in the span, one bar is no longer needed for flexure at 2.27 m from the support. Here, $V = 89.8$ kN and $V_{Rd,s} = 129.2$ kN with $\cot\theta = 2.5$. Thus, $\cot\theta = (V/V_{Rd,s}) \times 2.5 = 1.74$ is sufficient and the bar should extend beyond these points for a minimum distance $a_1 = z(\cot\theta)/2 = 0.45d\cot\theta$	At bottom of interior span, stop 1H25 at 1950 mm from gridlines 2 and 3
9.2.1.3 (2)	$a_1 = 0.45 \times 400 \times 1.74 = 313$ mm, $x - a_1 = 2270 - 313 = 1950$ mm say	

Example 1 **Calculation Sheet 11**

Reference	CALCULATIONS	OUTPUT
	At the top of the beam, 2H12 will be provided to support the links for the length of each span. With $d = 400$ mm, the following values are obtained:	
	$A_s f_{yk}/bdf_{ck} = 226 \times 500/(300 \times 400 \times 32) = 0.029$, $z/d = 0.95$ (max)	
	$M = 226 \times 0.87 \times 500 \times 0.95 \times 400 \times 10^{-6} = 37.3$ kNm	
	If V and M_s are the values at the support, the distance x from a support to a point where $M = 37.3$ kNm is given by $Vx - nx^2/2 = M_s - 37.3$	
	For the end span, for load cases 1 (after redistribution) and 2:	
	$V = 185.4$ kN, $M_s = 112.8$ kNm, $n = 67.5$ kN/m, giving $x = 0.45$ and 5.05 m	
	For the interior span, for load case 2:	
	$V = 128.9$ kN, $M_s = 172.3$ kNm, $n = 35.8$ kN/m, giving $x = 1.25$ and 5.95 m	
9.2.1.3 (2)	At these points, $\cot\theta = 2.5$, and the bars to be curtailed should extend for a further minimum distance $a_1 = 0.45 \times 400 \times 2.5 = 450$ mm. It is also necessary to ensure that the bars extend for a distance not less than $(a_1 + l_{bd})$ beyond the face of the support. For simplicity, $l_{bd} = l_{b,rqd}$ will be assumed, as the modification coefficients have only a minor effect. For the U-bars at an end support, $l_{b,rqd} = 900$ mm. Thus, since $l_{bd} > x = 450$ mm, the top leg should extend beyond the face of the support a distance $(a_1 + l_{bd}) = 450 + 900 = 1350$ mm. At an interior support, assuming poor	$a_1 = 450$ mm At end supports, extend upper leg of U-bars for 1350 mm from the face of support
8.4.3.2	bond conditions for the bars within the width of the web, $l_{b,rqd} = 50\phi$. For the bars in the flange, good bond conditions may be assumed and $l_{b,rqd} = 35\phi$.	
	Thus, the H25 bars should extend not less than $a_1 + 50\phi = 1700$ mm from the face of support, nor less than the following distances from the centre of support:	At interior supports, extend on each side from face of support, H25 for 1700 mm and B12 for 900 mm
	End span: $a_1 + 950 = 1400$ mm Interior span: $a_1 + 1250 = 1700$ mm	
	The B12 bars should extend beyond the face of support: $a_1 + 35\phi = 900$ mm say	
	Tying requirements (see *Reynolds*, Table 4.29)	
9.10.2.3 Table NA.1	The longitudinal reinforcement in the bottom of each span can be used to provide continuous internal ties. With $l_r = 7.2$ m and $F_t = (20 + 4n_0) \le 60$,	
	$F_{tie,int} = [(g_k + q_k)/7.5](l_r/5)F_t = (9.0/7.5)(7.2/5)(20 + 4 \times 6) = 76$ kN/m	
9.10.1 (4)	For beams at 4.8 m centres, with $\sigma_s = 500$ MPa, minimum area of reinforcement	
	$A_{s,min} = 4.8 \times 76 \times 1000/500 = 730$ mm^2 Use 2H25	2H25
8.7.3 (1)	At the supports, where the bars will be lapped, design lap length	
	$l_0 = \alpha_6 \times (35\phi) \times A_{s,req}/A_{s,prov} = 1.5 \times (35 \times 25) \times (730/982) = 1000$ mm say	Lap = 1000 mm

Bar Marks	Commentary on Bar Arrangement (Drawing 4)
01, 03	Bars in corners of links curtailed 50 mm from column face at each end to avoid clashing with column bars.
02	Loose bars positioned inside column bars. Bars lap 1000 mm with bars 01, and bars 03 in the adjacent span, to provide continuity of internal ties.
04	Bar curtailed 1950 mm (see calculation sheet 10) from centre line of column at each end of span.
05	Loose U-bars, shape code 13, positioned inside column bars. Upper leg extends 1350 mm beyond face of column to satisfy curtailment requirement (see calculation sheet 11) and lower leg laps 1000 mm with bar mark 01. Overall dimension of semi-circular bend provides tolerance for U-bar to fit inside links.
06, 09	Bars in corners of links curtailed 50 mm from column face.
07	Bars positioned inside column bars, and extending beyond centre line of column 1850 mm into end span and interior span (see calculation sheet 11).
08	Bars positioned in slab as shown on section B-B, and extending beyond centreline of column 1050 mm into end span and interior span.
10	Closed links, shape code 51, with 25 mm nominal cover. Spacing of links determined by requirements for shear reinforcement (see calculation sheet 8), and transverse reinforcement in lap zones of main bars. Where diameter of lapped bars $\phi \ge 20$ mm, transverse bars of total area not less than area of one lapped bar should be provided within outer thirds of lap zone (see clause 8.7.4.1). For full zone, with $\phi = 25$ mm and allowing for $A_{s,req} < A_{s,prov}$, total area of transverse bars $A_{st} = 1.5 \times 491 \times 730/982 = 548$ mm^2 (11H8-100).

Example: Reinforcement in Main Beams (Ground and Upper Floors)   **Drawing 4**

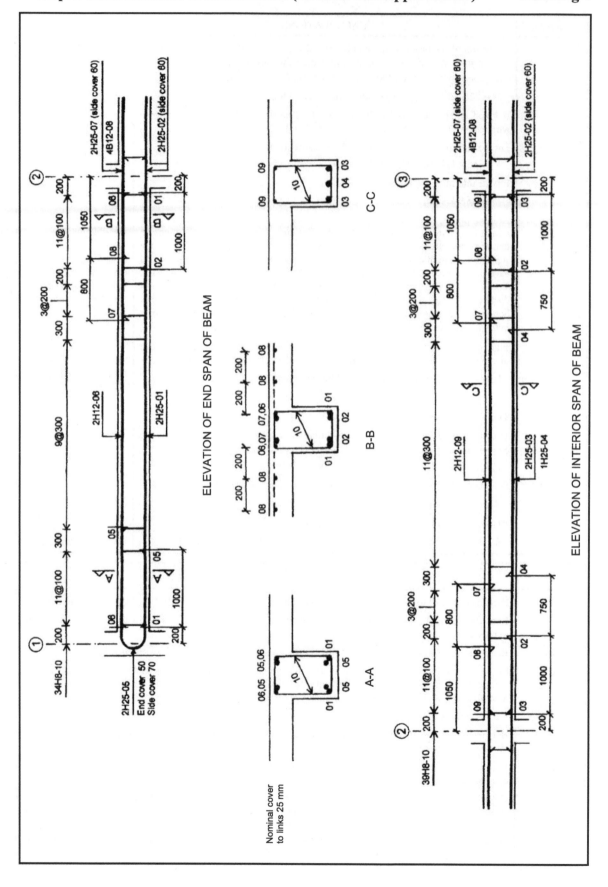

Example 1   Calculation Sheet 12

Reference	CALCULATIONS	OUTPUT
	EDGE BEAMS (GROUND AND UPPER FLOORS)	

Loading

In addition to self-weight, the beam will be designed to support a triangular area of floor slab, and say a load due to walling, cladding and windows of 5 kN/m. If the triangular area is taken to be formed by lines at 45° from the intersection of the faces of the main beams and the edge beam, area $= 0.5 \times 4.5 \times 2.25 = 5.0 \ m^2$.

The total design ultimate loads for each 4.8 m span are:

Beam and walling: $1.25 \times (0.3 \times 0.3 \times 25 + 5.0) \times 4.8 = 43.5$ kN (dead)
Slab: $1.25 \times 5.0 \times 5.0 + 1.5 \times 4.0 \times 5.0 = 31.3$ kN (dead) + 30.0 kN (live)

Analysis

Reference: 5.1.3 (1)P Table NA.1

Although the beam is part of a frame, it will be analysed as a continuous member on knife-edge supports. The simplified load cases allowed in the National Annex will be assumed. Bending moment and shear force coefficients for uniform and triangular loads are given in *Reynolds*, Tables 2.30, 2.31 and 2.32, respectively. Hence, for a continuous beam with five or more equal spans, the maximum values are as follows:

Location	Bending moment (kN m)
End span	$(0.078 \times 43.5 + 0.105 \times 31.3 + 0.135 \times 30) \times 4.8 = 51.5$
1st interior support	$(0.105 \times 43.5 + 0.132 \times 31.3 + 0.132 \times 30) \times 4.8 = 60.8$
Interior spans	$(0.046 \times 43.5 + 0.068 \times 31.3 + 0.117 \times 30) \times 4.8 = 36.7$
Other supports	$(0.079 \times 43.5 + 0.099 \times 31.3 + 0.099 \times 30) \times 4.8 = 45.6$

Location	Shear force (kN)
End support	$(0.395 \times 43.5 + 0.369 \times 31.3 + 0.434 \times 30) = 41.8$
1st interior support	$(0.605 \times 43.5 + 0.631 \times 31.3 + 0.649 \times 30) = 65.5$
ditto (interior span)	$(0.526 \times 43.5 + 0.532 \times 31.3 + 0.622 \times 30) = 58.2$
Other supports	$(0.500 \times 43.5 + 0.500 \times 31.3 + 0.614 \times 30) = 55.8$

Flexural design

At the supports, allowing for 16 mm bars with 62 mm cover, so that the bar at the inside face of the edge beam passes below the U-bars in the main beam,

$d = 450 - (62 + 16/2) = 380$ mm, $b = 300$ mm

In the spans, allowing for 35 mm cover with 8 mm links and 16 mm bars,

Reference: 5.3.2.1 (3)

$d = 400$ mm say, $b_{eff} = b_w + 0.2 \times 0.7l = 300 + 0.14 \times 4800 = 970$ mm

From Table A1, the required reinforcement can be determined as follows:

Location	$M/bd^2 f_{ck}$	z/d	A_s mm^2	Bars
End span (pinned end)	0.010	0.95	312	2H16
First interior support	0.044	0.95	387	2H16
Interior spans	0.008	0.95	222	2H12
Other interior supports	0.033	0.95	290	2H16

Reference: 5.3.2.1 (3)

At the interior supports, the tension flange is considered to extend beyond the side face of the beam for a distance given by $b_{eff,i} = 0.2 \times 0.3l = 0.06 \times 4800 = 290$ mm. The 2H16 will be positioned inside the links with an additional H12 in the flange.

Shear design

Minimum requirements for vertical links are given by

Reference: 9.2.2 (5)

$A_{sw}/s = (0.08\sqrt{f_{ck}}) b_w / f_{yk} = (0.08\sqrt{32}) \times 300/500 = 0.27$ mm^2/mm

Reference: 9.2.2 (6)

$s \leq 0.75d = 0.75 \times 380 = 285$ mm.

Example 1   **Calculation Sheet 13**

Reference	CALCULATIONS	OUTPUT
6.2.3 (3)	From *Reynolds*, Table 4.20, H8-275 links provides 0.36 mm²/mm which gives a design shear resistance: $V_{\text{Rd,s}} = (A_{\text{sw}}/s)\,f_{\text{ywd}}\,z\cot\theta = 0.36 \times 0.87 \times 500 \times 0.9 \times 380 \times 2.5 \times 10^{-3} = 133.9$ kN Comparison with the maximum shear forces shows that the provision of minimum links is sufficient in all cases.	H8-275 links
	Deflection Since the loading and the span are both less than those for the main beam, there is no need to check this requirement.	
	Cracking At the interior supports, $b_{\text{eff}} = 590$ mm, and the depth to the neutral axis for the uncracked section ignoring the effect of the reinforcement is given by $x = \dfrac{b_{\text{w}}h^2 + (b_{\text{f}}-b_{\text{w}})h_{\text{f}}^2}{2[b_{\text{w}}h + (b_{\text{f}}-b_{\text{w}})h_{\text{f}}]} = \dfrac{300 \times 450^2 + 290 \times 150^2}{2[300 \times 450 + 290 \times 150]} = 188$ mm ($>h_{\text{f}}$) For the flange, $k_{\text{c}} = 0.9(1 - h_{\text{f}}/2x) = 0.9 \times [1 - 150/(2 \times 188)] = 0.54$, $k = 0.80$ (for $b = 590$ mm), $A_{\text{ct}} = 590 \times 150 = 88.5 \times 10^3$ mm². Thus,	
7.3.2 (2)	$A_{\text{s,min}} = 0.54 \times 0.8 \times 3.0 \times 88.5 \times 10^3/500 = 230$ mm² (<515 mm² provided). In the spans, where $b_{\text{eff}} = 970$ mm, $x = \dfrac{300 \times 450^2 + 670 \times 150^2}{2[300 \times 450 + 670 \times 150]} = 160$ mm $(h - x) = 450 - 160 = 290$ mm For the web, $k_{\text{c}} = 0.4$, $k = 0.90$ (for $h = 450$ mm), $A_{\text{ct}} = 290 \times 300 = 87 \times 10^3$ mm² $A_{\text{s,min}} = 0.4 \times 0.9 \times 3.0 \times 87 \times 10^3/500 = 188$ mm² (<226 mm² provided). For the interior spans, the stress in the reinforcement under quasi-permanent load: $\sigma_{\text{s}} = (500/1.15)(224/226)(59.8 + 0.3 \times 20)/104.75 = 270$ MPa For $w_{\text{k}} = 0.3$ mm and $\sigma_{\text{s}} = 270$ MPa, maximum bar size (by interpolation):	
Table 7.2	$\phi_{\text{s}} = \phi^*_{\text{s}}\,(f_{\text{ct,eff}}/2.9)[k_{\text{c}}\,h_{\text{cr}}/2(h-d)] = 13 \times (3.0/2.9) \times [0.4 \times 290/(2 \times 50)] = 15$ mm	Check complies
	Detailing requirements	
9.2.1.1 (1)	Minimum area of longitudinal tension reinforcement: $A_{\text{s,min}} = 0.26(f_{\text{ctm}}/f_{\text{yk}})b_{\text{t}}d = 0.26 \times (3.0/500)\,b_{\text{t}}d = 0.00156bd \geq 0.0013\,b_{\text{t}}d$ For the interior support regions, $A_{\text{s,min}} = 0.00156 \times 590 \times 380 = 350$ mm² (<515 mm² provided) For the span regions, $A_{\text{s,min}} = 0.00156 \times 300 \times 400 = 188$ mm² (<226 mm² provided)	
	Curtailment of longitudinal tension reinforcement	
9.2.1.5 (1)	At the bottom of the beam, 2H16 in the end spans and 2H12 in the interior spans will be provided for the length of the span. At the supports where at least 25% of the area in the span is needed, 2H12 will be provided with a lap length of 300 mm. At the top of the beam, 2H12 will be provided to support the links for the length of each span. At the end supports where the columns provide partial fixity, 2H12 will be provided. At each support, the bars should extend beyond the face of the support far enough to provide a lap length. Assuming poor bond conditions, Projection from face of support $= (50 + 1.5 \times 50 \times 12) = 950$ mm	At all supports, extend top bars for 950 mm from face of support
	Tying requirements	
9.10.2.3 Table NA.1	The longitudinal reinforcement can be used to resist the peripheral tie force, $F_{\text{tie,per}} = 20 + 4n_0 = (20 + 4 \times 6) = 44$ kN	
9.10.1 (4)	Minimum area of reinforcement required with $\sigma_{\text{s}} = 500$ MPa, $A_{\text{s,min}} = 44 \times 1000/500 = 88$ mm² Use 1H12 The bar on the inside face at the top of the beam will form the peripheral tie.	1H12 as peripheral tie

Example 1: Reinforcement in Edge Beams (Ground and Upper Floors)   Drawing 5

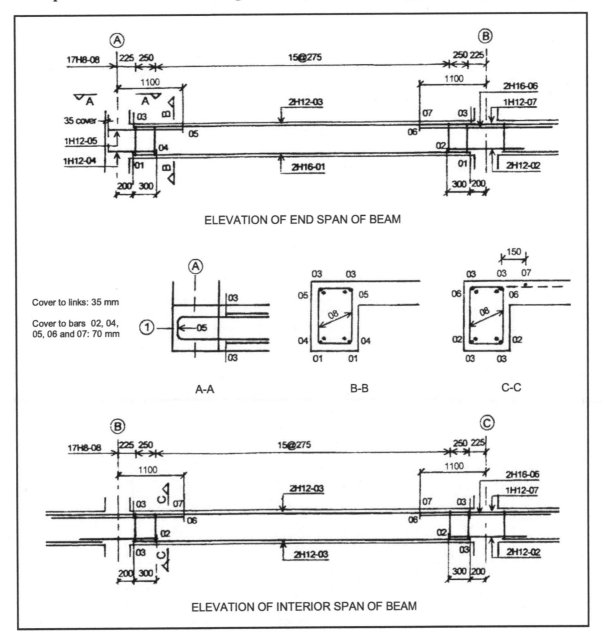

ELEVATION OF END SPAN OF BEAM

A-A   B-B   C-C

Cover to links: 35 mm

Cover to bars 02, 04, 05, 06 and 07: 70 mm

ELEVATION OF INTERIOR SPAN OF BEAM

Bar Marks	Commentary on Bar Arrangement (Drawing 5)
01	Bars in corners of links curtailed 50 mm from column face at each end to avoid clashing with column bars.
02	Loose bars positioned inside column bars and above bars in main beam. Bars lap 300 mm with bars 01, and bars 03 in adjacent span.
03	Bars in corners of links curtailed 50 mm from column face at each end to avoid clashing with column bars.
04	Loose U-bar, shape code 21, positioned inside column bars and above bottom leg of U-bars in main beam. Bar laps 300 mm with bars 01.
05	Loose U-bar, shape code 21, positioned inside column bars and below top leg of U-bars in main beam. Bar laps 900 mm with bars 03.
06	Loose bars positioned inside column bars and under bars in main beam. Bars lap 900 mm with bars 03.
07	Bar positioned in slab as shown in section C-C, and extending 1100 mm either side of column centreline.
08	Closed links, shape code 51, with 35 mm nominal cover.

Example 1   **Calculation Sheet 14**

Reference	CALCULATIONS	OUTPUT
	ACTIONS ON COLUMNS	

ACTIONS ON COLUMNS

For the columns on line B, the sub-frame analysis results shown on calculation sheets 5–7 give beam shears and column moments for three load cases, and apply at 2nd, 3rd and 4th floor levels. For simplicity, the same values will be used at lower floor levels, even though the storey heights result in sub-frame dimensions that are slightly different. At roof level, the sub-frame and the loading are significantly different, and another analysis is required. Loading details are as follows:

Characteristic loading for roof slab:

Slab and finishes: $(3.75 + 1.5) = 5.25$ kN/m^2 Imposed: 0.6 kN/m^2

Design ultimate load for roof slab: $n = 1.25 \times 5.25 + 1.5 \times 0.6 = 7.5$ kN/m^2

Design ultimate loads for first interior roof beam:

$1.1 \times 4.8 \times 7.5 + 2.8 = 42.4$ kN/m (max), 37.6 kN/m (min)

Loads per storey due to the self-weight of the columns:

Columns up to 1st floor: $1.25 \times 0.3 \times 0.3 \times 25 \times 3.55 = 10.0$ kN

Columns above 1st floor: $1.25 \times 0.3 \times 0.3 \times 25 \times 3.05 = 8.6$ kN

Reference: BS EN 1991-1-1 6.3.1.2 NA.2.

A reduction may be made in the total imposed floor load, according to the number of storeys being supported at the level considered. For up to five storeys, this load may be multiplied by $\alpha_n = 1.1 - n/10$, where n is the number of storeys.

EXTERNAL COLUMN B1

At each level, the load applied is the sum of the shear force for the main beam at line 1, and the uniform load only on the edge beam. Thus, at each floor, the load from the edge beam is $F = 43.5$ kN. At the roof, the load due to the self-weight of the beam and parapet is: $F = 1.25 \times (0.3 \times 0.3 + 0.15 \times 1.0) \times 25 \times 4.8 = 36.0$ kN

Reference: 5.8.8.3 (3)

The maximum moment and maximum coexistent load occur when load case 2 is applied at all levels (see below). Maximum moment and minimum coexistent load occur when load case 2 is applied at the level considered, and $1.0G_k$ is applied at levels above. This arrangement can be critical for values of $N_{Ed} < N_{bal}$ where

$N_{bal} = 0.4A_c f_{cd} = 0.4 \times 300 \times 300 \times 0.85 \times 32/1.5 \times 10^{-3} = 653$ kN

Values of axial load N (kN) and bending moment M (kNm)					
Loading	$1.25G_k + 1.5Q_k$			$1.5Q_k$	
Load case	1		2		1	2
Member	N	M	N	M	N	N
Roof beams	140.8	41.5	142.8	43.5	(11.8)	
Column	<u>8.6</u>		<u>8.6</u>			
	149.4	48.9	151.4	56.4		
4th floor beams	<u>217.7</u>		<u>228.9</u>		81.8	93.0
	367.1	48.9	380.3	56.4		
Column	<u>8.6</u>		<u>8.6</u>			
	375.7	48.9	388.9	56.4		
3rd floor beams	<u>217.7</u>		<u>228.9</u>		<u>81.8</u>	<u>93.0</u>
	593.4	48.9	617.8	56.4	163.6	186.0
Column	<u>8.6</u>		<u>8.6</u>			
	602.0	48.9	626.4	56.4		
2nd floor beams	<u>217.7</u>		<u>228.9</u>		<u>81.8</u>	<u>93.0</u>
	819.7	48.9	855.3	56.4	245.4	279.0
Column	<u>8.6</u>		<u>8.6</u>			
	828.3	48.9	863.9	56.4		
1st floor beams	<u>217.7</u>		<u>228.9</u>		<u>81.8</u>	<u>93.0</u>
	1046.0	48.9	1092.8	56.4	327.2	372.0
Column	<u>10.0</u>		<u>10.0</u>			
	1056.0	48.9	1102.8	56.4		
Grd. floor beam	<u>217.7</u>		<u>228.9</u>		<u>81.8</u>	<u>93.0</u>
Basement wall	1273.7	48.9	1331.7	56.4	409.0	465.0

Example 1   Calculation Sheet 15

Reference	CALCULATIONS	OUTPUT
	For the storey from ground to 1st floor, with load case 2 at ground floor level:	

For the storey from ground to 1st floor, with load case 2 at ground floor level:

$M_{bot} = 56.4$ kN m, $M_{top} = -0.5 M_{bot} = -28.2$ kN m

With load case 2 at levels above: $N_{Ed} = 1102.8 - 0.3 \times 372.0 = 991$ kN (max)

With $1.0 G_k$ at levels above: $N_{Ed} = [1046 - (327.2 + 11.8)]/1.25 = 574$ kN (min)

6.1 (4)   Minimum total design moment, with $e_0 = h/30 = 300/30 \geq 20$ mm:

$M_{min} = N_{Ed} e_0 = 991 \times 0.02 = 19.8$ kN m

Effective length and slenderness

5.8.3.2   Effective length for braced members in regular frames is given by

$l_0 = 0.5l\sqrt{(1+\alpha_1)(1+\alpha_2)}$ where, at joints 1 and 2,

$\alpha = k/(0.45 + k)$ and k is the relative flexibility of the joint

Guidance in assessing the values to use in this relationship is given in document PD 6687. A simplified method given in Concise Eurocode 2 will be used here. For condition 1 (monolithic connection to beams at least as deep as the overall depth of the column) at both top and bottom of column,

$l_0 = 0.75l = 0.75 \times 3.55 = 2.66$ m (for storeys above 1st floor, $l_0 = 2.29$ m)

5.2 (9)   First order moment from imperfections (simplified procedure):

$M_i = Nl_0/400 = 991 \times 2.66/400 = 6.6$ kN m

First order moments, including the effect of imperfections:

$M_{01} = -28.2 + 6.6 = -21.6$ kN m, $M_{02} = 56.4 + 6.6 = 63.0$ kN m

Radius of gyration of uncracked concrete section, $i = h/\sqrt{12} = 0.087$ m

5.8.3.2 (1)   Slenderness ratio $\lambda = l_0/i = 2.66/0.087 = 30.6$

5.8.3.1 (1)   Slenderness criterion, $\lambda_{lim} = 20(A \times B \times C)/\sqrt{n}$ where:

$n = N/A_c f_{cd} = N/(300^2 \times 0.85 \times 32/1.5) = 991/1632 = 0.61$

Taking $A = 0.7$, $B = 1.1$ and $C = 1.7 - M_{01}/M_{02} = 1.7 + 21.6/63 = 2.0$

$\lambda_{lim} = 20 \times 0.7 \times 1.1 \times 2.0/\sqrt{0.61} = 39.4 \ (>\lambda = 30.6)$

Since $\lambda < \lambda_{lim}$, second-order effects may be ignored and $M_{Ed} = M_{02} \ (>M_{min})$

Design of cross-section

Allowing 35 mm nominal cover, 8 mm links and 20 mm longitudinal bars, results in $d = 300 - (35 + 8 + 20/2) = 247$ mm, $d/h = 247/300 = 0.82$. Reinforcement can be determined from the design chart in Table A2 ($d/h = 0.8$) as follows:

$N_{Ed}/bhf_{ck} = (991 \text{ or } 583) \times 10^3/(300 \times 300 \times 32) = 0.35 \text{ or } 0.20$

$M_{Ed}/bh^2 f_{ck} = 63.0 \times 10^6/(300 \times 300^2 \times 32) = 0.073$

$A_s f_{yk}/bhf_{ck} = 0.03$, which gives $A_s = 0.03 \times 300 \times 300 \times 32/500 = 173$ mm^2

9.5.2 (2)   Minimum amount of longitudinal reinforcement:

$A_{s,min} = 0.1 N_{Ed}/f_{yd} = 0.1 \times 991 \times 10^3/(500/1.15) = 228$ mm^2 (4H12)

$\geq 0.002 A_c = 0.002 \times 300 \times 300 = 180$ mm^2

Similar calculations can be performed for the upper storeys, and all the results are summarised below:

Storey	N_{Ed} (kN max/min)	M_{Ed} (kN m)	$\dfrac{N_{Ed}}{bhf_{ck}}$	$\dfrac{M_{Ed}}{bh^2 f_{ck}}$	$\dfrac{A_s f_{yk}}{bhf_{ck}}$	A_s (mm^2)
4th floor–roof	151/110	57.3	0.05/0.04	0.067	0.17	980
3rd–4th floor	389/226	58.7	0.14/0.08	0.068	0.13	749
2nd–3rd floor	608/341	59.9	0.21/0.12	0.070	0.09	519
1st–2nd floor	808/457	61.0	0.28/0.16	0.071	0.06	346
Grd–1st floor	991/574	63.0	0.35/0.20	0.073	0.03	228

A reasonable arrangement, allowing for laps at the bottom of each storey, would be to provide 4H16 for the bottom three storeys and 4H20 for the top two storeys.

OUTPUT:

4H16 (Grd–3rd floor)
4H20 (3rd floor–roof)

Example 1   Calculation Sheet 16

Reference	CALCULATIONS	OUTPUT
PD 6687 BS EN 1990 A1.3.2 Table NA.A1.3	**Tying requirements** The UK Building Regulations require vertical ties for buildings within Classes 2B and 3, as defined in Approved Document A—structure. Class 2B includes offices greater than four storeys but not exceeding 15 storeys, for which: All columns and walls carrying vertical load should be tied continuously from the lowest to the highest level. The tie should be capable of carrying a tensile force equal to the load likely to be received by the column or wall from any one storey under the accidental design situation. The accidental design load is taken as: $G_k + \psi_1 Q_k$ where, for shopping areas, $\psi_1 = 0.7$ (Table 1.1) For the slab, accidental design load $= 5.0 + 0.7 \times 4.0 = 7.8$ kN/m² (max), 5.0 kN/m² (min) For the first interior main beam, accidental design load $= 1.1 \times 4.8 \times 7.8 + 0.3 \times 0.3 \times 25 = 43.5$ kN/m (max), 28.7 kN/m² (min) For the column, approximate accidental design load for load case 2 on main beam (calculation sheet 7), plus edge beam and walling (calculation sheet 12): $N_{Ad} = (43.5/67.5) \times 185.4 + 43.5/1.25 = 154.3$ kN Minimum area of reinforcement required with $\sigma_s = 500$ MPa, $A_{s,min} = 154.3 \times 10^3/500 = 309$ mm² (4H12 sufficient)	

INTERNAL COLUMN B2

The load from the main beam is the total shear force at line 2.

Values of axial load N (kN) and bending moment M (kNm)						
Loading	$1.25G_k + 1.5Q_k$				$1.5Q_k$	
Load case	1		3		1	3
Member	N	M	N	M	N	N
Roof beam	302.2	4.0	287.2	9.6	(33.9)	
Column	8.6		8.6			
	310.8	8.5	295.8	34.6		
4th floor beam	473.8		376.6		222.5	125.3
	784.6	8.5	672.4	34.6		
Column	8.6		8.6			
	793.2	8.5	681.0	34.6		
3rd floor beam	473.8		376.6		222.5	125.3
	1267.0	8.5	1057.6	34.6	445.0	250.6
Column	8.6		8.6			
	1275.6	8.5	1066.2	34.6		
2nd floor beam	473.8		376.6		222.5	125.3
	1749.4	8.5	1442.8	34.6	667.5	375.9
Column	8.6		8.6			
	1758.0	8.5	1451.4	34.6		
1st floor beam	473.8		376.6		222.5	125.3
	2231.8	8.5	1828.0	34.6	890.0	501.2
Column	10.0		10.0			
	2241.8	8.5	1838.0	34.6		
Grd. floor beam	473.8		376.6		222.5	125.3
	2715.6	8.5	2214.6	34.6	1112.5	626.5
Column	10.0		10.0			
Foundation	2725.6		2224.6			

The maximum moment occurs when load case 3 is applied at the level considered. Maximum coexistent load occurs when load case 1 is applied at levels above, and minimum coexistent load occurs when $1.0G_k$ is applied at levels above. The latter arrangement can be critical for values of $N_{Ed} < 653$ kN (see calculation sheet 14). The maximum load with a smaller coexistent moment results when load case 1 is applied at all levels.

Example 1   **Calculation Sheet 17**

Reference	CALCULATIONS	OUTPUT
	For the basement storey with load case 1 at all levels:	
	$N_{Ed} = 2725.6 - 0.4 \times 1112.5 = 2281$ kN	
6.1 (4)	Minimum total design moment, with $e_0 = h/30 = 300/30 \geq 20$ mm:	
	$M_{min} = N_{Ed} e_0 = 2281 \times 0.02 = 45.6$ kNm	
	For the basement storey with load case 3 at ground floor level:	
	$M_{top} = 34.6$ kNm, $M_{bot} = -0.5 M_{top} = -17.3$ kN m	
	$N_{Ed} = 376.6 + 2241.8 - 0.4 \times (125.3 + 890.0) = 2212$ kN (max)	
	$N_{Ed} = 376.6 + [2241.8 - (890.0 + 33.9)]/1.25 = 1431$ kN (min)	
	Effective length and slenderness	
	As for column B1, $l_0 = 2.66$ m and $\lambda = 30.6$	
5.2 (9)	First-order moment from imperfections (with load case 3 at ground floor level):	
	$M_i = N l_0 / 400 = 2212 \times 2.66/400 = 14.7$ kN m	
	First-order moments, including the effect of imperfections:	
	$M_{01} = -17.3 + 14.7 = -2.6$ kN m, $M_{02} = 34.6 + 14.7 = 49.3$ kN m	
5.8.3.1 (1)	Slenderness criterion: $\lambda_{lim} = 20(A \times B \times C)/\sqrt{n}$ where	
	$n = N/A_c f_{cd} = 2212/1632 = 1.36$ and	
	$A = 0.7$, $B = (1 + 2\omega)^{0.5}$, $C = 1.7 + 2.6/49.3 = 1.75$ where	
	$\omega = A_s f_{yd}/A_c f_{cd} = A_s \times 500/(1.15 \times 1632 \times 10^3) = A_s/3754$	
	Assuming 4H32, $\omega = 3217/3754 = 0.85$, $B = (1 + 2\omega)^{0.5} = 1.64$ and	
	$\lambda_{lim} = 20 \times 0.7 \times 1.64 \times 1.75/\sqrt{1.36} = 34.4$ $(>\lambda = 30.6)$	
	Since $\lambda < \lambda_{lim}$, second-order effects may be ignored and $M_{Ed} = M_{02}$ $(>M_{min})$	
	Design of cross-section	
	Although the nominal cover needed for durability is 25 mm, this will be increased to 35 mm to ensure a minimum cover to the H32 bars not less than the bar size.	
	Allowing 35 mm nominal cover, 8 mm links and 32 mm longitudinal bars, results in $d = 300 - (35 + 8 + 32/2) = 240$ mm say. Reinforcement can be determined from the design chart in Table A2 $(d/h = 0.8)$ as follows:	
	$N_{Ed}/bh f_{ck} = (2212$ or $1431) \times 10^3/(300 \times 300 \times 32) = 0.77$ or 0.50	
	$M_{Ed}/bh^2 f_{ck} = 49.3 \times 10^6/(300 \times 300^2 \times 32) = 0.057$	
	$A_s f_{yk}/bh f_{ck} = 0.42$, which gives $A_s = 0.42 \times 300 \times 300 \times 32/500 = 2419$ mm^2	
	Similar calculations can be performed for the upper storeys, and all the results are summarised below:	

Storey	N_{Ed} (kN max/min)	M_{Ed} (kN m)	$\dfrac{N_{Ed}}{bh f_{ck}}$	$\dfrac{M_{Ed}}{bh^2 f_{ck}}$	$\dfrac{A_s f_{yk}}{bh f_{ck}}$	A_s (mm^2)
4th floor–roof	311/221	36.7	0.11/0.08	0.043	0.04	231
3rd–4th floor	793/429	39.9	0.28/0.15	0.046	0	min
2nd–3rd floor	1231/637	42.8	0.43/0.22	0.050	0	min
1st–2nd floor	1625/845	45.4	0.57/0.29	0.053	0.17	980
Grd–1st floor	1975/1054	47.7	0.69/0.37	0.055	0.32	1844
Basement	2212/1431	49.3	0.77/0.50	0.057	0.42	2419

A reasonable arrangement would be to provide 4H32 for the bottom storey, 4H25 for the next storey, 4H20 for the next storey, and 4H16 for the top three storeys.

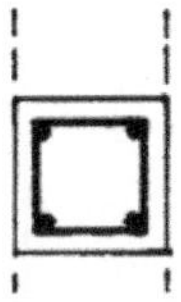

Tying requirements

Accidental design load (see calculation sheet 16) for load case 1 on main beam:

$N_{Ad} = (43.5/67.5) \times 473.8 = 305.4$ kN

$A_{s,min} = 305.4 \times 10^3/500 = 611$ mm^2 (4H16 sufficient)

OUTPUT column:

4H32 (Bottom storey)
4H25 (Grd–1st floor)
4H20 (1st–2nd floor)
4H16 (2nd floor–roof)

Example 1   **Calculation Sheet 18**

Reference	CALCULATIONS	OUTPUT
	CORNER COLUMN A1	
	From the calculations for column B1, it can be seen that the most critical condition occurs at the bottom of the top storey, with minimum load $1.0G_k$ at roof level and maximum design load at the 4th floor level.	
	Maximum loading for beams at 4th floor level:	

	Beam on line A kN/m	Beam on line 1 kN
Slab	$0.4 \times 4.8 \times 12.25 = 23.5$	$31.3 + 30.0 = 61.3$
Beam and walling	$43.5/4.8 = \underline{9.1}$	$= \underline{43.5}$
	32.6	104.8

<table>
<tr><td rowspan="21">5.2 (9)

5.8.9 (2)

5.8.9 (4)</td><td>Column moments for the frame on line A can be estimated from the results for the sub-frame on line B (see calculation sheets 5–7). Thus, for load case 2:</td><td></td></tr>
<tr><td>$M_z = (32.6/67.5) \times 56.4 = 27.3 \text{ kN m}$</td><td></td></tr>
<tr><td>Column moments for the frame on line 1 can be estimated on the assumption that the column and beam ends, remote from the junction, are fixed and that the beam possesses half its actual stiffness. Thus (see calculation sheet 5):</td><td></td></tr>
<tr><td>$K_{b,end} = 0.5 \times (2.278 \times 10^9)/4800 = 0.237 \times 10^6 \text{ mm}^3$</td><td></td></tr>
<tr><td>$K_{c,upper} = K_{c,lpwer} = 0.193 \times 10^6 \text{ mm}^3$</td><td></td></tr>
<tr><td>Beam fixed-end moment and resulting column moments:</td><td></td></tr>
<tr><td>$M_{FEM} = (0.104 \times 61.3 + 0.083 \times 43.5) \times 4.8 = 48.0 \text{ kN m}$</td><td></td></tr>
<tr><td>$M_y = [0.193/(2 \times 0.193 + 0.237)] \times 48.0 = 14.9 \text{ kN m}$</td><td></td></tr>
<tr><td>Minimum loading for beams at roof level:</td><td></td></tr>
<tr><td>

	Beam on line A kN/m	Beam on line 1 kN
Slab	$0.4 \times 4.8 \times 5.25 = 12.0$	(included in line A)
Beam and parapet	$28.8/4.8 = \underline{6.0}$	$36.0/1.25 = \underline{28.8}$
	18.0	28.8

</td><td></td></tr>
<tr><td>Minimum load at bottom of column (roof beams and weight of column):</td><td></td></tr>
<tr><td>$N_{Ed} = (18.0/67.5) \times 174.2 + 0.45 \times 28.8 + 8.6/1.25 = 66 \text{ kN}$</td><td></td></tr>
<tr><td>First-order moment from imperfections (simplified procedure):</td><td></td></tr>
<tr><td>$M_i = Nl_0/400 = 66 \times 2.66/400 = 0.5 \text{ kN m}$</td><td></td></tr>
<tr><td>Since imperfections need to be taken into account only in the direction where they will have the most unfavourable effect,</td><td></td></tr>
<tr><td>$M_{0z} = 27.3 + 0.5 = 27.8 \text{ kN m}, \ M_{0y} = 14.9 \text{ kN m}$</td><td></td></tr>
<tr><td>**Design of cross-section**</td><td></td></tr>
<tr><td>Assuming 4H16, the design resistance of the column for bending about either axis can be determined from the design chart in Table A2 as follows:</td><td></td></tr>
<tr><td>$A_s f_{yk}/bhf_{ck} = 804 \times 500/(300 \times 300 \times 32) = 0.14$</td><td></td></tr>
<tr><td>$N_{Ed}/bhf_{ck} = 66 \times 10^3/(300 \times 300 \times 32) = 0.023$</td><td></td></tr>
<tr><td>$M_{Ed}/bh^2f_{ck} = 0.059 \ (\text{for } d/h = 0.8)$</td><td></td></tr>
</table>

	Thus, $M_{Rdz} = M_{Rdy} = 0.059 \times 300 \times 300^2 \times 32 \times 10^{-6} = 51.0 \text{ kN m}$	
5.8.9 (4)	In the absence of a precise design for biaxial bending, a simplified criterion check for compliance may be made as follows:	
	$N_{Rd} = A_c f_{cd} + A_s f_{yd} = (300^2 \times 0.85 \times 32/1.5 + 804 \times 500/1.15) \times 10^{-3} = 1981 \text{ kN}$	
	$N_{Ed}/N_{Rd} = 66/1981 = 0.03.$ For values of $N_{Ed}/N_{Rd} \leq 0.1$, exponent $a = 1.0.$	
	$(M_{Edz}/M_{Rdz})^a + (M_{Edy}/M_{Rdy})^a = (27.8/51.0)^{1.0} + (14.9/51.0)^{1.0} = 0.84 \ (\leq 1.0)$	
	Since the criterion is met at the most critical condition, a reasonable arrangement would be to provide 4H16 for all storeys.	
	Tying requirements	
	From the calculations for column B1, it can be seen that 4H12 would be sufficient.	4H16 (Grd floor–roof)

Example 1   **Calculation Sheet 19**

Reference	CALCULATIONS	OUTPUT
	FIRE-RESISTANCE	
BS EN 1992-1-2	In Eurocode 2: Part 1.2, prescriptive rules for fire resistance are given for braced columns only. The tabulated data in Section 5.3 is of limited application, including restrictions on the first-order eccentricity under fire conditions. Limiting values of $e \leq 0.15b$ for Method A, and $e \leq 0.25b$ for Method B, apply. Further tabulated data, based on simplified calculation methods in Annex B, is given in Annex C for values of $e = 0.025b$ (≥ 10 mm), $0.25b$ (≤ 100 mm) and $0.5b$ (≤ 200 mm). In the following calculations, the effective length and first-order eccentricity under fire conditions are assumed to be equal to those at normal temperature.	
	Column B1	
	With 35 mm cover, 8 mm links and 16 mm longitudinal bars, axis distance of the main bars, $a = 35 + 8 + 16/2 = 50$ mm say	
	For the storey from ground to first floor, from calculation sheet 15:	
	$N_{Ed} = 991$ kN, $M_{Ed} = 63$ kN m, $e = 63/991 = 0.064$ m, $e/b = 0.064/0.3 = 0.21$,	
	$\lambda = 30$ say, $\omega = A_s f_{yd}/A_c f_{cd} = 804 \times (500/1.15)/(300^2 \times 0.85 \times 32/1.5) = 0.21$,	
	$n = N_{Ed}/(1 + \omega)A_c f_{cd} = 991 \times 10^3/(1.21 \times 300^2 \times 0.85 \times 32/1.5) = 0.50$	
BS EN 1992-1-2 5.3 Annex C	Since $0.15 < e/b < 0.25$, method A is not applicable but method B can be used. From Table 5.2b, for R 60 with $\omega = 0.1$ and $n = 0.5$, the minimum dimensions for column width/axis distance are 300/40. The same results can be obtained for the next storey. Annex C can be used with some difficulty for all but the top storey. For R 60 with $\lambda = 30$, $n = 0.3$ and $e/b = 0.5$, values of 200/40 and 500/50 are obtained with $\omega = 0.5$ and 0.1, respectively. Thus, with $\omega = 0.33$, values of 300/40 can be reasonably deduced, as indicated below.	
	<table><tr><th>Storey</th><th>N_{Ed}</th><th>M_{Ed}</th><th>e/b</th><th>ω</th><th>n</th><th>Column</th></tr><tr><td>4th floor–roof</td><td>151</td><td>58</td><td>1.28</td><td>0.33</td><td>0.07</td><td>NA</td></tr><tr><td>3rd–4th floor</td><td>389</td><td>59</td><td>0.50</td><td>0.33</td><td>0.18</td><td>300/40</td></tr><tr><td>2nd–3rd floor</td><td>608</td><td>61</td><td>0.33</td><td>0.21</td><td>0.30</td><td>300/40</td></tr><tr><td>1st–2nd floor</td><td>808</td><td>62</td><td>0.25</td><td>0.21</td><td>0.40</td><td>300/40</td></tr><tr><td>Grd–1st floor</td><td>991</td><td>63</td><td>0.21</td><td>0.21</td><td>0.50</td><td>300/40</td></tr></table>	
	For the top storey, if the column is considered as a beam and $N_{Ed} < N_{bal}$ is ignored, $A_s f_{yk}/bhf_{ck} = 0.21$, $A_s = 1210$ mm^2 and 4H20 are still sufficient. Then, for R 60 and $b_{min} = 300$ mm, the required minimum axis distance is 25 mm.	Column OK for 1 h fire resistance
	Column B2	
	With 35 mm cover, 8 mm links and 32 mm longitudinal bars, axis distance of the main bars, $a = 35 + 8 + 32/2 = 60$ mm say.	
	For the basement storey, from calculation sheet 17:	
	$N_{Ed} = 2212$ kN, $M_{Ed} = 50$ kN m, $e = 50/2212 = 0.023$ m, $e/b = 0.023/0.3 = 0.08$	
	$\omega = A_s f_{yd}/A_c f_{cd} = 3217 \times (500/1.15)/(300^2 \times 0.85 \times 32/1.5) = 0.86$	
	$n = N_{Ed}/(1 + \omega)A_c f_{cd} = 2212 \times 10^3/(1.86 \times 300^2 \times 0.85 \times 32/1.5) = 0.73$	
BS EN 1992-1-2 5.3.2 Equation 5.7	Since $e/b = 0.08$ (<0.15), $l_0 = 2.66$ m (<3.0 m) and $A_s/A_c = 0.036$ (<0.4), method A may be used. In this case, for R 90 and $\mu_{fi} = 0.7$, Table 5.2a indicates minimum dimensions for column width/axis distance = 350/53. Method A also provides an equation for R, in which specific values may be used for each parameter.	
	$R = 120[(R_{\eta fi} + R_a + R_l + R_b + R_n)/120]^{1.8}$ where, with $\mu_{fi} = n = 0.73$,	
	$R_{\eta fi} = 83(1 - \mu_{fi}) = 83 \times 0.27 = 22.4$, $R_a = 1.6 (a - 30) = 1.6 \times (60 - 30) = 48$	
	$R_l = 9.6 (5 - l_0) = 9.6 \times (5 - 2.66) = 22.4$, $R_b = 0.09b' = 0.09 \times 300 = 27$, $R_n = 0$	Basement storey OK for 1.5 h fire resistance
	Hence, $R = 120[(22.4 + 48 + 22.4 + 27 + 0)/120]^{1.8} = 120$ min (>1.5 h)	
	(Note: Minimum axis distance is $a = 50$ mm, which gives $R_a = 32$ and $R = 90$ min)	
	For the storeys above the basement, a fire resistance of 1 h is sufficient. Also, for the storeys above 1st floor, $l_0 = 2.29$ m. Using the equation for all but the top storey, even though $e/b = 0.17 > 0.15$ in one case, yields the following results:	

Example 1   **Calculation Sheet 20**

Reference	CALCULATIONS	OUTPUT

Storey	N_{Ed}	M_{Ed}	e/b	ω	n	R	Column
4th floor–roof	311	37	0.40	0.21	0.16	60	–
3rd–4th floor	793	40	0.17	0.21	0.40	60	300/25
2nd–3rd floor	1231	43	0.12	0.21	0.62	60	300/30
1st–2nd floor	1625	46	0.10	0.33	0.75	60	300/35
Grd–1st floor	1975	48	0.08	0.52	0.80	60	300/40
Basement	2212	50	0.08	0.86	0.73	90	300/50

For the top storey, if the column is considered as a beam and $N_{Ed} < N_{bal}$ is ignored, $A_s f_{yk}/bhf_{ck} = 0.13$, $A_s = 749$ mm^2 and 4H16 are still sufficient. Then, for R 60 and $b_{min} = 300$ mm, the required minimum axis distance is 25 mm.

OUTPUT: All storeys above basement are OK for 1 h fire resistance

Column A1

From the calculations for column B1, it is reasonable to conclude that column A1 is satisfactory for a fire resistance of 1 h.

OUTPUT: Column OK for 1 h fire resistance

Bar marks	Commentary on Bar Arrangement (Drawing 6)
	Details shown are for columns B1 (top of basement wall to 1st floor level) and B2 (top of basement floor to 1st floor level). Details for column A1 will be similar to column B1, except that two U-bars (bar mark 03) will be provided to restrain the outer longitudinal bars in each direction.
01	Bars (shape code 26) bearing on 75 mm kicker and cranked to fit alongside bars projecting from basement wall. Projection of starter bars = $1.5 \times 35 \times 16 + 75 = 925$ mm say. Crank to begin 75 mm from end of starter bar. Length of crank = 13ϕ and overall offset dimension = 2ϕ. Since 4H16 are also sufficient at the next level, projection of bars above first floor = 925 mm.
02, 06	Closed links (shape code 51), with 35 mm nominal cover, starting above kicker and stopping below beams at next floor level. The following requirements apply (see clause 9.5.3):
	Minimum diameter of links = 0.25 × maximum diameter of longitudinal bars ≥ 6 mm. Spacing should not exceed least of 20 × minimum diameter of longitudinal bars, lesser dimension of column cross-section or 400 mm. In regions within a distance equal to the larger dimension of column cross-section above or below a beam or slab, link spacing should not exceed 0.6 times the preceding values. Thus, the maximum link spacing is 300 mm generally, and 180 mm for a distance of 300 mm above or below a beam or slab.
	In the lap zones of the main bars, where the diameter of lapped bars $\phi \geq 20$ mm, transverse bars of total area not less than area of one lapped bar should be provided within outer thirds of lap zone (see clause 8.7.4.1). Thus, allowing for $A_{s,req} < A_{s,prov}$ (see calculation sheet 17), total area of transverse bars for the full lap zone should be not less than area of one lapped bar multiplied by 1.5 $A_{s,req}/A_{s,prov}$. For column B2, the following requirements apply:
	Fdn–1st floor: $A_{st} = 1.5 \times 2419/3217 \times 804 = 907$ mm^2 (12H10)
	Grd–1st floor: $A_{st} = 1.5 \times 1844/1963 \times 491 = 692$ mm^2 (14H8)
03	Open U-bar (shape code 21), instead of closed link, to restrain outer longitudinal bars.
04	Bars (similar to bar mark 01) cranked to fit alongside bars projecting from foundation. Projection of starter bars = $1.5 \times 35 \times 32 \times 2419/3217 + 75 = 1400$ mm say. Since 4H25 are sufficient at the next level, projection of bars above ground floor level = $1.5 \times 35 \times 25 + 75 = 1400$ mm say.
05	Bars (similar to bar mark 04). Since 4H20 are sufficient at the next level, projection of bars above 1st floor level = $1.5 \times 35 \times 20 + 75 = 1125$ mm.

Example 1: Reinforcement in Columns B1 and B2   **Drawing 6**

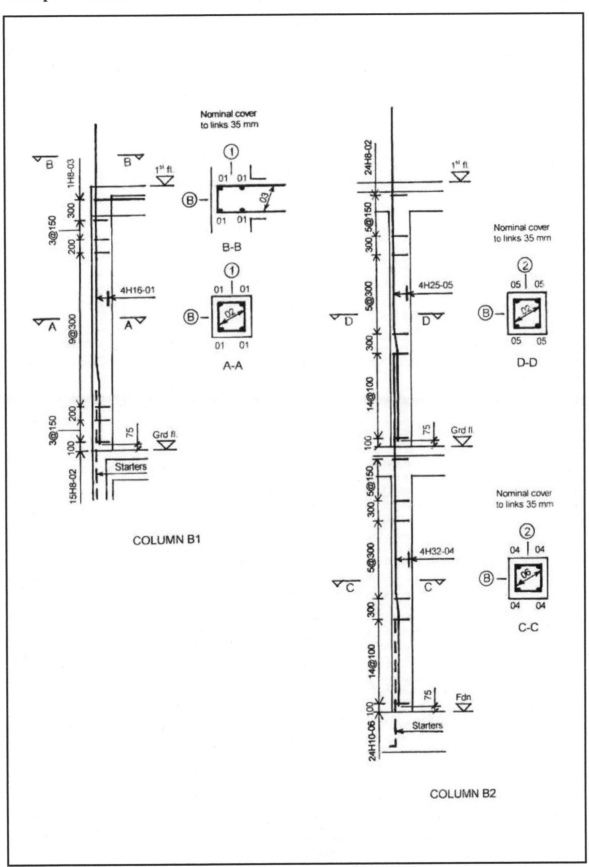

Example 1   **Calculation Sheet 21**

Reference	CALCULATIONS	OUTPUT
	FLAT SLAB FLOOR CONSTRUCTION	
	FLOOR SLABS (GROUND AND UPPER FLOORS)	
	For a solid slab without drops and a characteristic imposed load ≤ 5 kN/m^2, try a span/effective depth ratio of 36. Allowing for 25 mm cover and 16 mm bars gives an estimated thickness $= 7200/36 + (25 + 16/2) = 240$ mm say.	$h = 240$ mm
	Fire resistance	
BS EN 1992-1-2 5.7.4 (1) Tables 5.8 and 5.9	Allowing for the design to be based on more than 15% redistribution of moment, the axis distance should be taken as for a one-way simply supported slab. For the ground floor (minimum fire period 1.5 h), the required minimum dimensions (Table A2) are: Slab thickness: 200 mm Axis distance (to centre of lower bars): 30 mm Since the cover required for durability is 25 mm, the axis distance is sufficient.	Sufficient for 1.5 h fire period
	Loading	
	Details of the characteristic imposed loads, and the action combination options for the ultimate limit state, are given in calculation sheet 1.	
BS EN 1990 6.4.3.2 (3) Table A1.2(B)	<table><tr><td>Permanent load</td><td>kN/m^2</td><td>Variable load</td><td>kN/m^2</td></tr><tr><td>Self-weight of slab 0.240×25</td><td>$= 6.00$</td><td>Imposed</td><td>$= 2.5$</td></tr><tr><td>Finishes and services</td><td>$= \underline{1.25}$</td><td>Partitions</td><td>$= \underline{1.5}$</td></tr><tr><td></td><td>$g_k = \underline{7.25}$</td><td></td><td>$q_k = \underline{4.0}$</td></tr></table> Design ultimate load, $n = \xi\gamma_G G_k + \gamma_Q Q_k = 1.25 \times 7.25 + 1.5 \times 4.0 = 15.0$ kN/m^2	$g_k = 7.25$ kN/m^2 $q_k = 4.0$ kN/m^2 $n = 15.0$ kN/m^2
	Analysis	
	Several methods of analysis are available, including the use of simplified moment coefficients, equivalent frame analysis, finite element analysis, grillage analysis and yield-line methods. Equivalent frame analysis will be used in this example.	
	The structure is divided, into two orthogonal directions, into frames consisting of columns and strips of slab. For vertical loading, the effective stiffness of the slab strips may be based on the width between centrelines of panels. Each frame will be simplified to a series of sub-frames, consisting of the slab at one level together with the columns above and below the slab.	
	The equivalent frame model overestimates the restraint provided at edge columns by assuming a continuous line support. A reasonable allowance for this inaccuracy is to take the stiffness of the edge columns as 0.7 times the actual value (a similar approach is taken in The Concrete Society Technical Report No. 64).	
	For large columns ($h >$ span/10 or 500 mm), an additional moment transfer due to the different shear forces at the opposite faces of the column should be considered. The Concrete Society Technical Report No. 64 suggests the additional moment is taken as $h/3 \times$ (difference in shear forces). In this case, hogging moments in the slab greater than those at a distance $h/3$ from the centre of the column may be ignored. The column size in this example does not warrant such modifications.	
I.1.2 (5) Figure I.2	In the absence of edge beams that are adequately designed for torsion, the moment transferred to an edge or corner column should be limited to $M_{t,max} = 0.17b_e\, d^2 f_{ck}$ where b_e is the effective width of the moment transfer strip.	
	For an edge column, $b_e = c + y$ where c is the width of the column, and y is the distance from the innermost face of the column to the edge of the slab. Referring to drawing 2, $y = 800$ mm, which gives $b_e = 400 + 800 = 1200$ mm.	
5.1.3 (1)P Table NA.1	Although a design based on the single load case of all spans carrying the design maximum load is permitted for slabs, the case of alternate spans carrying design maximum load with other spans carrying only design permanent load will also be considered. The analysis of sub-frames where the columns above and below the floor are identical (i.e., 2nd, 3rd and 4th floors) will be shown.	
	An analysis and subsequent slab design for flexure, deflection and cracking will be carried out for the sub-frames on lines B and 2, followed by a design for shear at an internal column (B2) and an edge column (A2).	

Example 1

Reference	CALCULATIONS	OUTPUT
	SUB-FRAME ON LINE B	

Dimensions of sub-frame

Since the sub-frame and the load cases are symmetrical about the centreline, an analysis can be carried out for one half of the sub-frame by taking the stiffness of the central span as half the actual value.

$$I_s = 7200 \times 240^3/12 = 8.29 \times 10^9 \text{ mm}^4 \qquad I_c = 400 \times 400^3/12 = 2.13 \times 10^9 \text{ mm}^4$$

$$K_{s,end} = I_s/6000 = 1.38 \times 10^6 \text{ mm}^3 \qquad K_{s,int} = I_s/7200 = 1.15 \times 10^6 \text{ mm}^3$$

$$K_{c,upper} = K_{s,lower} = I_c/3500 = 0.61 \times 10^6 \text{ mm}^3$$

Distribution factors for unit moment applied at an end joint are:

$$D_s = 1.38/(1.38 + 2 \times 0.7 \times 0.61) = 0.618, \quad D_c = (1 - 0.618)/2 = 0.191$$

Distribution factors for unit moment applied at an interior joint are:

$$D_{s,end} = 1.38/(1.38 + 0.5 \times 1.15 + 2 \times 0.61) = 1.38/3.175 = 0.435$$

$$D_{s,int} = 0.5 \times 1.15/3.175 = 0.181, \quad D_c = 0.61/3.175 = 0.192$$

Maximum and minimum design loads for a full panel width of 7.2 m, assuming shear force coefficients in the orthogonal directional of 0.6 for each span, are:

$$1.2 \times 7.2 \times 15 = 129.6 \text{ kN/m (max)} \qquad 1.2 \times 7.2 \times 9 = 77.8 \text{ kN/m (min)}$$

Fixed-end moments due to maximum load on beams are:

$$M_{end} = 129.6 \times 6^2/12 = 388.8 \text{ kN m} \qquad M_{int} = 129.6 \times 7.2^2/12 = 559.9 \text{ kN m}$$

Fixed-end moments due to minimum load on beams are:

$$M_{end} = 77.8 \times 6^2/12 = 233.4 \text{ kN m} \qquad M_{int} = 77.8 \times 7.2^2/12 = 336.1 \text{ kN m}$$

An analysis similar to that on calculation sheet 6 yields the following results:

Load Case	Location and Member	End Support	End Span	Interior Support	Interior Support	Interior Span
No.	Bending Moment (kN m) in Members for Load Case					
1	Slab	−144.0	262.5	529.4	−550.0	289.8
	Upper column	72.0		10.3		
	Lower column	72.0		10.3		
2	Slab	−164.2	288.7	441.2	−369.6	134.6
	Upper column	82.1		−35.8		
	Lower column	82.1		−35.8		
3	Slab	−66.2	134.7	405.8	−510.6	329.2
	Upper column	33.1		52.4		
	Lower column	33.1		52.4		
No.	Shear Force (kN) in Members for Load Case					
1	Slab	324.6		453.0	466.6	
2	Slab	342.6		435.0	280.1	
3	Slab	176.8		290.0	466.6	

Example 1　　　　　　　　　　　　　　　　　　　**Calculation Sheet 23**

Reference	CALCULATIONS	OUTPUT
	Allowing for some redistribution of moment, the maximum hogging moments in the slab will be taken as 164.2 kN m at the end support for load cases 1 and 2, and 441.2 kN m at the interior support for load cases 1 and 3. As a result, the maximum sagging moment in the end span for load case 1 will be the same as that for load case 2. In the interior span, the maximum sagging moment for load cases 1 and 3 will increase in order to maintain equilibrium, as follows: $M = 129.6 \times 7.2^2 /8 - 441.2 = 398.6$ kN m **Flexural design**	
I.1.2 (3) Figure I.1 Table I.1	The panels should be notionally divided into column and middle strips, and the bending moments for the full panel width apportioned within specified limits. On lines B and E, the width of the column strip will be taken as $7200/2 = 3600$ mm. The hogging moments at the internal columns will be allocated in the proportions: 75% on column strips, 25% on middle strips. The sagging moments in the spans will be allocated in the proportions: 55% on column strips, 45% on middle strips.	Column strip width 3600 mm
	According to the values of $M/bd^2 f_{ck}$ where the strip width $b = 3600$ mm, except at the edge columns where $b_e = 1200$ mm, appropriate values of $z/d \leq 0.95$ and A_s can be determined (Table A1) and suitable bars selected (Table A9).	
	Allowing for 25 mm cover and 16 mm bars in each direction, for the second layer of bars, $d = 240 - (25 + 16 + 16/2) = 190$ mm say. At an edge column, $M_{t,max} = 0.17 b_e d^2 f_{ck} = 0.17 \times 1200 \times 190^2 \times 32 \times 10^{-6} = 235.6$ kN m	$d = 190$ mm
	At the end supports:	
	For the moment transfer strip: $M = 164.2$ kN m (<235.6 kN m) $M/bd^2 f_{ck} = 164.2 \times 10^6 /(1200 \times 190^2 \times 32) = 0.119$　　$z/d = 0.881$ $A_s = M/(0.87 f_{ck} z) = 164.2 \times 10^6 /(0.87 \times 500 \times 0.881 \times 190) = 2255$ mm^2 (12H16)	At end supports, bars in moment transfer strip 12H16 (T)
	At the interior supports:	
	For the column strip: $M = 0.75 \times 441.2 = 330.9$ kN m $M/bd^2 f_{ck} = 330.9 \times 10^6 /(3600 \times 190^2 \times 32) = 0.080$　　$z/d = 0.924$ $A_s = 330.9 \times 10^6 /(0.87 \times 500 \times 0.924 \times 190) = 4333$ mm^2 (22H16)	At interior supports, bars in column strip 22H16 (T)
	For the middle strip: $M = 0.25 \times 441.2 = 110.3$ kN m $M/bd^2 f_{ck} = 110.3 \times 10^6 /(3600 \times 190^2 \times 32) = 0.027$　　$z/d = 0.95$ (max) $A_s = 110.3 \times 10^6 /(0.87 \times 500 \times 0.95 \times 190) = 1405$ mm^2 (18H10)	bars in middle strip 18H10 (T)
5.5 (4) Table NA.1	Redistribution of moment gives $\delta = 441.2/550 = 0.8$. This is allowable without an explicit check on the rotation capacity, provided $x/d \leq (\delta - 0.4) = 0.4$. For the column strip, $z/d = 0.924$ so that $x/d = 2.5(1 - z/d) = 0.190$ (<0.4).	
	In the end spans:	
	For the column strip: $M = 0.55 \times 288.7 = 158.8$ kN m　　$M/bd^2 f_{ck} = 0.038$ $A_s = 158.8 \times 10^6 /(0.87 \times 500 \times 0.95 \times 190) = 2023$ mm^2 (18H12)	In the end spans, bars in column strip 18H12 (B)
	For the middle strip: $M = 0.45 \times 288.7 = 129.9$ kN m　　$M/bd^2 f_{ck} = 0.031$ $A_s = 129.9 \times 10^6 /(0.87 \times 500 \times 0.95 \times 190) = 1655$ mm^2 (15H12)	bars in middle strip (15H12)
	In the interior span:	
	For the column strip: $M = 0.55 \times 398.6 = 219.3$ kN m　　$M/bd^2 f_{ck} = 0.053$ $A_s = 219.3 \times 10^6 /(0.87 \times 500 \times 0.95 \times 190) = 2793$ mm^2 (14H16)	In the interior spans, bars in column strip 14H16 (B)
	For the middle strip: $M = 0.45 \times 398.6 = 179.4$ kN m $A_s = 179.4 \times 10^6 /(0.87 \times 500 \times 0.95 \times 190) = 2285$ mm^2 (12H16)	bars in middle strip (12H16)
	Deflection	
7.4.1 (6)	Deflection requirements may be met by limiting the span-effective depth ratio. For the interior span, the actual span/effective depth = $7200/190 = 38$.	
	The characteristic load for a full panel width is given by $g_k + q_k = 1.2 \times 7.2 \times 11.25 = 97.2$ kN/m	

Example 1   **Calculation Sheet 24**

Reference	CALCULATIONS	OUTPUT
	Taking account of the moment redistribution in the analysis, the service stress in the bottom reinforcement for the full panel width (i.e., total for column and middle strips) under the characteristic load is given approximately by $\sigma_s = (f_{yk}/\gamma_s)(M_{elastic}/M_{design})(A_{s,req}/A_{s,prov})[(g_k + q_k)/n]$ $= (500/1.15)(329.2/398.6)(5078/5228)(97.2/129.6) = 262$ MPa From *Reynolds*, Table 4.21, limiting l/d = basic ratio $\times \alpha_s \times \beta_s$ where: For $100A_s/bd = 100 \times 5078/(7200 \times 190) = 0.37 < 0.1f_{ck}^{0.5} = 0.1 \times 32^{0.5} = 0.56$, $\alpha_s = 0.55 + 0.0075f_{ck}/(100A_s/bd) + 0.005f_{ck}^{0.5}[f_{ck}^{0.5}/(100A_s/bd) - 10]^{1.5}$ $= 0.55 + 0.0075 \times 32/0.37 + 0.005 \times 32^{0.5} \times (32^{0.5}/0.37 - 10)^{1.5} = 1.54$ (Note: The value of α_s can also be obtained from *Reynolds*, Table 4.21, for the given values of $f_{ck} = 32$ MPa and $100A_s/bd = 0.37$) $\beta_s = 310/\sigma_s = 310/262 = 1.18$	
7.4.2 Table NA.5	For a flat slab, basic ratio = 24. Since the span does not exceed 8.5 m, there is no need to modify this value and hence Limiting $l/d = 24 \times \alpha_s \times \beta_s = 24 \times 1.54 \times 1.18 = 43.6$ (> actual $l/d = 38$)	Check complies
	Cracking	
7.3.2 (2)	Minimum area of reinforcement required in tension zone for crack control: $A_{s,min} = k_c k f_{ct,eff} A_{ct}/\sigma_s$ Taking values of $k_c = 0.4$, $k = 1.0$, $f_{ct,eff} = f_{ctm} = 0.3f_{ck}^{(2/3)} = 3.0$ MPa (for general design purposes), $A_{ct} = bh/2$ (for plain concrete section) and $\sigma_s \leq f_{yk} = 500$ MPa $A_{s,min} = 0.4 \times 1.0 \times 3.0 \times 1000 \times (240/2)/500 = 288$ mm²/m (Note: A value for $100A_{s,min}/A_{ct} = 0.24$ can be obtained from *Reynolds*, Table 4.23, giving $A_{s,min} = 0.0024\,A_{ct} = 0.0024 \times 1000 \times 240/2 = 288$ mm²/m)	Minimum tension reinforcement H10-250
BS EN 1990 Table NA.A1.1	The quasi-permanent load, where $\psi_2 = 0.3$ is obtained from the National Annex to the Eurocode (Table 1.1), is given by $g_k + \psi_2 q_k = 1.2 \times 7.2 \times (7.25 + 0.3 \times 4.0) = 73.0$ kN/m	
7.3.3 (2)	The crack width criterion can be satisfied by limiting either the bar size or the bar spacing. For the top of the slab, it is reasonable to ignore any requirement based on appearance, since the surface of the slab will not be visible below the finishes. The service stress in the bottom reinforcement under the quasi-permanent load is given approximately pro rata to the stress under the characteristic load as $\sigma_s = (73.0/97.2) \times 262 = 197$ MPa	
Table 7.2 Table 7.3	For $w_k = 0.3$ mm and $\sigma_s = 200$ MPa, the recommended maximum values are either $\phi_s^* = 25$ mm or bar spacing = 250 mm (*Reynolds*, Table 4.24). Maximum bar size: $\phi_s = \phi_s^*(f_{ct,eff}/2.9)[k_c h_{cr}/2(h-d)]$ $= 25 \times (3.0/2.9) \times [0.4 \times 120/(2 \times 50)] = 12$ mm	Maximum bar size 12 mm or maximum bar spacing 250 mm
	Detailing requirements	
9.3.1.1 (1)	Minimum area of longitudinal tension reinforcement (*Reynolds*, Table 4.28): $A_{s,min} = 0.26(f_{ctm}/f_{yk})b_t d = 0.26 \times (3.0/500)\,b_t d = 0.00156bd \geq 0.0013\,b_t d$ $= 0.00156 \times 1000 \times 190 = 297$ mm²/m	Minimum tension reinforcement H10-250
9.3.1.1 (2)	Minimum area of secondary reinforcement (20% of principal reinforcement): $A_{s,min} = 0.2 \times 2793/3.6 = 155$ mm²/m (for interior span column strip)	
9.3.1.1 (3)	Maximum spacing of principal reinforcement in area of maximum moment: $2h = 480 \leq 250$ mm. Elsewhere: $3h = 720$ mm ≤ 400 mm Maximum spacing of secondary reinforcement in area of maximum moment: $3h = 720 \leq 400$ mm. Elsewhere: $3.5h = 840$ mm ≤ 450 mm	
9.4.1 (2)	At internal columns, top reinforcement in the column strip should be placed with two-thirds of the required area concentrated in the central half of the strip.	

Example 1   Calculation Sheet 25

Reference	CALCULATIONS	OUTPUT
	A suitable arrangement would be to provide 25H16 arranged in groups as follows: 17H16 in the central half, and 4H16 in each outer quarter.	
BS EN 1992-1-2 5.7.4 (2)	For fire ratings of REI90 and above, top reinforcement that is continuous over the full span should be provided in the column strips. The reinforcement area should be at least 20% of that required at the internal columns for the full panel width.	
	$A_{s,min} = 0.2 \times 5738 = 1148$ mm^2 (15H10 required in ground floor slab)	
	An arrangement of 18H10-200 at all levels would be suitable.	
	In the spans, for ease of construction, the bar spacing for the bottom reinforcement will be made uniform for the full panel width. Thus, bars provided in the middle strip will be the same as those required in the column strip. Suitable arrangements would be 36H12-200 in the end spans and 36H16-200 in the interior span.	
9.4.1 (3)	Bottom reinforcement (≥ 2 bars) in each orthogonal direction should be provided to pass through all internal columns.	
9.4.2 Figure 9.9	At the edge columns, the reinforcement required for moment transfer should be placed within the effective width $b_e = 1200$ mm (see calculation sheet 21). An arrangement of 13H16-100 would be suitable.	
	Between the moment transfer zones, at a free edge, nominal reinforcement should be provided in the form of U bars in the vertical plane with legs, of length equal to $0.15 \times$ span, perpendicular to the edge of the slab. Bars parallel to the edge of the slab should be placed in the corners of the U bars and distributed along the top and bottom faces of the slab.	
	Curtailment of longitudinal tension reinforcement	
	In the absence of an elastic moment envelope covering all appropriate load cases, the simplified curtailment rules for one-way continuous slabs will be used in each orthogonal direction (see calculation sheet 4).	
	Tying requirements (see *Reynolds*, Table 4.29)	
	The principal reinforcement in the bottom of each span can be utilised to provide continuous internal ties. With $l_r = 7.2$ m and $F_t = (20 + 4n_0) \leq 60$,	
	$F_{tie,int} = [(g_k + q_k)/7.5](l_r/5)F_t = (11.25/7.5)(7.2/5)(20 + 4 \times 6) = 95$ kN/m	
	Minimum area of reinforcement required with $\sigma_s = 500$ MPa (i.e., $\gamma_s = 1.0$)	
	$A_{s,min} = 95 \times 1000/500 = 190$ mm^2/m < minimum for normal design situations.	
	If all bars are lapped at same position, design lap length (*Reynolds*, Table 4.31):	
	$l_0 = \alpha_6 l_{bd} \geq l_{0,min} = 200$ mm, where $\alpha_6 = 1.5$ for > 50% bars lapped	
	For accidental design situations, $\gamma_c = 1.2$ (*Reynolds*, Table 4.1), and the value of l_{bd} will be taken as that determined for normal design situations. Thus,	Use minimum tension reinforcement with lap length = 550 mm
	$l_0 = \alpha_6 \times (35\phi) = 1.5 \times (35 \times 10) = 550$ mm say	
	SUB-FRAME ON LINE 2	
	The support conditions at C and D are difficult to model. All the walls enclosing the central core of the building are stiffened by return walls, while some walls are also perforated by openings. Since the walls are 200 mm thick and the column dimensions are 400 × 400, the stiffness of a column is equivalent to a wall length equal to 8 × 400 = 3200 mm (approximately half the length of the wall on line C).	
	For ease of construction, the connection between the slab and the walls will utilise bent-out bars cast into the wall, for which proprietary systems are generally used. The bars that can be provided are typically limited to a maximum of H16-150.	
	For ease of analysis, the stiffness of the wall on line C will be considered the same as that of two edge columns on line A. If the resulting moment at C is beyond the capacity of a proprietary reinforcement system, this moment will be reduced and the span moment increased accordingly. An alternative approach suggested in The Concrete Society Technical Report No. 64 would be to consider a pinned support at C. The wall will be designed to resist the maximum slab moment that can be generated by the chosen set of bent-out bars.	

Example 1   **Calculation Sheet 26**

Reference	CALCULATIONS	OUTPUT

Dimensions of simplified sub-frame

The properties of the members are:

$I_s = 6600 \times 240^3/12 = 7.60 \times 10^9 \text{ mm}^4$ $I_c = 400 \times 400^3/12 = 2.13 \times 10^9 \text{ mm}^4$

$K_s = I_s/7200 = 1.06 \times 10^6 \text{ mm}^3$ $K_c = I_c/3500 = 0.61 \text{ mm}^3$

Distribution factors for unit moment applied at an end joint are:

$D_s = 1.06/(1.06 + 2 \times 0.7 \times 0.61) = 0.554$, $D_c = (1 - 0.554)/2 = 0.223$

Distribution factors for unit moment applied at the interior joint are:

$D_s = 1.06/(2 \times 1.06 + 2 \times 0.61) = 1.06/3.34 = 0.318$, $D_c = 0.61/3.34 = 0.182$

Assuming shear force coefficients for the spans in the orthogonal directional of 0.60 for the end span and 0.50 for the interior span, loaded width for sub-frame on line 2 is $(0.6 \times 6.0 + 0.5 \times 7.2) = 7.2$ m. Thus, the maximum and minimum loads for the full panel width are:

$7.2 \times 15 = 108 \text{ kN/m (max)}$ $7.2 \times 9.0 = 64.8 \text{ kN/m (min)}$

Fixed-end moments due to maximum and minimum load on slab are:

$M_{max} = 108 \times 7.2^2/12 = 466.6 \text{ kN m}$, $M_{min} = 64.8 \times 7.2^2/12 = 280 \text{ kN m}$

The following results are obtained for load case 1 (maximum load on both spans), load case 2 (maximum load on span AB, minimum load on span BC) and load case 3 (minimum load on span AB, maximum load on span BC).

Load Case	Location and Member	Support A	Span AB	Support B	Support B	Span BC	Support C
No.		Bending Moment (kN m) in Members for Load Case					Moment
1	Slab	−208.2	311.4	595.9	−595.9	311.4	208.2
	Upper column	104.1		0			−104.1
	Lower column	104.1		0			−104.1
2	Slab	−226.8	332.3	524.3	−429.1	168.0	106.2
	Upper column	113.4		−47.6			−53.1
	Lower column	113.4		−47.6			−53.1
3	Slab	−106.2	168.0	429.1	−524.3	332.3	226.8
	Upper column	53.1		47.6			−113.4
	Lower column	53.1		47.6			−113.4
No.		Shear Force (kN) in Members for Load Case					Shear
1	Slab	335.0		442.6	442.6		335.0
2	Slab	347.5		430.1	278.1		188.5
3	Slab	188.5		278.1	430.1		347.5

Allowing for some redistribution of moment, the maximum hogging moments in the slab will be taken as 226.8 kN m at supports A and C, and 524.3 kN m at support B. As a result, the maximum sagging moments in the spans will remain unchanged for load cases 2 and 3, but will increase to 332.3 kN m for load case 1.

Example 1   **Calculation Sheet 27**

Reference	CALCULATIONS	OUTPUT
	Flexural design The width of the column strip on lines 2 and 3, and the middle strip for the panel between lines 2 and 3, will be taken as $7200/2 = 3600$ mm. Allocation of the panel moments between column and middle strips is specified in calculation sheet 23. Assuming the bars are in the first layer, and allowing for 25 mm cover and 16 mm bars, $d = 240 - (25 + 16/2) = 206$ mm say. At an edge column, $M_{t,max} = 0.17b_e d_2 f_{ck} = 0.17 \times 1200 \times 206^2 \times 32 \times 10^{-6} = 277$ kN m (>226.8 kN m) Calculations similar to those in calculation sheet 23, yield the following results:	$d = 206$ mm

Location	Strip	M(kN m)	$M/bd^2 f_{ck}$	z/d	A_s	Bars
Support A	Transfer	226.8	0.139	0.857	2954	15H16
Span A-B	column	182.8	0.038	0.95	2148	19H12
(and B-C)	middle	149.5	0.031	0.95	1756	16H12
Support B	column	393.2	0.081	0.923	4754	24H16
	middle	131.1	0.027	0.95	1540	20H10
Support C	total	226.8	0.024	0.95	2665	34H10

Reference	CALCULATIONS	OUTPUT
7.4.1 (6) 7.4.2 Table NA.5	**Deflection** The deflection requirements may be met by limiting the span/effective depth ratio, where the actual span/effective depth $= 7200/206 = 35$ From *Reynolds*, Table 4.21, limiting l/d = basic ratio $\times \alpha_s \times \beta_s$ where: For $100A_s/bd = 100 \times 3904/(7200 \times 206) = 0.27 < 0.1f_{ck}^{0.5} = 0.1 \times 32^{0.5} = 0.56$, $\alpha_s = 0.55 + 0.0075f_{ck}/(100A_s/bd) + 0.005f_{ck}^{0.5}[f_{ck}^{0.5}/(100A_s/bd) - 10]^{1.5}$ $\quad = 0.55 + 0.0075 \times 32/0.27 + 0.005 \times 32^{0.5} \times (32^{0.5}/0.27 - 10)^{1.5} = 2.46$ The service stress in the bottom reinforcement for the full panel width under the characteristic load is given approximately by $\sigma_s = (f_{yk}/\gamma_s)(A_{s,req}/A_{s,prov})[(g_k + q_k)/n]$ $\quad = (500/1.15)(3904/3958)(11.25/15) = 322$ MPa $\beta_s = 310/\sigma_s = 310/322 = 0.96$ Limiting $l/d = 24 \times \alpha_s \times \beta_s = 24 \times 2.46 \times 0.96 = 56.6$ ($>$actual $l/d = 35$) **Other considerations** The requirements for cracking, detailing, curtailment and tying are similar to those for the sub-frame on line B (see calculation sheets 24 and 25). The following bar arrangements would be suitable: At support A (transfer strip): 15H16 in zone of width 1200 mm At support B (column strip): 17H16 in central half, 4H16 in each outer quarter (middle strip): 19H10-200 At support C (full panel): 36H10-200 (included with wall reinforcement) In both spans (full panel): 36H12-200	
	SUB-FRAME ON LINE A The loading comprises the shear force from the span in the orthogonal direction plus the load resulting from a 600 mm wide edge strip of slab, and a uniform load of 5 kN/m to cover walling, cladding and windows. Total design loads, assuming a shear force coefficient of 0.4 for the span in the orthogonal direction are: $(0.4 \times 7.2 + 0.6) \times 15.0 + 1.25 \times 5.0 = 58.5$ kN/m (max), 37.6 kN/m (min) Width of strip from edge of slab to centre of panel $= 600 + 3600 = 4200$ mm. The analysis of the sub-frame will be similar to that for the sub-frame on line B, except that the column/slab stiffness ratio is greater. However, since the maximum load is only 45% of that for the sub-frame on line B, it will be sufficient to provide a similar layout of reinforcement in the slab.	

Example 1

Calculation Sheet 28

Reference	CALCULATIONS	OUTPUT
	SHEAR DESIGN	
	Shear stresses in the slab are checked on control perimeters that are constructed so as to minimise the length of the perimeter. The basic control perimeter is taken at distance $2d$ from the column perimeter, where d is taken as the mean effective depth for the reinforcement in two orthogonal directions. At the column perimeter, the maximum shear stress should not exceed $v_{Rd,max}$. With $\alpha_{cc} = 1.0$,	
6.4.5 (3)		
6.2.2 (6)	$v_{Rd,max} = 0.5vf_{cd} = 0.5 \times 0.6(1 - f_{ck}/250) \times (\alpha_{cc} f_{ck}/1.5) = 0.2(1 - f_{ck}/250)f_{ck}$	
6.4.3 (3)	The column reaction is taken as the greater of the values obtained from an analysis in two orthogonal directions. The maximum shear stress on a control perimeter is taken as $\beta \times$ mean shear stress, where β is a factor to be determined. For structures where lateral stability does not depend on frame action, and adjacent span lengths differ by no more than 25%, approximate values for β may be used.	
6.4.3 (6)		
	Column B2	Basic control perimeter for internal column
	For the top reinforcement in the column strips in the two orthogonal directions, the mean effective depth is $d = (190 + 206)/2 = 198$ mm.	
6.4.2 (1) Figure 6.13	The width of the basic control perimeter, taken at distance $2d = 400$ mm say from the face of the column, is $b = c + 2 \times 2d = 400 + 2 \times 400 = 1200$ mm	
6.4.3 (3) Equation 6.43	For an internal rectangular column, where the loading is eccentric about one axis only, $\beta = 1 + 1.8M/Vb$ may be used. Thus, the following values are obtained:	
	From the analyses on calculation sheets 22 and 26, respectively, the results for the sub-frame on line B are critical. From these, the following values are obtained:	
	Load case 1: $V = 453.0 + 466.6 = 919.6$ kN, $M = 10.3 + 10.3 = 20.6$ kN m	
	$\beta = 1 + 1.8 \times 20.6/(919.6 \times 1.2) = 1.04$, $\beta V = 956.4$ kN	
	Load case 3: $V = 290.0 + 466.6 = 756.6$ kN, $M = 52.4 + 52.4 = 104.8$ kN m	
	$\beta = 1 + 1.8 \times 104.8/(756.6 \times 1.2) = 1.21$, $\beta V = 915.5$ kN	
6.4.3 (6) Figure 6.21N	Alternatively, since lateral stability does not depend on frame action, and adjacent span lengths do not differ by more than 25%, $\beta = 1.15$ could be used in both cases. The value of βV determined for load case 1 is used in the following calculations.	
	Thus, at the column perimeter	
6.4.5 (3)	$v = \beta V/u_0 d = 1.04 \times 919.6 \times 10^3 / (4 \times 400 \times 198) = 3.02$ MPa	
Equation 6.53	$v_{Rd,max} = 0.2(1 - f_{ck}/250)f_{ck} = 0.2 \times (1 - 32/250) \times 32 = 5.58$ MPa $(>v)$	
	The length of the basic control perimeter is $u_1 = 4 \times 400 + 2\pi \times 400 = 4113$ mm.	
	$v = \beta V/u_1 d = 1.04 \times 919.6 \times 10^3 / (4113 \times 198) = 1.18$ MPa	
6.4.4 (1)	The punching shear resistance is assessed on the basis of the mean value, for the two orthogonal directions, of the tension reinforcement in a slab width equal to the column width plus $3d$ each side $= 400 + 6 \times 198 = 1600$ mm say. The central half of each column strip is 1800 mm wide and contains 17H16-100. Thus,	
	$\rho_1 = A_{sl}/b_w d = 3418/(1800 \times 198) = 0.0096$ and, with $k = 2.0$ for $d \leq 200$ mm:	
Equation 6.47	$v_{Rd,c} = (0.18k/\gamma_c)(100\rho_1 f_{ck})^{1/3} = (0.18 \times 2.0/1.5)(0.96 \times 32)^{1/3} = 0.75$ MPa	
6.2.2 (1)	$v_{min} = 0.035k^{3/2}f_{ck}^{1/2} = 0.035 \times 2^{3/2} \times 32^{1/2} = 0.56$ MPa	
6.4.3 (2) 6.4.5 (1)	Since $v > v_{Rd,c}$, shear reinforcement is required. With the effective design strength of the reinforcement $f_{ywd,ef} = 250 + 0.25d = 300$ MPa say, the area required in one perimeter of vertical shear reinforcement, placed at the maximum radial spacing $s_r = 0.75d = 150$ mm say, is given by	
	$A_{sw} = (v - 0.75v_{Rd,c})u_1 s_r/(1.5f_{ywd,ef})$	
	$= (1.18 - 0.75 \times 0.75) \times 4113 \times 150/(1.5 \times 300) = 847$ mm^2 (12H10 say)	
6.4.5 (4)	The length of the control perimeter at which $v = v_{Rd,c}$ is given by	
	$u_{out} = \beta V/(v_{Rd,c} d) = 1.04 \times 919.6 \times 10^3 / (0.75 \times 198) = 6440$ mm	
	The distance of this control perimeter from the face of the column is given by	

Example 1 {style="display:inline"}

Reference	CALCULATIONS	OUTPUT
	$(u_{out} - 4c)/2\pi = (6440 - 4 \times 400)/2\pi = 770$ mm	
	The distance of the final perimeter of reinforcement from the control perimeter at which $v = v_{Rd,c}$ should not exceed $1.5d = 300$ mm say.	Shear reinforcement on four perimeters with 12H10 on each one.
	Thus, 4 reinforcement perimeters spaced at $s_r = 150$ mm, with the first perimeter at 100 mm from the column perimeter, would be suitable.	
	Column B1	
6.4.2 (4) Figure 6.15	Since the slab extends 400 mm beyond the outer face of the column, the length of the basic control perimeter at distance $2d = 400$ mm say from the column face is	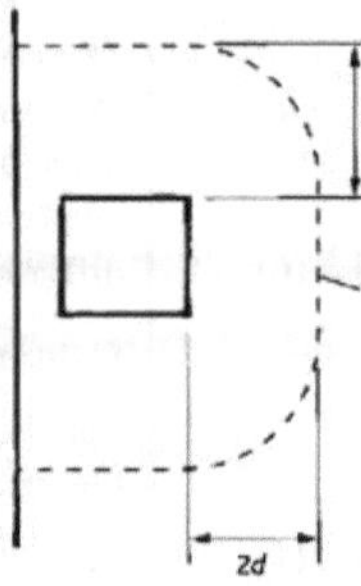
	$u_1 = 2 \times 400 + 3 \times 400 + \pi \times 400 = 3256$ mm	
6.4.3 (6) Figure 6.21N	Since lateral stability does not depend on frame action, and adjacent span lengths do not differ by more than 25%, $\beta = 1.4$ may be used.	
	From the results of the analysis for the sub-frame on line B (calculation sheet 22), the maximum shear force (for load case 2) is 342.6 kN. To this must be added the load resulting from a 600 mm wide edge strip of slab plus 5 kN/m due to walling, cladding and windows. Thus, the total shear force transferred to the column is	
	$V = 342.6 + 1.2 \times 7.2 \times (0.6 \times 15.0 + 1.25 \times 5.0) = 474.4$ kN	
6.4.3 (3) Equation 6.38	Maximum shear stress, with $\beta = 1.4$, along the basic control perimeter is	Basic control perimeter for edge column
	$v = \beta V/u_1 d = 1.4 \times 474.4 \times 10^3 / (3256 \times 198) = 1.03$ MPa	
6.4.3 (4) Figure 6.20a	Alternatively, in cases where there is no eccentricity parallel to the slab edge, and the eccentricity perpendicular to the slab edge is toward the interior, the punching shear force may be taken as uniformly distributed along an equivalent (reduced) control perimeter (see adjacent figure). The length of the reduced perimeter is	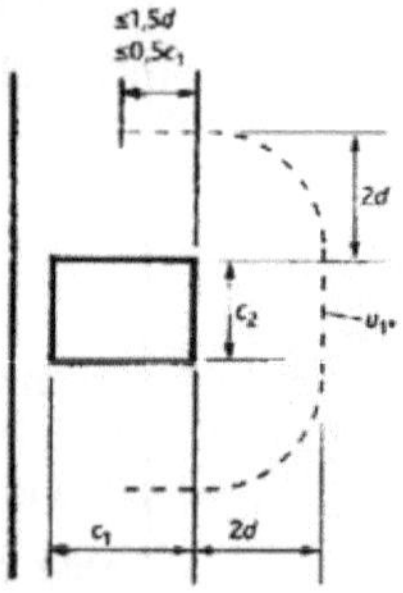
	$u_{1*} = 2 \times 400 + \pi \times 400 = 2056$ mm	
	Uniform shear stress along reduced control perimeter is	
	$v = V/u_{1*} d = 474.4 \times 10^3 / (2056 \times 198) = 1.17$ MPa	
	The maximum shear stress obtained for the basic control perimeter will be used in the following calculations, that is, $v = 1.03$ MPa	
	In the direction perpendicular to the slab edge, the moment transfer strip contains 13H16. Ignoring the nominal reinforcement provided outside this zone, for a slab width equal to the column width plus $3d$ each side = 1600 mm say,	Equivalent (reduced) control perimeter for edge column
	$\rho_1 = A_{sl}/b_w d = 2614/(1600 \times 198) = 0.0082$	
	In the direction parallel to the slab edge, $\rho_1 = 0.0096$ as for column B2. Thus, the mean value is $\rho_1 = 0.0089$, for which $v_{Rd,c} = 0.24 \times (0.89 \times 32)^{1/3} = 0.73$ MPa.	
6.4.3 (2)	Since $v > v_{Rd,c}$, shear reinforcement is required and, with $s_r = 0.75d = 150$ mm:	
	$A_{sw} = (v - 0.75 v_{Rd,c}) u_1 s_r /(1.5 f_{ywd,ef})$	
	$= (1.03 - 0.75 \times 0.73) \times 3256 \times 150/(1.5 \times 300) = 524$ mm^2 (8H10 say)	
6.4.5 (4)	The length of the control perimeter at which $v = v_{Rd,c}$ is given by	
	$u_{out} = \beta V/(v_{Rd,c} d) = 1.4 \times 474.4 \times 10^3 / (0.73 \times 198) = 4595$ mm	
	The distance of this control perimeter from the face of the column is given by	
	$(u_{out} - 2 \times 400 - 3c)/\pi = (4595 - 800 - 1200)/\pi = 826$ mm	Shear reinforcement on four perimeters with 8H10 on each one.
	Thus, 4 reinforcement perimeters spaced at $s_r = 150$ mm, with the first perimeter at 100 mm from the column perimeter, would be suitable.	
	Column A2	
	From the results of the analysis for the sub-frame on line 2 (calculation sheet 26), the maximum shear force (for load case 2) is 347.5 kN. Thus, the total shear force transferred to the column is	
	$V = 347.5 + 7.2 \times (0.6 \times 15.0 + 1.25 \times 5.0) = 457.3$ kN	Shear reinforcement on four perimeters with 8H10 on each one.
	Since this value is only slightly less than that for column B1, a similar layout of shear reinforcement will be required.	

Example 1   Calculation Sheet 30

Reference	CALCULATIONS	OUTPUT
	Column A1	
6.4.2 (4) Figure 6.15	Since the slab extends 400 mm beyond the outer face of the column, the length of the basic control perimeter at distance $2d = 400$ mm say from the column face is $u_1 = 2 \times 400 + 2 \times 400 + (\pi/2) \times 400 = 2228$ mm	
6.4.3 (6) Figure 6.21N	Since lateral stability does not depend on frame action, and adjacent span lengths do not differ by more than 25%, $\beta = 1.5$ may be used.	
	For the sub-frame on line A (calculation sheet 27), the maximum design load is 58.5 kN/m. Considering the additional edge loading for the sub-frame on line 1, and assuming shear force coefficients of 0.45 for both sub-frames, the total shear force transferred to the column is $V = 0.45 \times [58.5 \times 6.0 + (0.6 \times 15.0 + 1.25 \times 5.0) \times 7.2] = 207.4$ kN	Basic control perimeter for corner column
6.4.3 (4) Equation 6.38	Maximum shear stress, with $\beta = 1.5$, along the basic control perimeter is $v = \beta V / u_1 d = 1.5 \times 207.4 \times 10^3 / (2228 \times 198) = 0.70$ MPa	
	Suppose that each moment transfer strip contains 11H16. Ignoring the nominal reinforcement provided outside these zones, for a width taken from the edge of the slab to a position $3d$ beyond the inner face of the column = 1400 mm say: $\rho_1 = 2212/(1400 \times 198) = 0.0080$, $v_{Rd,c} = 0.24 \times (0.80 \times 32)^{1/3} = 0.70$ MPa	Shear reinforcement not required
6.4.3 (2)	Since v does not exceed $v_{Rd,c}$, there is no need to provide shear reinforcement.	

Bar Marks	Commentary on Bar Arrangement (Drawings 7, 8 and 9)
	For the bottom bars, a spacing of 200 mm in each direction suits the column layout and ensures that two bars pass through each column. The preferred arrangement would be to use alternate long and short bars with the long bars, being lapped on the column lines, providing continuous internal ties. However, this would result in H12-400 at the ends of the spans, which is less than the minimum reinforcement requirements. Instead, an arrangement is used in which the spacing remains at 200 mm throughout, but the bar diameters are changed.
	For the top bars required at the column positions, it is recommended that the curtailment position of alternate bars should be staggered at distances from the face of the column of $0.2 \times$ span and $0.3 \times$ span respectively. However, the layout is already intricate enough without this further complication.
01	Bars (shape code 21) providing at least 50% of area needed in span A−B. Length of top leg not less than $0.15 \times$ span = 1200 mm say. Bottom leg laps with bar 02, where lap length = $1.5 \times 35\phi = 550$ mm say.
02	Straight bars providing 100% of area needed in span A−B. Bar curtailed at distance from centre of columns not more than $0.2 \times$ span = 1400 mm say.
03	Straight bars providing at least 50% of area needed in span A−B. Bar laps 550 mm with bar 02.
04	Bars (shape code 21) providing at least 50% of area needed in span 1−2. Length of top leg = 1200 mm say. Bottom leg laps 550 mm with bar 05.
05	Straight bars providing 100% of area needed in span 1−2, and at least 50% of area needed in span 2−3. Bar curtailed at distance not more than $0.2 \times$ span = 1200 mm from centre of columns on line 1. Bar laps with bar 05, where lap length = $1.5 \times 35\phi = 650$ mm say.
06	Straight bars providing 100% of area needed in span 2−3. Bar curtailed at distance from centre of columns not more than $0.2 \times$ span = 1400 mm say,
07	Bars (shape code 11, minimum end projection) extending $0.3 \times$ span = 1800 mm beyond face of column.
08	Straight bars providing minimum reinforcement and lapping 550 mm with bars 07 and 10.
09, 14	Straight bars providing minimum reinforcement and lapping 550 mm with bars 10.
10, 11	Straight bars extending not less than $0.3 \times$ span = 2200 mm say beyond faces of column.
12	Straight bars providing minimum reinforcement and lapping 550 mm with bars 10 and 13.
13	Bars (shape code 11, minimum end projection) extending $0.3 \times$ span = 2200 mm say beyond face of column.
15	Straight bars in corners of links and extending $35\phi = 350$ mm beyond last link.
16	Links (shape code 22) anchored around bars in inner layers.
17	Bars (shape code 11, minimum end projection) extending $35\phi = 350$ mm beyond last link.

Example 1: Bottom Reinforcement in Flat Slab Floor   **Drawing 7**

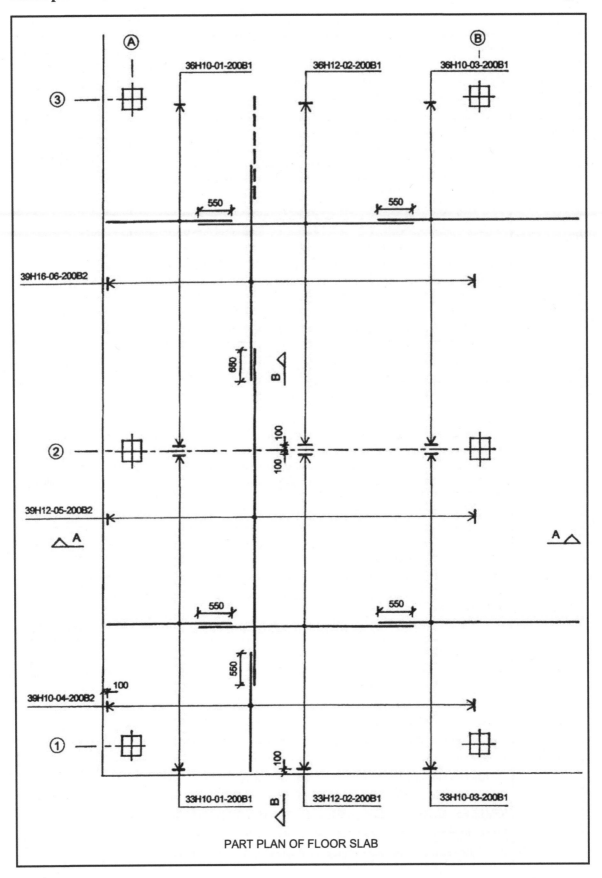

PART PLAN OF FLOOR SLAB

Example 1: Top Reinforcement in Flat Slab Floor   **Drawing 8**

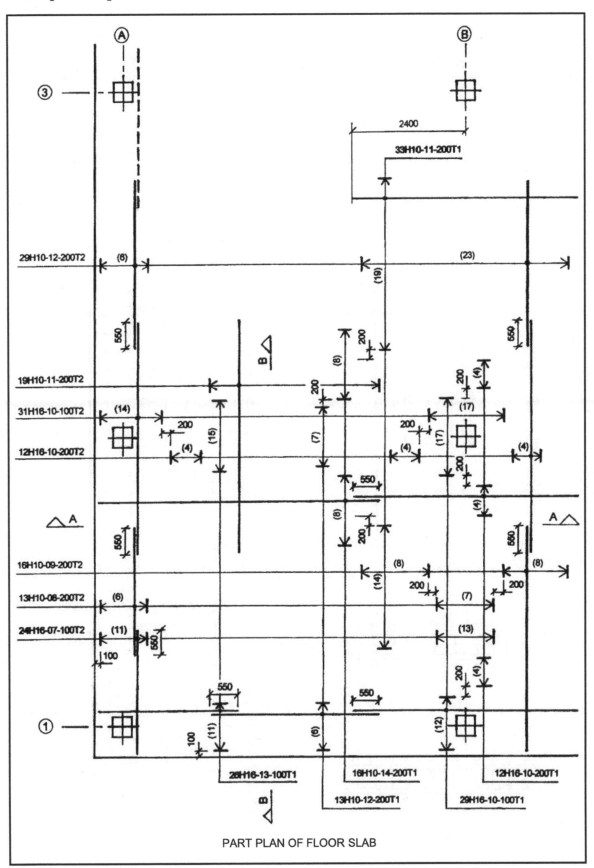

PART PLAN OF FLOOR SLAB

Example 1: Cross-Sections and Shear Reinforcement Flat Slab Floor   Drawing 9

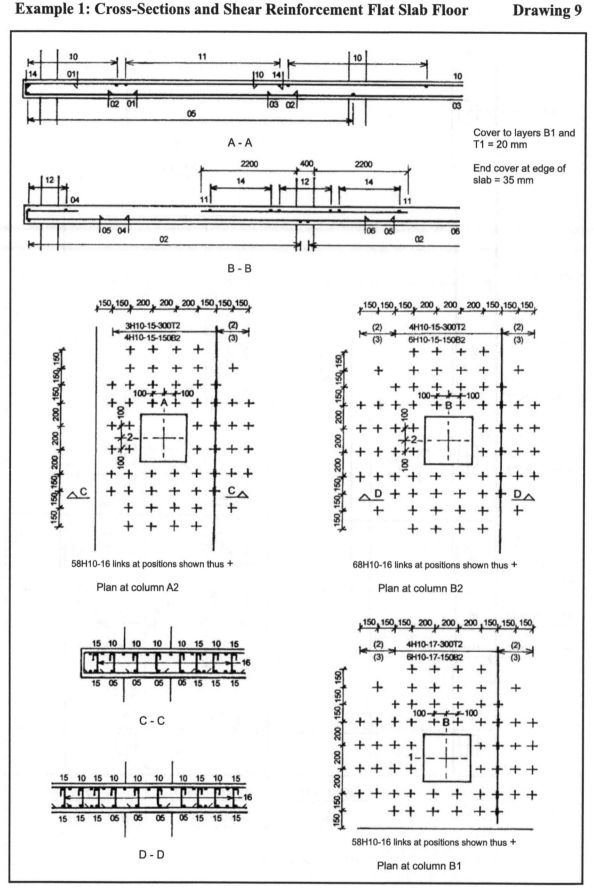

A - A

Cover to layers B1 and
T1 = 20 mm

End cover at edge of
slab = 35 mm

B - B

58H10-16 links at positions shown thus +

Plan at column A2

68H10-16 links at positions shown thus +

Plan at column B2

C - C

D - D

58H10-16 links at positions shown thus +

Plan at column B1

Example 1   **Calculation Sheet 31**

Reference	CALCULATIONS	OUTPUT
	SLAB WITH DROP PANELS	
	The necessity for shear reinforcement at the internal and edge columns could be avoided by introducing drop panels. The increased slab stiffness at the columns will increase the hogging moments in the slab, but the effect will be small in this example and can be offset by increasing the moment redistribution at these points. In the following calculations, the moments and shear forces obtained for the slab without drop panels, and the same tension reinforcement, will be assumed. Since the effective depth of the reinforcement is increased at the drop panel, the moment of resistance will be increased locally. Taking the depth of the drop panel below the slab soffit as 140 mm, the mean effective depth $d_d = 140 + 198 = 338$ mm.	$d_d = 338$ mm
	Column B2	
	The length of the basic control perimeter, taken at distance $2d_d = 675$ mm say from the face of the column, is: $u_1 = 4 \times 400 + 2\pi \times 675 = 5841$ mm.	
6.4.3 (3) Equation 6.38	The maximum shear stress, with $\beta = 1.04$, along the basic control perimeter is $v = \beta V / u_1 d = 1.04 \times 919.6 \times 10^3 / (5841 \times 338) = 0.49$ MPa	
6.4.4 (1)	For a slab width equal to the column width plus $3d_d$ each side = 2430 mm say, that contains 17H16 + 4H16 as provided in the slab without drop panels,	
Equation 6.47	$\rho_1 = A_{sl}/b_w d = 4223/(2430 \times 338) = 0.0051$ and, with $k = 1.77$ for $d = 338$ mm: $v_{Rd,c} = (0.18 \times 1.77/1.5)(0.51 \times 32)^{1/3} = 0.54$ MPa $(>v)$	
	Distance from edge of drop panel to control perimeter for slab at which $v = v_{Rd,c}$ should not exceed $2d$. Thus, length of side of drop panel (see calculation sheet 28) should be not less than $(6440 - 4\pi d)/4 = (6440 - 4\pi \times 198)/4 = 1200$ mm say.	Size of drop panel $1200 \times 1200 \times 380$ deep overall
	Column B1	
	The length of the basic control perimeter is: $u_1 = 5 \times 400 + \pi \times 675 = 4120$ mm	
6.4.3 (3) Equation 6.38	The maximum shear stress, with $\beta = 1.4$, along the basic control perimeter is $v = \beta V / u_1 d = 1.4 \times 474.4 \times 10^3 / (4120 \times 338) = 0.48$ MPa	
	In the direction perpendicular to the slab edge, for a slab width of 2430 mm that contains 13H16 + 6H10, $\rho_1 = 3085/(2430 \times 338) = 0.0037$. In the direction parallel to the slab edge, $\rho_1 = 0.0051$ as for column B2. The mean value is $\rho_1 = 0.0044$, for which $v_{Rd,c} = (0.18 \times 1.77/1.5)(0.44 \times 32)^{1/3} = 0.51$ MPa $(>v)$	Size of drop panel $1200 \times 1200 \times 380$ deep overall
	For $v = v_{Rd,c}$ at basic control perimeter for slab, length of side of drop panel (see calculation sheet 29) should be not less than $(4595 - 2\pi \times 198)/3 = 1200$ mm say.	

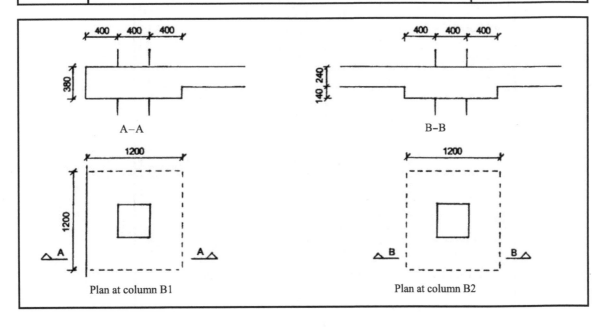

Example 1

Calculation Sheet 32

Reference	CALCULATIONS	OUTPUT
	ACTIONS ON COLUMNS	

For the columns on line B, the sub-frame analysis results shown in calculation sheet 22 given beam shears 2nd, 3rd and 4th floor levels. For simplicity, the same values will be used at lower floor levels, even though the storey heights result in sub-frame dimensions that are slightly different. At the roof level, the sub-frame and the loading are significantly different, and another analysis is required. Loading details are as follows:

Characteristic loading for roof slab:

Slab and finishes: $(6.0 + 1.5) = 7.5$ kN/m^2 — Imposed: 0.6 kN/m^2

Design ultimate load for roof slab: $n = 1.25 \times 7.5 + 1.5 \times 0.6 = 10.3$ kN/m^2

Loads per storey due to the self-weight of the columns:

Columns up to 1st floor: $1.25 \times 0.4 \times 0.4 \times 25 \times 3.76 = 18.8$ kN

Columns above 1st floor: $1.25 \times 0.4 \times 0.4 \times 25 \times 3.26 = 16.3$ kN

BS EN 1991-1-1 6.3.1.2 NA.2.

A reduction may be made in the total imposed floor load, according to the number of storeys being supported at the level considered. For up to five storeys, this load may be multiplied by $\alpha_n = 1.1 - n/10$, where n is the number of storeys.

EDGE COLUMN B1

At each level, the load applied is the shear force from the sub-frame on line B (see calculation sheet 22) plus the edge loading. At each floor, the load due to the edge slab and walling $= 1.2 \times 7.2 \times (0.6 \times 15.0 + 1.25 \times 5.0) = 131.8$ kN. At the roof, the additional load $= 1.2 \times 7.2 \times (0.6 \times 10.3 + 1.25 \times 0.15 \times 1.0 \times 25) = 93.9$ kN.

The maximum moment and maximum coexistent load occur when load case 2 is applied at all levels (see below). Maximum moment and minimum coexistent load occur when load case 2 is applied at the level considered, and $1.0G_k$ is applied at levels above. This arrangement can be critical for values of $N_{Ed} < N_{bal}$ where:

$N_{bal} = 0.4A_c f_{cd} = 0.4 \times 400 \times 400 \times 0.85 \times 32/1.5 \times 10^{-3} = 1160$ kN

Values of axial load N (kN) and bending moment M (kN m)						
Loading	$1.25G_k + 1.5Q_k$				$1.5Q_k$	
Load case	1		2		1	2
Member	N	M	N	M	N	N
Roof slab	308.7	61.7	311.8	64.2	(18.8)	
Column	<u>16.3</u>		<u>16.3</u>			
	325.0	72.0	328.1	82.1		
4th floor slab	<u>456.4</u>		<u>474.4</u>		129.8	147.8
	781.4	72.0	802.5	82.1		
Column	<u>16.3</u>		<u>16.3</u>			
	797.7	72.0	818.8	82.1		
3rd floor slab	<u>456.4</u>		<u>474.4</u>		<u>129.8</u>	<u>147.8</u>
	1254.1	72.0	1293.2	82.1	259.6	295.6
Column	<u>16.3</u>		<u>16.3</u>			
	1270.4	72.0	1309.5	82.1		
2nd floor slab	<u>456.4</u>		<u>474.4</u>		<u>129.8</u>	<u>147.8</u>
	1726.8	72.0	1783.9	82.1	389.4	443.4
Column	<u>16.3</u>		<u>16.3</u>			
	1743.1	72.0	1800.2	82.1		
1st floor slab	<u>456.4</u>		<u>474.4</u>		<u>129.8</u>	<u>147.8</u>
	2199.5	72.0	2274.6	82.1	519.2	591.2
Column	<u>18.8</u>		<u>18.8</u>			
	2218.3	72.0	2293.4	82.1		
Grd. floor slab	<u>456.4</u>		<u>474.4</u>		<u>129.8</u>	<u>147.8</u>
Basement wall	2674.7	72.0	2767.8	82.1	649.0	739.0

For the storey from ground to 1st floor, with load case 2 at ground floor level:

$M_{bot} = 82.1$ kN m, $M_{top} = -0.5M_{bot} = -41.1$ kN m

Example 1

Calculation Sheet 33

Reference	CALCULATIONS	OUTPUT
	With load case 2 at levels above: $N_{Ed} = 2293.4 - 0.3 \times 591.2 = 2116$ kN (max)	
	With $1.0 G_k$ at levels above: $N_{Ed} = [2199.5 - (519.2 + 18.8)]/1.25 = 1329$ kN (min)	
6.1 (4)	Minimum total design moment, with $e_0 = h/30 = 300/30 \geq 20$ mm:	
	$M_{min} = N_{Ed}\, e_0 = 2116 \times 0.02 = 42.4$ kN m	
	Effective length and slenderness	
	Using the simplified method in Concise Eurocode 2, with condition 2 (monolithic connection to members shallower than the overall depth of the column) at both top and bottom of the column,	
	$l_0 = 0.85l = 0.85 \times 3.76 = 3.2$ m (for storeys above 1st floor, $l_0 = 2.77$ m)	
5.2 (9)	First-order moment from imperfections (simplified procedure):	
	$M_i = Nl_0/400 = 2116 \times 3.2/400 = 17.0$ kN m	
	First-order moments, including the effect of imperfections:	
	$M_{01} = -41.1 + 17.0 = -24.1$ kNm, $M_{02} = 82.1 + 17.0 = 99.1$ kN m	
	Radius of gyration of uncracked concrete section, $i = h/\sqrt{12} = 0.115$ m	
5.8.3.2 (1)	Slenderness ratio $\lambda = l_0/i = 3.2/0.115 = 27.8$	
5.8.3.1 (1)	Slenderness criterion, $\lambda_{lim} = 20(A \times B \times C)/\sqrt{n}$ where:	
	$n = N/A_c f_{cd} = N/(400^2 \times 0.85 \times 32/1.5) = 2116/2901 = 0.73$	
	Taking $A = 0.7$, $B = 1.1$ and $C = 1.7 - M_{01}/M_{02} = 1.7 + 24.1/99.1 = 1.94$	
	$\lambda_{lim} = 20 \times 0.7 \times 1.1 \times 1.94/\sqrt{0.73} = 34.9$ ($> \lambda = 27.8$)	
	Since $\lambda < \lambda_{lim}$, second order effects may be ignored and $M_{Ed} = M_{02}$ ($\geq M_{min}$)	
	Design of cross-section	
	Allowing 35 mm nominal cover, 8 mm links and 32 mm longitudinal bars, results in $d = 400 - (35 + 8 + 32/2) = 340$ mm say, $d/h = 340/400 = 0.85$. Reinforcement can be determined from the design chart in Table A3 as follows:	
	$N_{Ed}/bhf_{ck} = (2116 \text{ or } 1329) \times 10^3/(400 \times 400 \times 32) = 0.42$ or 0.26	
	$M_{Ed}/bh^2 f_{ck} = 99.1 \times 10^6/(400 \times 400^2 \times 32) = 0.049$, $A_s f_{yk}/bhf_{ck} = 0$	
9.5.2 (2)	Minimum amount of longitudinal reinforcement:	
	$A_{s,min} = 0.1 N_{Ed}/f_{yd} = 0.1 \times 2116 \times 10^3/(500/1.15) = 487$ mm^2 (4H16)	
	$\geq 0.002 A_c = 0.002 \times 400 \times 400 = 320$ mm^2	
	Similar calculations for the other storeys provide results as summarised below.	

Storey	N_{Ed}(kN max/min)	M_{Ed} (kN m)	$\dfrac{N_{Ed}}{bhf_{ck}}$	$\dfrac{M_{Ed}}{bh^2 f_{ck}}$	$\dfrac{A_s f_{yk}}{bhf_{ck}}$	A_s (mm^2)
4th floor–roof	328/245	84.4	0.07/0.05	0.042	0.07	717
3rd–4th floor	819/519	87.8	0.16/0.10	0.043	0	320
2nd–3rd floor	1284/794	91.0	0.25/0.16	0.045	0	320
1st–2nd floor	1722/1068	94.0	0.34/0.21	0.046	0	396
Grd–1st floor	2116/1329	99.1	0.42/0.26	0.049	0	487

Reference	CALCULATIONS	OUTPUT
	Tying requirements	
BS EN 1990 A1.3.2 Table NA.A1.3	For the slab, the accidental design load $(G_k + \psi_1 Q_k)$	
	$= 7.25 + 0.7 \times 4.0 = 10$ kN/m^2 (max), 7.25 kN/m^2 (min)	
	For the column, approximate accidental design load for load case 2 on sub-frame on line B (calculation sheet 22), plus edge loading (calculation sheet 32):	
	$N_{Ad} = (10/15) \times 342.6 + 1.2 \times 7.2 \times (0.6 \times 10 + 5.0) = 323.5$ kN	
	Minimum area of reinforcement required with $\sigma_s = 500$ MPa,	
	$A_{s,min} = 323.5 \times 10^3/500 = 647$ mm^2 (4H16 sufficient)	
	An arrangement of 4H16 at all storeys is sufficient to meet both normal structural and tying requirements.	4H16

Example 1   Calculation Sheet 34

Reference	CALCULATIONS	OUTPUT

INTERNAL COLUMN B2

The load from the sub-frame on line B is the total shear force at line 2.

Loading	$1.25G_k + 1.5Q_k$				$1.5Q_k$	
Load case	1		3		1	3
Member	N	M	N	M	N	N
Roof slab	639.6	5.0	614.6	13.7	(55.9)	
Column	<u>16.3</u>		<u>16.3</u>			
	655.9	10.3	630.9	52.4		
4th floor slab	<u>919.6</u>		<u>756.6</u>		367.8	204.5
	1575.5	10.3	1387.5	52.4		
Column	<u>16.3</u>		<u>16.3</u>			
	1591.8	10.3	1403.8	52.4		
3rd floor slab	<u>919.6</u>		<u>756.6</u>		<u>367.8</u>	<u>204.5</u>
	2511.4	10.3	2160.4	52.4	735.6	409.0
Column	<u>16.3</u>		<u>16.3</u>			
	2527.7	10.3	2176.7	52.4		
2nd floor slab	<u>919.6</u>		<u>756.6</u>		<u>367.8</u>	<u>204.5</u>
	3447.3	10.3	2933.3	52.4	1103.4	613.5
Column	<u>16.3</u>		<u>16.3</u>			
	3463.6	10.3	2949.6	52.4		
1st floor slab	<u>919.6</u>		<u>756.6</u>		<u>367.8</u>	<u>204.5</u>
	4383.2	10.3	3706.2	52.4	1471.2	818.0
Column	<u>18.8</u>		<u>18.8</u>			
	4402.0	10.3	3725.0	52.4		
Grd floor slab	<u>919.6</u>		<u>756.6</u>		<u>367.8</u>	<u>204.5</u>
	5321.6	10.3	4481.6	52.4	1839.0	1022.5
Column	<u>18.8</u>		<u>18.8</u>			
Foundation	5340.4		4500.4			

Values of axial load N (kN) and bending moment M (kN m)

The maximum moment occurs when load case 3 is applied at the level considered. Maximum coexistent load occurs when load case 1 is applied at levels above, and minimum coexistent load occurs when $1.0G_k$ is applied at levels above. The latter arrangement can be critical for values of $N_{Ed} < 1160$ kN. The maximum load with a smaller coexistent moment results when load case 1 is applied at all levels.

For the basement storey with load case 1 at all levels:

$N_{Ed} = 5340.4 - 0.4 \times 1839.0 = 4605$ kN

6.1 (4)	Minimum total design moment, with $e_0 = h/30 = 400/30 \geq 20$ mm:

$M_{min} = N_{Ed} e_0 = 4605 \times 0.02 = 92.1$ kN m

For the basement storey with load case 3 at ground floor level:

$M_{top} = 52.4$ kN m, $M_{bot} = -0.5M_{top} = -26.2$ kN m

$N_{Ed} = 756.6 + 4402.0 - 0.4 \times (204.5 + 1471.2) = 4488$ kN (max)

$N_{Ed} = 756.6 + [4402.0 - (1471.2 + 55.9)]/1.25 = 3057$ kN (min)

Effective length and slenderness

As for column B1, $l_0 = 3.2$ m and $\lambda = 27.8$

5.2 (9)	First-order moment from imperfections (with load case 3 at ground floor level):

$M_i = Nl_0/400 = 4488 \times 3.2/400 = 35.9$ kN m

First-order moments, including the effect of imperfections:

$M_{01} = -26.2 + 35.9 = 9.7$ kN m, $M_{02} = 52.4 + 35.9 = 88.3$ kN m

5.8.3.1 (1)	Slenderness criterion: $\lambda_{lim} = 20(A \times B \times C)/\sqrt{n}$ where

$n = N/A_c f_{cd} = 4488/2901 = 1.55$ and

$A = 0.7$, $B = (1 + 2\omega)^{0.5}$, $C = 1.7 - 9.7/88.3 = 1.59$ where

Example 1

Reference	CALCULATIONS	OUTPUT
	$\omega = A_s f_{yd}/A_c f_{cd} = A_s \times 500/(1.15 \times 2901 \times 10^3) = A_s/6672$	
	Assuming 6H32, $\omega = 4825/6672 = 0.72$, $B = (1 + 2\omega)^{0.5} = 1.56$ and	
	$\lambda_{lim} = 20 \times 0.7 \times 1.56 \times 1.59/\sqrt{1.55} = 27.9$ $(>\lambda = 27.8)$	
	Since $\lambda < \lambda_{lim}$, second-order effects may be ignored and $M_{Ed} = M_{02}(\geq M_{min})$.	
	Design of cross-section	
	Although the nominal cover needed for durability is 25 mm, this will be increased to 35 mm to ensure a minimum cover to the H32 bars not less than the bar size.	
	Since $M_{min} > M_{02}$, the critical condition occurs with load case 1 at all levels. The reinforcement can be determined from the design chart in Table A3 as follows.	
	$N_{Ed}/bhf_{ck} = 4605 \times 10^3/(400 \times 400 \times 32) = 0.90$	
	$M_{Ed}/bh^2 f_{ck} = 92.1 \times 10^6/(400 \times 400^2 \times 32) = 0.045$	
	$A_s f_{yk}/bhf_{ck} = 0.52$, which gives $A_s = 0.52 \times 400 \times 400 \times 32/500 = 5325$ mm^2	
	A reasonable arrangement would be to provide 8H32, with a bar at each corner of the column and a bar at the mid-point of each side. For the upper storeys, critical conditions occur with load case 3 at the level considered and either maximum load (case 1) or minimum load ($1.0G_k$) at levels above, as summarised below.	

Storey	N_{Ed} (kN max/min)	M_{Ed} (kN m)	$\dfrac{N_{Ed}}{bhf_{ck}}$	$\dfrac{M_{Ed}}{bh^2 f_{ck}}$	$\dfrac{A_s f_{yk}}{bhf_{ck}}$	A_s (mm^2)
4th floor–roof	656/483	57.0	0.13/0.09	0.028	0	320
3rd–4th floor	1592/922	63.4	0.31/0.18	0.031	0	366
2nd–3rd floor	2454/1376	69.4	0.48/0.27	0.034	0	565
1st–2nd floor	3243/1831	74.9	0.63/0.36	0.037	0.19	1946
Grd–1st floor	3814/2285	82.9	0.75/0.45	0.041	0.34	3482
Basement	4605	92.1	0.90	0.045	0.52	5325

Reference	CALCULATIONS	OUTPUT
9.5.2 (2)	Minimum amount of longitudinal reinforcement (2nd–3rd floor):	
	$A_{s,min} = 0.1N_{Ed}/f_{yd} = 0.1 \times 2454 \times 10^3/(500/1.15) = 565$ mm^2 (4H16)	
9.5.2 (3)	Maximum amount of longitudinal reinforcement:	
	$A_{s,max} = 0.04A_c = 0.04 \times 400 \times 400 = 6400$ mm^2 (8H32)	
	Tying requirements	
BS EN 1990 A1.3.2 Table NA.A1.3	Accidental design load applied to column for load case 1 on sub-frame on line B:	
	$N_{Ad} = (10/15) \times 919.6 = 613.0$ kN	
	Minimum area of reinforcement required with $\sigma_s = 500$ MPa,	
	$A_{s,min} = 613 \times 10^3/500 = 1226$ mm^2 (4H20 sufficient)	8H32 (Bottom storey) 8H25 (Grd–1st floor) 4H25 (1st–2nd floor) 4H20 (2nd floor–roof)
	A reasonable arrangement would be to provide 8H32 for the bottom storey, 8H25 for the next storey, 4H25 for the next storey and 4H20 for the top three storeys.	
	CORNER COLUMN A1	
	From the calculations for column B1, it can be seen that the most critical condition occurs at the bottom of the top storey, with minimum load $1.0G_k$ at roof level and maximum design load at 4th floor level.	
	Maximum loading for sub-frames at 4th floor level:	

	Frame on line A kN/m	Frame on line 1 kN/m
Slab	$0.4 \times 7.2 \times 15.0 = 43.2$	$0.4 \times 6.0 \times 15.0 = 36.0$
Edge strip and walling	$0.6 \times 15 + 1.25 \times 5.0 = \underline{15.3}$	$\underline{15.3}$
	58.5	51.3

Reference	CALCULATIONS	OUTPUT
	Maximum column moments for the sub-frames on lines A and 1 can be taken pro rata to the results for the sub-frames on lines B and 2 (for load case 2), as follows:	
	Frame on line A, $M_z = (58.5/129.6) \times 82.1 = 37.0$ kN m	
	Frame on line 1, $M_y = (51.3/108.0) \times 113.4 = 53.9$ kN m	

Example 1   Calculation Sheet 36

Reference	CALCULATIONS	OUTPUT
	Minimum loading for sub-frames at roof level:	
		Frame on line A	kN/m	Frame on line 1	kN/m
	Slab $\quad$ $0.4 \times 7.2 \times 7.25 = 20.9$ $\quad$ (included in line A)	
	Edge strip & parapet $\quad$ $0.6 \times 7.25 + 0.15 \times 25 = \underline{8.1}$ $\qquad$ $\underline{8.1}$	
	$\qquad\qquad\qquad\qquad\qquad\qquad$ $\underline{29.0}$ $\qquad\qquad\qquad$ $\underline{8.1}$	
	Minimum load at bottom of column (roof slab and weight of column):	
	$N_{Ed} = 0.45 \times (29.0 \times 6.0 + 8.1 \times 7.2) + 16.3/1.25 = 118 \text{ kN}$	
5.2 (9)	First-order moment from imperfections (simplified procedure):	
	$M_i = Nl_0/400 = 118 \times 2.77/400 = 0.8 \text{ kN m}$	
5.8.9 (2)	Since imperfections need to be taken into account only in the direction where they will have the most unfavourable effect,	
	$M_{0z} = 37.0 \text{ kN m}, \ M_{0y} = 53.9 + 0.8 = 54.7 \text{ kN m}$	
	Design of cross-section	
	Assuming 4H20, the design resistance of the column for bending about either axis can be determined from the design chart in Table A3 as follows:	
	$A_s f_{yk}/bhf_{ck} = 1257 \times 500/(400 \times 400 \times 32) = 0.12$	
	$N_{Ed}/bhf_{ck} = 118 \times 10^3/(400 \times 400 \times 32) = 0.023$	
	$M_{Ed}/bh^2 f_{ck} = 0.050 \ (\text{for } d/h = 0.85)$	
	Thus, $M_{Rdz} = M_{Rdy} = 0.05 \times 400 \times 400^2 \times 32 \times 10^{-6} = 102.4 \text{ kN m}$	
5.8.9 (4)	In the absence of a precise design for biaxial bending, a simplified criterion check for compliance may be made as follows:	
	$N_{Rd} = A_c f_{cd} + A_s f_{yd} = (400^2 \times 0.85 \times 32/1.5 + 1257 \times 500/1.15) \times 10^{-3} = 3448 \text{ kN}$	
	$N_{Ed}/N_{Rd} = 118/3448 = 0.034.$ For values of $N_{Ed}/N_{Rd} \leq 0.1$, exponent $a = 1.0$.	
	$(M_{Edz}/M_{Rdz})^a + (M_{Edy}/M_{Rdy})^a = (37.0/102.4)^{1.0} + (54.7/102.4)^{1.0} = 0.90 \ (\leq 1.0)$	
	Tying requirements	
	From the calculations for column B1, it can be seen that 4H16 would be sufficient. A reasonable arrangement would be to provide 4H20 for the top storey, and 4H16 for the lower storeys.	4H16 (Grd—4th floor) 4H20 (4th floor—roof)
	FIRE RESISTANCE	
	The columns can be assumed to meet the requirements, since 300×300 columns were sufficient for the beam and slab construction (calculation sheets 19–20).	

Bar Marks	Commentary on Bar Arrangement (Drawing 10)
01	Bars (shape code 26) bearing on 75 mm kicker and cranked to fit alongside bars projecting from basement wall. Projection of starter bars $= 1.5 \times 35 \times 16 + 75 = 925 \text{ mm}$ say. Crank to begin 75 mm from end of starter bar. Length of crank $= 13\phi$ and overall offset dimension $= 2\phi$. Since 4H16 are also sufficient at the next level, projection of bars above first floor $= 925 \text{ mm}$.
02, 05	Closed links (shape code 51), with 35 mm nominal cover, starting above kicker and stopping below slab at next floor level. See bar commentary in calculation sheet 20 for details of code requirements. For main bars of 16 mm diameter, link spacing should not exceed $20 \times 16 = 320 \text{ mm}$ generally, or $0.6 \times 320 = 192 \text{ mm}$ for a distance of 400 mm above or below the slab. For larger diameter main bars, maximum link spacing values are 400 mm and 240 mm, respectively. For column B2, required areas of transverse bars in lap zones are: Fdn–Grd floor: $A_{st} = 1.5 \times 5325/6434 \times 804 = 998 \text{ mm}^2$ (13H10) Grd–1st floor: $\ A_{st} = 1.5 \times 491 = 737 \text{ mm}^2$ (15H8)
03	Bars (similar to bar mark 01) cranked to fit alongside bars projecting from foundation. Projection of starter bars $= 1.5 \times 35 \times 32 \times 5325/6434 + 75 = 1500 \text{ mm}$ say. Since 8H25 are sufficient at the next level, projection of bars above ground floor level $= 1.5 \times 35 \times 25 + 75 = 1400 \text{ mm}$ say.
04	Bars (similar to bar mark 03). Projection of bars above first floor level $= 1400 \text{ mm}$.

Example 1: Reinforcement in Columns B1 and B2   **Drawing 10**

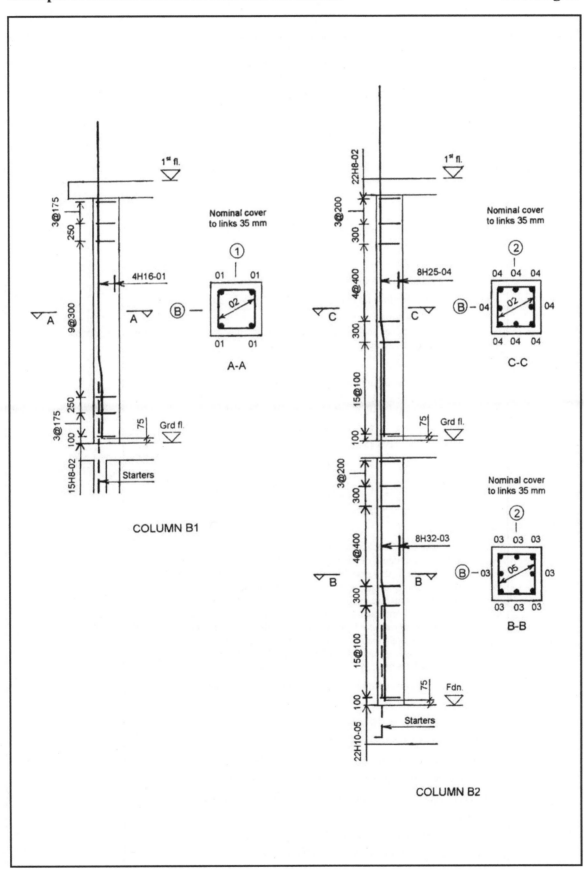

COLUMN B1

COLUMN B2

Example 1   **Calculation Sheet 37**

Reference	CALCULATIONS	OUTPUT
	INTEGRAL BEAM AND RIBBED SLAB FLOOR CONSTRUCTION	
	FLOOR SLABS (GROUND AND UPPER FLOORS)	
	For the end span of a ribbed one-way continuous slab and a characteristic imposed load ≤ 5 kN/m^2, try a span/effective depth ratio of $0.8 \times 40 = 32$. Since the span exceeds 7 m and supports partitions liable to be damaged by excessive deflection, this value should be multiplied by 7/span giving $32 \times 7/9.6 = 24$ say.	
	Allowing for 25 mm bars with 35 mm nominal cover gives an estimated overall depth = $9600/24 + (35 + 25/2) = 450$ mm say. If a 350 mm deep trough mould is used with a 100 mm thick flange, overall depth = 450 mm.	
	Fire resistance	
BS EN 1991-1-2 5.7.5 (1) Table 5.6 Table 5.8	Allowing for the design to be based on no more than 15% moment redistribution, the slab may be taken as continuous. For the ground floor (minimum fire period 1.5 h), the required minimum dimensions are:	
	Flange thickness: 100 mm Axis distance (to centre of bars): 20 mm	
	Rib width: 150 mm Axis distance (to centre of bars in one layer): 35 mm	
	Axis distance to side of rib for corner bars: $(35 + 10) = 45$ mm	
	If the finishes included not less than 25 mm non-combustible material, the flange thickness could be reduced to 75 mm. With 25 mm nominal cover for durability, the axis distances to the main bars will be sufficient.	Sufficient for 1.5 h fire period
	Loading	
	Details of the characteristic imposed loads, and the action combination options for the ultimate limit state, are given in calculation sheet 1. The volume of concrete for the chosen section is 0.210 m^3 per square metre of floor area.	
	Permanent load kN/m^2 Variable load kN/m^2	
	Self-weight of slab 0.210×25 $= 5.25$ Imposed $= 2.5$ Finishes and services $= \underline{1.25}$ Partitions $= \underline{1.5}$ $g_k = \underline{6.50}$ $q_k = \underline{4.0}$	$g_k = 6.5$ kN/m^2 $q_k = 4.0$ kN/m^2 $n = 14.1$ kN/m^2
	Design ultimate load = $\xi\gamma_G G_k + \gamma_Q Q_k = 1.25 \times 6.5 + 1.5 \times 4.0 = 14.1$ kN/m^2	
	Design permanent load = $\xi\gamma_G G_k = 1.25 \times 6.5 = 8.1$ kN/m^2	
	Analysis	
	The effective span could be taken as the distance between the faces of the integral beams plus the overall depth of the slab, provided the beams are designed to resist the resulting torsion. In this example, the slab will be designed to span between the centres of the beams assuming that the supports provide no rotational restraint. The following two load cases will be considered: (1) both spans carrying design ultimate load, and (2) one span carrying design ultimate load with the other span carrying only design permanent load. The effect of the ribbed slab becoming solid towards the ends of each span is to increase the elastic hogging moment at the internal support. However, since this effect is small and can be offset by moment redistribution, it will be ignored in the analysis.	
	The elastic moments and corresponding shears can be calculated as follows:	
	Load case 1	
	Hogging moment at interior support 2:	
	$M = 0.125 \times 14.1 \times 9.6^2 = 162.5$ kN m/m	
	Shear force at end supports: $V = 0.5 \times 14.1 \times 9.6 - 162.5/9.6 = 50.8$ kN/m	
	Shear force at interior support: $V = 14.1 \times 9.6 - 50.8 = 84.6$ kN/m	
	Maximum sagging moment: $M = 0.5 \times 50.8^2/14.1 = 91.5$ kN m/m	
	Load case 2	
	Hogging moment at interior support 2:	
	$M = 0.0625 \times (14.1 + 8.1) \times 9.6^2 = 127.9$ kN m/m	

Example 1

Reference	CALCULATIONS	OUTPUT
	Shear force at end support 1: $V = 0.5 \times 14.1 \times 9.6 - 127.9/9.6 = 54.4$ kN/m	
	Maximum sagging moment in span 1–2: $M = 0.5 \times 54.4^2/14.1 = 105.0$ kN m/m	
	Shear force at end support 3: $V = 0.5 \times 8.1 \times 9.6 - 127.9/9.6 = 25.6$ kN/m	
	Maximum sagging moment in span 2–3: $M = 0.5 \times 25.6^2/8.1 = 40.5$ kN m/m	

Multiplying these values by 0.6, applicable to a rib spacing of 600 mm, gives the following results that will now be used to design the section.

Load Case	Member		Span 1–2			Span 2–3	
	Support/Span	1	Span	2	2	Span	3
1	Moment kN m	0	54.9	−97.5	−97.5	54.9	0
	Shear kN	30.5		50.8	50.8		30.5
2	Moment kN m	0	63.0	−76.8	−76.8	24.3	0
	Shear kN	32.7		48.6	31.3		15.4

Flexural design

The section is solid at the supports and flanged in the spans. Allowing for 25 mm cover, 8 mm links and 25 mm main bars, $d = 450 - (25 + 8 + 25/2) = 400$ mm say.

According to the values of M/bd^2f_{ck}, where $b = 600$ mm, appropriate values of z/d and A_s can be determined (Table A1) and suitable bars selected (Table A9). These values will be valid for the span section, provided the neutral axis depth does not exceed the flange thickness, that is $z/d = (1 - 0.4\, x/d) \geq (1 - 0.4 \times 75/400) = 0.925$.

At the interior support, for load case 1:

$\quad M/bd^2f_{ck} = 97.5 \times 10^6/(600 \times 400^2 \times 32) = 0.032 \qquad z/d = 0.95$ (maximum)

$\quad A_s = M/(0.87f_{yk}z) = 97.5 \times 10^6/(0.87 \times 500 \times 0.95 \times 400) = 590$ mm^2 (3H16)

Alternatively, 1H25 giving M_u (491/590) × 97.5 = 81.1 kN m could be provided. The resulting moment redistribution would be $100 \times (97.5 - 81.1)/97.5 = 17\%$. The span moment obtained for load case 2 would still exceed that for load case 1.

For the span section, for load case 2:

$\quad M/bd^2f_{ck} = 63.0 \times 10^6/(600 \times 400^2 \times 32) = 0.021 \qquad z/d = 0.95$ (maximum)

$\quad A_s = 63.0 \times 10^6/(0.87 \times 500 \times 0.95 \times 400) = 381$ mm^2 (1H25)

Shear design

The critical section for shear in the ribbed portion of the slab will be taken at the face of the beam at the interior support. For hogging, the neutral axis depth can be determined by iteration, where b is taken as 125 mm initially, as follows:

$\quad b = 125 + x \tan 10^\circ$, $A_s f_{yk}/bdf_{ck} = 603 \times 500/(b \times 400 \times 32) = 23.5/b$

$\quad b = 125$ mm, $A_s f_{yk}/bdf_{ck} = 0.188$, $x/d = 0.361$, $x = 144$ mm

$\quad b = 125 + 144 \tan 10^\circ = 150$ mm, $A_s f_{yk}/bdf_{ck} = 0.157$, $x/d = 0.301$, $x = 120$ mm

$\quad b = 125 + 120 \tan 10^\circ = 146$ mm, $A_s f_{yk}/bdf_{ck} = 0.161$, $x/d = 0.309$, $x = 123$ mm

The value of b_w is taken as the smallest width of the section in the tensile area, and is given by $b_w = 125 + 2x \tan 10^\circ = 125 + 2 \times 123 \tan 10^\circ = 168$ mm

At 600 mm from the centre of the interior support, for load case 1:

$\quad V = 50.8 - 0.6 \times 14.1 \times 0.6 = 45.7$ kN

$\quad v = V/b_w d = 45.7 \times 10^3/(168 \times 400) = 0.68$ MPa

6.2.2 (1)
Table NA.1

The design shear strength of a flexural member without shear reinforcement is given by

$$v_c = \left(\frac{0.18k}{\gamma_c}\right)\left(\frac{100 A_{sl} f_{ck}}{b_w d}\right)^{1/3} \geq v_{min} = 0.035 k^{3/2} f_{ck}^{1/2}$$

where $\quad k = 1 + \sqrt{\dfrac{200}{d}} \leq 2.0, \quad \left(\dfrac{100 A_{sl}}{b_w d}\right) \leq 2.0 \quad$ and $\quad \gamma_c = 1.5$

OUTPUT column:

$d = 400$ mm

At interior support, provide 3H16 per rib at top of slab

In each span, provide 1H25 per rib at bottom of slab

Example 1   **Calculation Sheet 39**

Reference	CALCULATIONS	OUTPUT
	With $k = 1.7$ (for $d = 400$ mm), $v_{min} = 0.035 \times 1.7^{3/2} \times 32^{1/2} = 0.44$ MPa	
	If the tension bars are distributed uniformly across the flange to provide H16-200, only 1H16 (over the rib) will be taken into account for shear resistance.	
	$100A_{sl}/b_w d = 100 \times 203/(168 \times 400) = 0.30$	
	$v_c = (0.18 \times 1.7/1.5)(0.30 \times 32)^{1/3} = 0.43 \geq v_{min} = 0.44$ MPa	
	Since $v > v_c$, shear reinforcement is required.	
	Minimum requirements for vertical links are given by	
	$A_{sw}/s = (0.08\sqrt{f_{ck}})\, b_w /f_{yk} = (0.08\sqrt{32}) \times 168/500 = 0.15$ mm^2/mm	Provide H6-300 links in each rib, from face of supporting beam to end of top bar
	$s \leq 0.75d = 0.75 \times 400 = 300$ mm.	
	Using H6-300 links provides 0.19 mm^2/mm giving a design shear resistance:	
	$V_{Rd,s} = (A_{sw}/s)\, f_{ywd} z \cot\theta = 0.19 \times 0.87 \times 500 \times 0.9 \times 400 \times 2.5 \times 10^{-3} = 74.4$ kN	
	In the sagging region, b_w is the width of the section at the level of the H25 bar at the bottom of the rib. At the point of contra-flexure, for load case 2, $V = 32.7$ kN.	
	$v = V/b_w d = 32.7 \times 10^3/(142 \times 400) = 0.58$ MPa	
	$100A_{sl}/b_w d = 100 \times 491/(142 \times 400) = 0.86$	
	$v_c = (0.18 \times 1.7/1.5)(0.86 \times 32)^{1/3} = 0.61$ MPa	
6.2.1 (4)	Since $v < v_c$, no shear reinforcement is required within the sagging region, except at the end supports. Here, the bar at the bottom of the rib will lap with an H16 bar and minimum links will be provided over the length of the lap.	
	Deflection	
7.4.1 (6)	Deflection requirements may be met by limiting the span/effective depth ratio. For each span, the actual span/effective depth ratio = 9600/400 = 24.	
	The characteristic load: $g_k + q_k = 6.5 + 4.0 = 10.5$ kN/m	
	The reinforcement stress under the characteristic load is given approximately by	
	$\sigma_s = (f_{yk}/\gamma_s)(A_{s,req}/A_{s,prov})[(g_k + q_k)/n]$	
	$= (500/1.15)(381/491)(10.5/14.1) = 252$ MPa	
7.4.2 Table NA.5 PD 6687	From *Reynolds*, Table 4.21, limiting l/d = basic ratio $\times \alpha_s \times \beta_s$ where:	
	With bd taken as $bh_f + b_w (d - h_f)$, where b_w is taken as the average width of the rib above the level of the bottom reinforcement,	
	$bd = 600 \times 100 + (125 + 2 \times 200\tan 10^o) \times 350 = 128.4 \times 10^3$,	
	$100A_s/bd = 100 \times 491/(128.4 \times 10^3) = 0.38 < 0.1f_{ck}^{0.5} = 0.1 \times 32^{0.5} = 0.56$	
	$\alpha_s = 0.55 + 0.0075f_{ck}/(100A_s/bd) + 0.005f_{ck}^{0.5}[f_{ck}^{0.5}/(100A_s/bd) - 10]^{1.5}$	
	$= 0.55 + 0.0075 \times 32/0.38 + 0.005 \times 32^{0.5} \times (32^{0.5}/0.38 - 10)^{1.5} = 1.48$	
	$\beta_s = 310/\sigma_s = 310/252 = 1.23$	
	For an end span of a continuous slab, basic ratio = 26. For flanged sections with values of b/b_w greater than 3, the basic ratio should be multiplied by 0.8. For slabs with spans exceeding 7 m, supporting partitions liable to be damaged by excessive deflections, the basic ratio should be multiplied by 7/span.	
	Limiting $l/d = 26 \times 0.8 \times 7.0/9.6 \times \alpha_s \times \beta_s = 15.1 \times 1.48 \times 1.23 = 27.5$ (>24)	Check complies
	Cracking	
7.3.2 (2)	Minimum area of reinforcement required in tension zone for crack control:	
	$A_{s,min} = k_c k f_{ct,eff} A_{ct}/\sigma_s$	
	Taking values of $k_c = 0.4$, $k = 1.0$, $f_{ct,eff} = f_{ctm} = 0.3f_{ck}^{(2/3)} = 3.0$ MPa (for general design purposes), $A_{ct} = bh/2$ (for solid section at support) and $\sigma_s \leq f_{yk} = 500$ MPa	
BS EN 1990 Table NA.A1.1	$A_{s,min} = 0.4 \times 1.0 \times 3.0 \times 600 \times (450/2)/500 = 324$ mm^2 (<603 mm^2)	
	The quasi-permanent load, where $\psi_2 = 0.3$ is obtained from the National Annex to the Eurocode (Table 1.1), is given by:	

Example 1   **Calculation Sheet 40**

Reference	CALCULATIONS	OUTPUT
	$g_k + \psi_2 q_k = 6.5 + 0.3 \times 4.0 = 7.7$ kN/m^2 The service stress in the reinforcement under the quasi-permanent load is given approximately by $\sigma_s = (f_{yk}/\gamma_s)(A_{s,req}/A_{s,prov})[(g_k + \psi_2 q_k)/n]$ $\quad = (500/1.15)(590/603)(7.7/14.1) = 232$ MPa	
7.3.3 (2) Table 7.2 Table 7.3	The crack width criterion can be satisfied by limiting either the bar size or the bar spacing. Although the top surface of the slab will not be visible below the finishes, the deemed-to-satisfy criteria for $w_k = 0.3$ mm will be checked. The recommended maximum values, by interpolation, are bar spacing = 210 mm or $\phi^*_s = 18$ mm. The maximum bar size is then given by $\phi_s = \phi^*_s (f_{ct,eff}/2.9)[k_c h_{cr}/2(h-d)] = 18 \times (3.0/2.9) \times 0.4 \times 225/(2 \times 50) = 16$ mm If the bars are distributed uniformly to provide H16-200, it can be inferred that w_k will be less than 0.3 mm.	
	Detailing requirements	
9.2.1.1 (1)	Minimum area of longitudinal tension reinforcement (*Reynolds*, Table 4.28): $A_{s,min} = 0.26(f_{ctm}/f_{yk})b_t d = 0.26 \times (3.0/500) b_t d = 0.00156\, b_t d \geq 0.0013\, b_t d$ where b_t is the mean width of the tension zone For the sagging region, $b_t = 125 + 350 \tan 10° = 187$ mm and $A_{s,min} = 0.00156 \times 187 \times 400 = 117$ mm^2 (<491 mm^2 provided) For the hogging region, the depth of the tension zone for the uncracked section: $h_{cr} = \dfrac{b_w h^2 + (b_f - b_w)h_f^2}{2[b_w h + (b_f - b_w)h_f]} = \dfrac{187 \times 450^2 + 413 \times 100^2}{2[187 \times 450 + 413 \times 100]} = 168$ mm $b_t = (600 \times 100 + 236 \times 68)/168 = 452$ mm $A_{s,min} = 0.00156 \times 452 \times 400 = 282$ mm^2 (<603 mm^2 provided)	
9.2.1.2 (1)	At an end support where partial fixity occurs, top reinforcement to resist at least 25% of the maximum moment in the end span should be provided. $A_{s,min} = 0.25 \times 381 = 96$ mm^2 (1H16)	
9.2.1.5 (1) 9.2.1.5 (2)	At the bottom of each span, at least 25% of the area provided in the span should continue to the supports and be provided with an anchorage length beyond the face of the support not less than 10ϕ. $A_{s,min} = 0.25 \times 491 = 123$ mm^2 (1H16)$\qquad l_{b,min} = 10 \times 16 = 160$ mm At an end support, the tensile force is given by $F = (a_l/z)V$, with $a_l = d$ and $z = 0.9d$. With $V = 32.7$ kN and $A_s = 201$ mm^2 $F = 32.7/0.9 = 36.3$ kN/m and $\sigma_s = V/A_s = 36.3 \times 10^3/201 = 181$ MPa	At end supports, provide U-bars in the vertical plane (1H16 per rib)
8.4.3 (2) 8.4.4	For good bond, $f_{ck} = 32$ MPa and $\sigma_s = 435$ MPa, $l_{b,rqd} = 35\phi$ (*Reynolds*, Table 4.30) The tabulated value may be multiplied by $\sigma_s/435$, where $\sigma_s = 181$ MPa, giving $l_{b,rqd} = (181/435) \times 35 \times 16 = 233$ mm $\geq l_{b,min} = 10\phi = 10 \times 16 = 160$ mm	
	Curtailment of longitudinal tension reinforcement If V_1 and V_2 are values at the end and interior supports, respectively, the distance x from the interior support to a point of contra-flexure is given by $V_2 - nx = V_1$. Load case 1 gives the following values: $V_1 = 30.5$ kN, $V_2 = 50.8$ kN, $n = 0.6 \times 14.1 = 8.46$ kN/m, $x = 2.4$ m The top bar in each rib will be extended beyond this point for a further minimum distance $a_l = 0.45d \cot\theta = 0.45 \times 400 \times 2.5 = 450$ mm Load case 2, with minimum load on the span, gives the following values: $V_1 = 15.4$ kN, $V_2 = 31.3$ kN, $n = 0.6 \times 8.1 = 4.86$ kN/m, $x = 3.3$ m	At interior support, curtail top bars at 2.85 m from centre of support

Example 1   **Calculation Sheet 41**

Reference	CALCULATIONS	OUTPUT
	If the flange is reinforced with A252 fabric, the hogging moment of resistance and shear resistance in the region where $x > 2.4$ m, are as follows:	Provide A252 fabric reinforcement in flange
	$M_u = A_s(0.87f_{yk}z) = (0.6 \times 252) \times (0.87 \times 500 \times 0.95 \times 410) \times 10^{-6} = 25.6$ kN m	
	$V_u = v_{min}\,b_w d = 0.44 \times 142 \times 410 \times 10^{-3} = 25.6$ kN	
	These resistances are sufficient to cater for the values that occur for load case 2, with minimum load on the span, in the region between $x = 2.4$ m and $x = 3.3$ m.	
	Tying requirements (see *Reynolds*, Table 4.29)	
9.10.2.3 Table NA.1	The longitudinal reinforcement in the bottom of each rib can be used to provide continuous internal ties. With $l_r = 9.6$ m and $F_t = (20 + 4n_0) \leq 60$,	At interior support, provide 1H16 bar at bottom of each rib
	$F_{tie,int} = [(g_k + q_k)/7.5](l_r/5)F_t = (10.5/7.5)(9.6/5)(20 + 4 \times 6) = 118.3$ kN/m	
9.10.1 (4)	With $\sigma_s = 500$ MPa, minimum area of reinforcement	
	$A_{s,min} = 118.3 \times 600/500 = 142$ mm^2 (1H16)	At each end of each span, provide 850 mm laps between bars in bottom of each rib
	Design lap length to develop full tensile resistance of bar	
8.7.3 (1)	$l_0 = \alpha_6 \times (35\phi) = 1.5 \times (35 \times 16) = 850$ mm say	

Bar Marks	Commentary on Bar Arrangement (Drawing 11)
01	Bar at bottom of longitudinal ribs, with 35 mm nominal cover, curtailed 50 mm from face of integral beam at each end.
02	U-bar, shape code 21, with legs of equal length. Lower leg positioned above bar mark 01, with 850 mm lap. Dimension in vertical plane to provide tolerance for U-bar to fit inside links.
03	Bar positioned above bar mark 01, with 850 mm lap in each span.
04	Top bars at spacing of 200 mm, extending beyond centreline of interior beam 2850 mm into each span.
05	Closed links, shape code 33, at maximum spacing permitted by requirements for shear reinforcement (see calculation sheet 39). No additional requirements for transverse reinforcement in lap zones, since diameter of lapped bars is less than 20 mm (see clause 8.7.4.1).
06	U-bar, shape code 21, with legs of equal length. Lower leg laps 650 mm with bar mark 07.
07	Bar at bottom of transverse rib, crossing over bar mark 01 in longitudinal ribs.
-	A252 fabric reinforcement in 2.4 m wide sheets with 300 mm laps. Sheets to intermesh with transverse wires in layer 2, and longitudinal wires in alternate sheets in layers 1 and 3, respectively.

Example 1: Reinforcement in Trough Slab Floor

Drawing 11

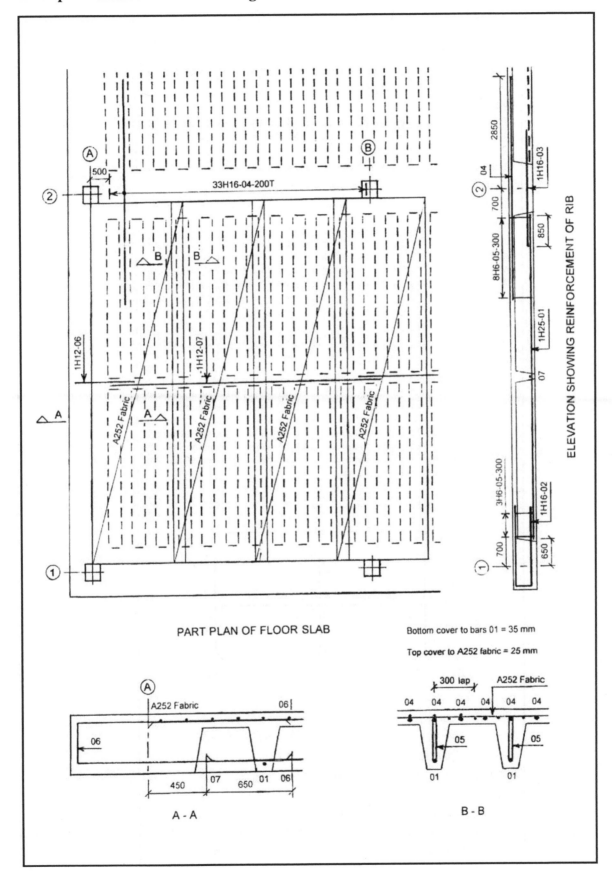

PART PLAN OF FLOOR SLAB

ELEVATION SHOWING REINFORCEMENT OF RIB

Bottom cover to bars 01 = 35 mm

Top cover to A252 fabric = 25 mm

A - A

B - B

Example 1   **Calculation Sheet 42**

Reference	CALCULATIONS	OUTPUT
BS EN 1992-1-2 5.6.3 (2) Table 5.6	**INTEGRAL BEAM ON LINE 2** The beam dimensions, which were assumed in the design of the ribbed slab, are 450 mm deep and 1200 mm wide. **Fire resistance** Allowing for the design to be based on no more than 15% moment redistribution, the beam may be taken as continuous. For the ground floor (minimum fire period 1.5 h), the required minimum dimensions are: Beam width: ≥ 250 mm. Axis distance (to centre of bars in one layer): 25 mm Axis distance to side of beam for corner bars: $(25 + 10) = 35$ mm Since the cover required for durability is 25 mm, the axis distances are sufficient.	Sufficient for 1.5 h fire period

Loading

For the ribbed slab, the maximum design load is 14.1 kN/m^2 and the minimum design load is 8.1 kN/m^2. The extra volume of concrete in the solid portion of the slab is $(0.45 - 0.21) = 0.24$ m^3 per square metre of floor area. The loads on the beam taking shear force coefficients for the slab of 0.625 for each span, are:

		(max)		(min)
Slab	$1.25 \times 9.6 \times 14.1 =$	169.2	$1.25 \times 9.6 \times 8.1 =$	97.2
Extra	$1.25 \times 1.2 \times 0.24 \times 25 =$	9.0	$=$	9.0
		178.2 kN/m		106.2 kN/m

Analysis

The beams are continuous over two spans. Design moments and shears will be derived from an elastic analysis of a sub-frame consisting of the beam at one level together with the columns above and below. The analysis of a sub-frame where the columns above and below the floor are identical will be shown. The support conditions at C and D are difficult to model. For ease of analysis, the stiffness of the wall on line C will be taken same as that of the column on line A. Since the wall is 200 mm thick and the column is 400 × 400, this is equivalent to taking a wall length equal to 8 × 400 = 3200 mm.

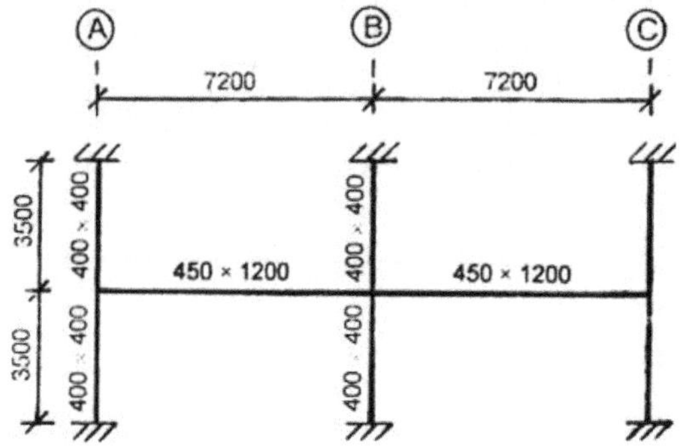

Dimensions of simplified sub-frame (2nd, 3rd and 4th floors)

The properties of the members are:

$I_b = 1200 \times 450^3 / 12 = 9.11 \times 10^9$ mm^4, $K_b = I_b / 7200 = 1.27 \times 10^6$ mm^3

$I_c = 400 \times 400^3 / 12 = 2.13 \times 10^9$ mm^4, $K_c = I_c / 3500 = 0.61$ mm^3

$I_w = 3200 \times 200^3 / 12 = 2.13 \times 10^9$ mm^4, $K_w = I_w / 3500 = 0.61$ mm^3

Distribution factors for unit moment applied at an end joint are:

$D_b = 1.27 / (1.27 + 2 \times 0.61) = 0.510$, $D_c = (1 - 0.510)/2 = 0.245$

Distribution factors for unit moment applied at the interior joint are:

$D_b = 1.27 / (2 \times 1.27 + 2 \times 0.61) = 1.27 / 3.76 = 0.338$, $D_c = 0.61/3.76 = 0.162$

Fixed-end moments due to maximum and minimum load on slab are:

Example 1   **Calculation Sheet 43**

Reference	CALCULATIONS	OUTPUT
	$M_{max} = 178.2 \times 7.2^2/12 = 769.8$ kN m, $M_{min} = 106.2 \times 7.2^2/12 = 458.8$ kN m	

The following results are obtained for load case 1 (maximum load on both spans), load case 2 (maximum load on span AB, minimum load on span BC), and load case 3 (minimum load on span AB, maximum load on span BC).

Load Case	Location and Member	Support A	Span AB	Support B	Support B	Span BC
No.	Bending Moment (kN m) in Members for Load Case					
1	Beam	−377.2	502.1	966.1	−966.1	502.1
	Upper column	188.6		0		
	Lower column	188.6		0		
2	Beam	−412.4	538.7	840.0	−701.8	266.4
	Upper column	206.2		−69.1		
	Lower column	206.2		−69.1		
3	Beam	−189.6	266.4	701.8	−840.0	538.7
	Upper column	94.8		69.1		
	Lower column	94.8		69.1		

No.	Shear Force (kN) in Members for Load Case			
1	Beam	559.8	723.3	723.3
2	Beam	582.2	700.9	453.5
3	Beam	311.2	453.5	700.9

Allowing for some moment redistribution, the maximum hogging moments in the beam will be taken as 412.4 kN m at supports A and C, and 840.0 kN m at support B. As a result, the maximum sagging moments in the spans will remain unchanged for load cases 2 and 3, but will increase to 538.7 kN m for load case 1.

Flexural design

At the top of the beam, allowing for 25 mm cover, A252 fabric, 16 mm transverse bars and 25 mm longitudinal bars,

$d = 450 - (25 + 2 \times 8 + 16 + 25/2) = 380$ mm

5.5 (4)
Table NA.1

At the interior support, the ratio of design moment to maximum elastic moment is $\delta = 840/966.1 = 0.87$, and the ductility criterion $x/d \le (\delta - 0.4) = 0.47$ applies.

$M/bd^2 f_{ck} = 840 \times 10^6/(1200 \times 380^2 \times 32) = 0.152$

From Table A1, $A_s f_{yk}/bdf_{ck} = 0.208$ and $x/d = 0.399$ (<0.47).

$A_s = 0.208 \times 1200 \times 380 \times 32/500 = 6070$ mm^2 (13H25)

At the end supports, where $M = 412.4$ kN m:

$M/bd^2 f_{ck} = 0.075$, $A_s f_{yk}/bdf_{ck} = 0.093$, $A_s = 2715$ mm^2 (6H25)

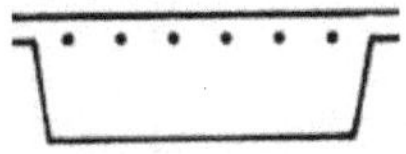

At the bottom of the beam, allowing for 25 mm cover with 8 mm links and 25 mm longitudinal bars, $d = 450 - (25 + 8 + 25/2) = 400$ mm say.

For the spans, where the effective flange width is given by:

5.3.2.1 (3)

$b_{eff} = b_w + 2 \times 0.2 \times 0.7l = 1200 + 0.28 \times 7200 = 3216$ mm

$M/bd^2 f_{ck} = 538.7 \times 10^6/(3216 \times 400^2 \times 32) = 0.033$ (<0.054), $z/d = 0.95$ (max)

$A_s = M/0.87 f_{yk} z = 538.7 \times 10^6/(0.87 \times 500 \times 0.95 \times 400) = 3259$ mm^2 (11H20)

Shear design

Since the load is uniformly distributed, the critical section for shear can be taken at distance d from the face of support, that is, 580 mm from the centre of column.

At interior support B, the critical shear value is:

$V = 723.3 - 178.2 \times 0.58 = 620$ kN

The required inclination of the concrete strut (defined by $\cot\theta$), to obtain the least amount of shear reinforcement, can be shown to depend on the following factor:

OUTPUT column:

13H25
At interior support

6H25
At end support

Spans
11H20

Example 1   Calculation Sheet 44

Reference	CALCULATIONS	OUTPUT
	$v_w = V/[b_w z (1 - f_{ck}/250) f_{ck}$ which, with $V = 620$ kN, gives	
	$v_w = 620 \times 10^3/[1200 \times 0.9 \times 380 \times (1 - 32/250) \times 32] = 0.054$	
	From *Reynolds*, Table 4.18, for vertical links and values of $v_w < 0.138$, $\cot\theta = 2.5$ can be used. The area of links required is then given by	
	$A_{sw}/s = V/f_{ywd} z \cot\theta$	
	$= 620 \times 10^3/(0.87 \times 500 \times 0.9 \times 380 \times 2.5) = 1.67$ mm^2/mm	At interior support
9.2.2 (6)	Maximum longitudinal spacing of vertical links, $s_{max} = 0.75d = 285$ mm	
9.2.2 (8)	Maximum transverse spacing of legs, $s_{t,max} = 0.75d = 285$ mm (≤ 600 mm)	
	6H8-175 gives $A_{sw}/s = 302/175 = 1.72$ mm^2/mm (3 sets of links)	H8-175 (3 sets)
	Minimum requirements for vertical links are given by	Minimum links
	$A_{sw}/s = (0.08\sqrt{f_{ck}}) b_w/f_{yk} = (0.08\sqrt{32}) \times 1200/500 = 1.09$ mm^2/mm	H8-200 (3 sets)
	6H8-200 links gives $A_{sw}/s = 302/200 = 1.51$ mm^2/mm	Note. Link spacing chosen to suit spacing
	$V_{Rd,s} = (A_{sw}/s) f_{ywd} z \cot\theta = 1.51 \times 0.87 \times 500 \times 0.9 \times 380 \times 2.5 \times 10^{-3} = 561.6$ kN	of bars in ribbed slab
	Distance from support B at which $V_{Rd,s} = 561.6$ kN is sufficient is given by	
	$(V - V_{Rd,s})/n = (723.3 - 561.6)/178.2 = 0.91$ m	
	At support A, critical shear is $V = 582.2 - 178.2 \times 0.58 = 479$ kN ($< V_{Rd,s}$)	
	Deflection (See *Reynolds*, Table 4.21)	
7.4.1 (6)	Deflection requirements may be met by limiting the span/effective depth ratio. The actual span/effective depth ratio = $7200/400 = 18$	
	The characteristic load is given by	
	$g_k + q_k = 1.25 \times 9.6 \times 10.5 + 1.2 \times 0.24 \times 25 = 133.2$ kN/m	
	The reinforcement stress under the characteristic load is given approximately by	
	$\sigma_s = (f_{yk}/\gamma_s)(A_{s,req}/A_{s,prov})[(g_k + q_k)/n]$	
	$= (500/1.15)(3259/3456)(133.2/178.2) = 307$ MPa	
7.4.2 Table NA.5 PD 6687	From *Reynolds*, Table 4.21, limiting l/d = basic ratio $\times \alpha_s \times \beta_s$ where:	
	With bd taken as $b_{eff} h_f + b_w(d - h_f) = 3216 \times 100 + 1200 \times 300 = 681.6 \times 10^3$,	
	$100 A_s/bd = 100 \times 3259/(681.6 \times 10^3) = 0.48 < 0.1 f_{ck}^{0.5} = 0.1 \times 32^{0.5} = 0.56$	
	$\alpha_s = 0.55 + 0.0075 f_{ck}/(100 A_s/bd) + 0.005 f_{ck}^{0.5}[f_{ck}^{0.5}/(100 A_s/bd) - 10]^{1.5}$	
	$= 0.55 + 0.0075 \times 32/0.48 + 0.005 \times 32^{0.5} \times (32^{0.5}/0.48 - 10)^{1.5} = 1.12$	
	$\beta_s = 310/\sigma_s = 310/307 = 1.01$	
	For an end span of a continuous beam, the basic ratio = 26. For beams with spans exceeding 7 m, supporting partitions liable to be damaged by excessive deflections, the basic ratio should be multiplied by 7/span. For flanged sections, the basic ratio should be multiplied by $(11 - b/b_w)/10 = (11 - 3216/1200)/10 = 0.83 \geq 0.8$.	
	Limiting $l/d = 26 \times 0.83 \times 7.0/7.2 \times \alpha_s \times \beta_s = 21 \times 1.12 \times 1.01 = 23.7$ (>18)	Check complies
	Cracking	
7.3.2 (2)	Minimum area of reinforcement required in tension zone for crack control:	
	$A_{s,min} = k_c k f_{ct,eff} A_{ct}/\sigma_s$	
	For the hogging region at the interior support, the tension flange is considered to extend beyond the side face of the beam for a distance given by	
5.3.2.1 (3)	$b_{eff,i} = 0.2 \times 0.15(l_1 + l_2) = 0.03 \times (7200 + 7200) = 400$ mm say.	
	For the uncracked section, the depth of the tension zone ignoring the effect of the reinforcement is given by	
	$h_{cr} = \dfrac{b_w h^2 + (b_f - b_w) h_f^2}{2[b_w h + (b_f - b_w) h_f]} = \dfrac{1200 \times 450^2 + 800 \times 100^2}{2[1200 \times 450 + 800 \times 100]} = 202$ mm ($> h_f$)	

Example 1   **Calculation Sheet 45**

Reference	CALCULATIONS	OUTPUT
	$A_{ct} = 2000 \times 100 + 1200 \times 102 = 322.4 \times 10^3$ mm^2, $k = 0.65$ (since $b \geq 800$ mm), $k_c = 0.9 \times (152/202) \times (200/322.4) = 0.42 \geq 0.5$. Hence, $A_{s,min} = 0.5 \times 0.65 \times 3.0 \times 322.4 \times 10^3/500 = 629$ mm^2 (<6380 mm^2 provided)	
BS EN 1990 Table NA.A1.1	The quasi-permanent load, where $\psi_2 = 0.3$ is obtained from the National Annex to the Eurocode (Table 1.1), is given by $g_k + \psi_2 q_k = 1.25 \times 9.6 \times (6.5 + 0.3 \times 4.0) + 9.0/1.25 = 99.6$ kN/m Taking account of the moment redistribution in the analysis, the service stress in the reinforcement under the quasi-permanent load is given approximately by $\sigma_s = (f_{yk}/\gamma_s)(M_{elastic}/M_{design})(A_{s,req}/A_{s,prov})[(g_k + \psi_2 q_k)/n]$ $= (500/1.15)(966.1/840.0)(6070/6380)(99.6/178.2) = 266$ MPa	
7.3.3 (2)	The crack width criterion can be satisfied by limiting either the bar size or the bar spacing. For the top of the beam, it is reasonable to ignore any requirement based on appearance, since the surface of the beam will not be visible below the finishes.	
Table 7.2 Table 7.3	Nevertheless, for $w_k = 0.3$ mm and $\sigma_s = 270$ MPa, the recommended maximum values, by interpolation, are bar spacing 160 mm or $\phi^*_s = 13$ mm. The maximum bar size is then given by: $\phi_s = \phi^*_s (f_{ct,eff}/2.9) [h_{cr}/4(h-d)] = 13 \times (3.0/2.9) \times [202/(4 \times 70)] = 10$ mm The reinforcement comprises 25 mm bars at 90 mm centres approximately. Thus, although there is no specific requirement to be satisfied, it can be inferred that w_k will be less than 0.3 mm. (Note: This is a rather dubious means of compliance in cases such as this where the cover is large). In the sagging regions, with no redistribution, the stress in the reinforcement under the quasi-permanent loading is given approximately by $\sigma_s = (500/1.15)(3259/3456)(99.6/178.2) = 230$ MPa	
Table 7.2 Table 7.3	For $w_k = 0.3$ mm, the maximum values, by interpolation, are: $\phi^*_s = 18$ mm, or bar spacing 210 mm. The bar size is 20 mm and the maximum bar spacing is 180 mm.	
	Detailing requirements	
9.2.1.1 (1)	Minimum area of longitudinal tension reinforcement (*Reynolds*, Table 4.28): $A_{s,min} = 0.26(f_{ctm}/f_{yk})b_t d = 0.26 \times (3.0/500) b_t d = 0.00156 b d \geq 0.0013 b_t d$ For the hogging region at the interior support, $h_{cr} = 202$ mm, $b_t = (2000 \times 100 + 1200 \times 102)/202 = 1596$ $A_{s,min} = 0.00156 \times 1596 \times 380 = 946$ mm^2 For the sagging regions, $A_{s,min} = 0.00156 \times 1200 \times 400 = 749$ mm^2	
9.2.1.5 (1) 9.2.1.5 (2)	At the bottom of each span, at least 25% of the area provided in the span should continue to the supports and be provided with an anchorage length beyond the face of the support of not less than 10ϕ. In the final detail, 6 bars are made effectively continuous for the whole length of the beam. $l_{b,min} = 10 \times 25 = 250$ mm	$l_{b,min} = 250$ mm
Figure 8.2	For the bars at the top of the beam, poor bond conditions are assumed. Hence, from *Reynolds*, Table 4.30, with $f_{ck} = 32$ MPa, $l_{b,rqd} = (A_{s,req}/A_{s,prov}) \times 50\phi \geq l_{b,min}$. Thus, at the end supports	$l_{b,rqd} = 1150$ mm
8.4.3 (2)	$l_{b,rqd} = (A_{s,req}/A_{s,prov}) \times 50\phi = (2715/2945) \times 50 \times 25 = 1150$ mm	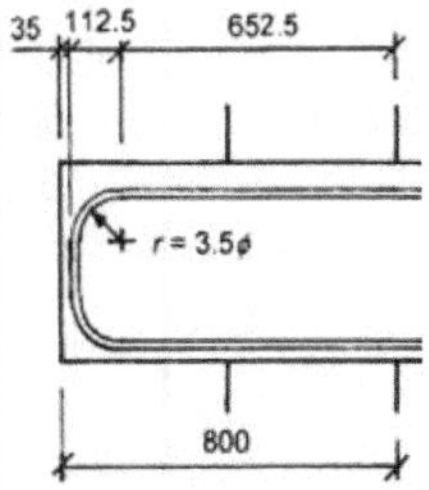
8.3 (3)	The minimum radius of bend of the bars depends on the value of a_b/ϕ, where a_b is taken as half the centre-to-centre distance between the bars. In the final detail, the bar spacing is about 200 mm, so that $a_b = 0.5 \times 200 = 100$ mm. From *Reynolds*, Table 4.31, with $f_{ck} = 32$ MPa and $a_b/\phi = 100/25 = 4$, $r_{min} = 7.1\phi$. This value can be reduced by allowing for $A_{s,req} < A_{s,prov}$, and taking into account the stress reduction in the bar between the edge of the support and the start of the bend. Thus, if a standard U-bar (shape code 21) is used, $r = 3.5\phi$, and distance from edge	

Example 1   **Calculation Sheet 46**

Reference	CALCULATIONS	OUTPUT
	of support to start of bend = 800 - (35 + 4.5 × 25) = 652.5 mm. Reduced value of r_{min} = (2715/2945)(1 − 652.5/1150) × 7.1ϕ = 2.9ϕ (<3.5ϕ) **Curtailment of longitudinal tension reinforcement** (see *Reynolds*, Table 4.32) Since the shear reinforcement consists of 3 sets of 2-leg links, a minimum of six longitudinal bars will be provided. The resistance moment provided by 6H20 at the bottom of the beam can be determined as follows: $M = A_s(0.87f_{yk})z = 1885 × 0.87 × 500 × 0.95 × 400 × 10^{-6} = 311.6$ kN m At the end support, M_s = 412.4 kN m and distance x from the support to a point where M = 311.6 kN m is given by: $Vx − nx^2/2 − 412.4 = 311.6$ kN m For load cases 1 (after redistribution) and 2: V = 582.2 kN and n = 178.2 kN/m giving the equation: $0.5x^2 − 3.27x + 4.06 = 0$ solutions of which are x = 1.67 m and 4.87 m	
9.2.1.3 (2)	Thus, of the 11H20 required in the spans, 5 bars are no longer needed for flexure at 1.67 and 4.87 m from the end support. Here V = 284.6 kn and $V_{Rd,s}$ = 446.3 kN with $\cot\theta$ = 2.5. Thus $\cot\theta = (V/V_{Rd,s}) × 2.5 = 1.60$ is sufficient and the bars should extend beyond these points for a minimum distance $a_1 = z(\cot\theta)/2 = 0.45d\cot\theta$ $a_1 = 0.45 × 400 × 1.6 = 288$ mm $x − a_1 = 1670 − 288 = 1350$ mm say, $x + a_1 = 4870 + 288 = 5200$ mm say At the top of the beam, with 6H12 supporting the links, the resistance moment: $M = 679 × 0.87 × 500 × 0.95 × 380 × 10^{-6} = 106.6$ kN m If V and M_s are the values at the end support, the distance x from the support to a point where M = 106.6 kN m is given by: $Vx − nx^2/2 = M_s − 106.6$ For load cases 1 (after redistribution) and 2: V = 582.2 kN, M_s = 412.4 kN m, n = 178.2 kN/m, giving x = 0.58 and 5.96 m For load case 3: V = 311.2 kN, M_s = 189.6 kN m, n = 106.2 kN/m, giving x = 0.28 and 5.58 m	At bottom of each span, stop 5H20 at 1350 mm from end support and 2000 mm from interior support
9.2.1.3 (2)	At these points, $\cot\theta$ = 2.5, and the bars to be curtailed should extend for a further distance not less than $a_1 = 0.45 × 380 × 2.5 = 430$ mm. It is also necessary to ensure that the bars extend for a distance not less than $(a_1 + l_{bd})$ beyond the face of the support. For simplicity, $l_{bd} = l_{b,rqd}$ will be assumed, as the modification coefficients have only a minor effect. For the U-bars at the end support, $l_{b,rqd}$ = 1150 mm. Thus, the critical distance is $(a_1 + l_{bd})$ = 430 + 1150 = 1600 mm say from face of support. At the interior support, the bars to be curtailed should extend for a distance not less than $a_1 + 50\phi$ from face of support, nor less than $a_1 + (l−x)$ from the centre of support. Thus, of the 13H25 bars required at the centre of support, 7 bars could be curtailed at $(a_1 + l_{bd})$ = 430 + 1250 = 1680 mm from face of support, with the other 6 bars curtailed at $a_1 + (l−x)$ = 430 + (7200 − 5580) = 2050 mm from the centre of support.	At the end supports, extend upper leg of U-bars for 1600 mm from face of support At interior support, extend all bars for 2050 mm from the centre of support
9.10.2.3 Table NA.1	**Tying requirements** (see *Reynolds*, Table 4.29) The longitudinal reinforcement at the bottom of each span can be used to provide continuous internal ties. With l_r = 7.2 m and $F_t = (20 + 4n_0) ≤ 60$, $F_{tie,int} = [(g_k)/7.5](l_r/5)F_t = (9.0/7.5)(7.2/5)(20 + 4 × 6) = 76$ kN/m	
9.10.1 (4)	For beams at 9.6 m centres, with σ_s = 500 MPa, minimum area of reinforcement $A_{s,min}$ = 9.6 × 76 × 1000/500 = 1460 mm^2 Use 6H20	6H20 continuous at bottom of beam
8.7.3 (1)	At the supports, where the bars will be lapped, design lap length $l_0 = \alpha_6 × (35\phi) × A_{s,req}/A_{s,prov}$ = 1.5 × (35 × 20) × (1460/1885) = 850 mm say	850 mm lap

Example 1

Calculation Sheet 47

Reference	CALCULATIONS	OUTPUT
	INTEGRAL BEAM ON LINE 1	
	The loading comprises the shear force from the ribbed slab, plus the load resulting from the 600 mm wide edge strip of slab and extra volume of concrete in the solid portion of the slab, and a load of 5 kN/m to cover walling, cladding and windows.	
	Total design load, assuming a shear force coefficient of 0.4 for the slab span is	
	$(0.4 \times 9.6 + 0.6) \times 14.1 + 1.25 \times (1.2 \times 0.24 \times 25 + 5.0) = 78$ kN/m (max)	
	The analysis of the sub-frame on line 1 will be similar to that for the sub-frame on line 2, except that the beam is continuous over five spans. The design load is 44% of that for the beam on line 2, giving pro-rata reinforcement as follows:	
	Interior support: $A_s = 0.44 \times 6070 = 2670$ mm^2 (6H25) End support: $A_s = 0.44 \times 2715 = 1195$ mm^2 (6H16) 7.2 m spans: $A_s = 0.44 \times 3259 = 1434$ mm^2 (6H20)	6H25 interior support 6H16 end supports 6H20 spans
	For shear links, providing H8-200 (three sets) throughout will suffice.	
9.10.2 Table NA.1	The longitudinal reinforcement at the bottom of each span can be used to provide continuous internal and peripheral ties, where $F_{tie,per} = F_t = 44$ kN.	6H16 continuous at bottom of beam
	$A_{s,min} = [(4.8 + 0.6) \times 76 + 44] \times 1000/500 = 909$ mm^2 Use 6H16	
	$l_0 = 1.5 \times (35 \times 16) \times (909/1206) = 650$ mm say	650 mm lap

Bar Marks	Commentary on Bar Arrangement (Drawing 12)
01, 08	Bars in corners of links curtailed 50 mm from column face at each end.
02	Loose U-bars, shape code 21. Upper leg extends $(a_1 + l_{bd}) = 430 + 50 \times 16 = 1250$ mm say beyond face of column, to satisfy curtailment requirement, and lower leg laps 650 mm with bar mark 01. Overall dimension of vertical leg provides tolerance for U-bar to fit inside links.
03	Loose bars lapping 650 mm with bars 01 to provide continuity of internal ties.
04, 12	Bars extending into each span 2050 mm beyond centreline of column.
05	Bars in corner of links lapping 300 mm with bars 02 and 04.
06, 07	Closed links, shape code 51, in sets of one 06 and two 07. Spacing of links determined by requirements for shear reinforcement (see calculation sheet 44), and transverse reinforcement in lap zones of main bars. Where diameter of lapped bars $\phi \geq 20$ mm, transverse bars of total area not less than area of one lapped bar should be provided within outer third of lap zone (see clause 8.7.4.1). For the beam on line 2, allowing for $A_{s,req} < A_{s,prov}$, total area of transverse bars for full lap zone is $A_{st} = 1.5 \times 314 \times 1460/1885 = 365$ mm^2 (8H8). The transverse reinforcement provided to the longitudinal bars comprises 4H8 to the outer two bars and 8H8 to the inner four bars, which is considered to be a reasonable arrangement.
09	Bars curtailed 1350 mm from centreline of column A and 2000 mm from centreline of column B (see calculation sheet 46).
10	Loose U-bars, shape code 21. Upper leg extends 1600 mm beyond face of column to satisfy curtailment requirement (see calculation sheet 46) and lower leg laps 850 mm with bar mark 01. Overall dimension of vertical leg provides tolerance for U-bar to fit inside links.
11	Loose bars lapping 850 mm with bars 08 to provide continuity of internal ties.
13	Bars in corner of links lapping 300 mm with bars 10 and 12.

Example 1: Reinforcement in Integral Beams on Lines 1 and 2   **Drawing 12**

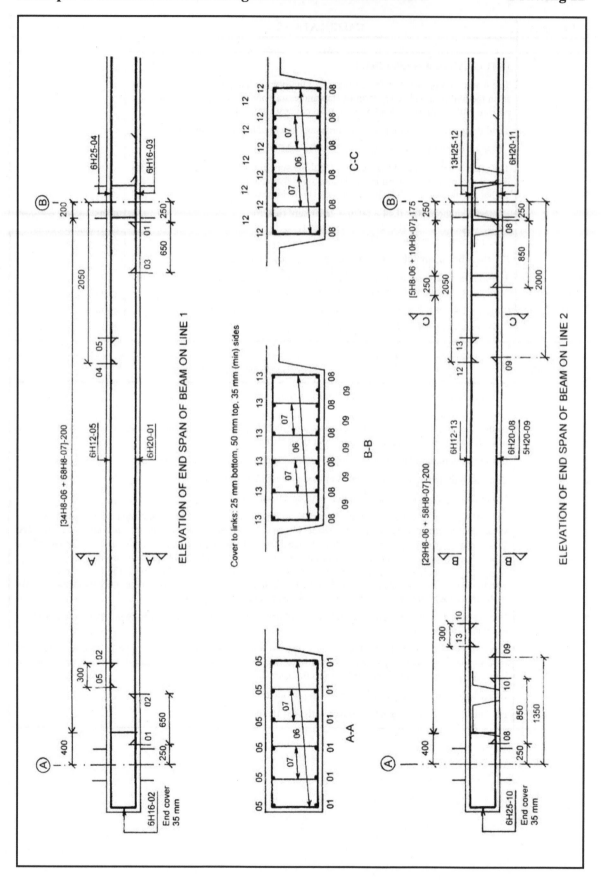

Example 1

Calculation Sheet 48

Reference	CALCULATIONS	OUTPUT
	ACTIONS ON COLUMNS	

ACTIONS ON COLUMNS

For the columns on line 2, the sub-frame analysis results shown in calculation sheet 43 give beam shears and column moments for three load cases, and apply at 2nd, 3rd and 4th floor levels. For simplicity, the same values will be used at lower floor levels, even though the storey heights result in sub-frame dimensions that are slightly different. At the roof level, the sub-frame and the loading are significantly different, and another analysis is required. Loading details are as follows:

Characteristic loading for roof slab:

  Slab and finishes: $(5.25 + 1.5) = 6.75$ kN/m^2 Imposed: 0.6 kN/m^2

Design ultimate load for roof slab: $n = 1.25 \times 6.75 + 1.5 \times 0.6 = 9.4$ kN/m^2

Loads per storey due to the self-weight of the columns:

  Columns up to 1st floor: $1.25 \times 0.4 \times 0.4 \times 25 \times 3.55 = 17.8$ kN

  Columns above 1st floor: $1.25 \times 0.4 \times 0.4 \times 25 \times 3.05 = 15.3$ kN

BS EN 1991-1-1 6.3.1.2 NA.2

A reduction may be made in the total imposed floor load, according to the number of storeys being supported at the level considered. For up to five storeys, this load may be multiplied by $\alpha_n = 1.1 - n/10$, where n is the number of storeys.

EDGE COLUMN A2

At each level, the load applied is the shear force from the sub-frame on line 2 (see calculation sheet 43) plus force F due to the edge loading on line A.

  Floor: $F = 1.2 \times 9.6 \times [0.6 \times 14.1 + 1.25 \times (1.0 \times 0.24 \times 25 + 5.0)] = 255.9$ kN

  Roof: $F = 1.2 \times 9.6 \times [0.6 \times 9.4 + 1.25 \times (0.24 + 0.15) \times 25] = 205.4$ kN

The maximum moment and maximum coexistent load occur when load case 2 is applied at all levels (see below). Maximum moment and minimum coexistent load occur when load case 2 is applied at the level considered, and $1.0G_k$ is applied at levels above. This arrangement can be critical for values of $N_{Ed} < N_{bal}$ where:

$N_{bal} = 0.4A_c f_{cd} = 0.4 \times 400 \times 400 \times 0.85 \times 32/1.5 \times 10^{-3} = 1160$ kN

Values of axial load N (kN) and bending moment M (kN m)						
Loading	$1.25G_k + 1.5Q_k$				$1.5Q_k$	
Load case	1		2		1	2
Member	N	M	N	M	N	N
Roof slab	569.8	170.5	573.8	175.3	(32.3)	
column	<u>15.3</u>		<u>15.3</u>			
	585.1	188.6	589.1	206.2		
4th floor slab	<u>815.7</u>		<u>838.1</u>		226.2	248.6
	1400.8	188.6	1427.2	206.2		
Column	<u>15.3</u>		<u>15.3</u>			
	1416.1	188.6	1442.5	206.2		
3rd floor slab	<u>815.7</u>		<u>838.1</u>		<u>226.2</u>	<u>248.6</u>
	2231.8	188.6	2280.6	206.2	452.4	497.2
Column	<u>15.3</u>		<u>15.3</u>			
	2247.1	188.6	2295.9	206.2		
2nd floor slab	<u>815.7</u>		<u>838.1</u>		<u>226.2</u>	<u>248.6</u>
	3062.8	188.6	3134.0	206.2	678.6	745.8
Column	<u>15.3</u>		<u>15.3</u>			
	3078.1	188.6	3149.3	206.2		
1st floor slab	<u>815.7</u>		<u>838.1</u>		<u>226.2</u>	<u>248.6</u>
	3893.8	188.6	3987.4	206.2	904.8	994.4
Column	<u>17.8</u>		<u>17.8</u>			
	3911.6	188.6	4005.2	206.2		
Grd. floor slab	<u>815.7</u>		<u>838.1</u>		<u>226.2</u>	<u>248.6</u>
Basement wall	4727.3	188.6	4843.3	206.2	1131.0	1243.0

For the storey from ground to 1st floor, with load case 2 at ground floor level:

Example 1   Calculation Sheet 49

Reference	CALCULATIONS	OUTPUT
	$M_{bot} = 188.6$ kN m, $M_{top} = -0.5M_{bot} = -94.3$ kN m	
	With load case 2 at levels above: $N_{Ed} = 4005.2 - 0.3 \times 994.4 = 3707$ kN (max)	
	With $1.0G_k$ at levels above: $N_{Ed} = [3893.8 - (904.8 + 32.3)]/1.25 = 2365$ kN (min)	
6.1 (4)	Minimum total design moment, with $e_0 = h/30 = 300/30 \geq 20$ mm:	
	$\quad M_{min} = N_{Ed}e_0 = 3707 \times 0.02 = 74.2$ kN m	
	Effective length and slenderness	
	Using the simplified method in Concise Eurocode 2, with condition 1 (monolithic connection to beams at least as deep as the overall depth of the column) at both top and bottom of the column,	
	$\quad l_0 = 0.75l = 0.75 \times 3.55 = 2.66$ m (for storeys above 1st floor, $l_0 = 2.29$ m)	
5.2 (9)	First-order moment from imperfections (simplified procedure):	
	$\quad M_i = Nl_0/400 = 3707 \times 2.66/400 = 24.7$ kN m	
	First-order moments, including the effect of imperfections:	
	$\quad M_{01} = -94.3 + 24.7 = -69.6$ kN m, $M_{02} = 188.6 + 24.7 = 213.3$ kN m	
	Radius of gyration of uncracked concrete section, $i = h/\sqrt{12} = 0.115$ m	
5.8.3.2 (1)	Slenderness ratio $\lambda = l_0/i = 2.66/0.115 = 23.2$	
5.8.3.1 (1)	Slenderness criterion, $\lambda_{lim} = 20(A \times B \times C)/\sqrt{n}$ where:	
	$\quad n = N/A_c f_{cd} = N/(400^2 \times 0.85 \times 32/1.5) = 3707/2901 = 1.28$	
	Taking $A = 0.7$, $B = 1.1$ and $C = 1.7 - M_{01}/M_{02} = 1.7 + 69.6/213.3 = 2.02$	
	$\quad \lambda_{lim} = 20 \times 0.7 \times 1.1 \times 2.02/\sqrt{1.28} = 27.5 \ (> \lambda = 23.2)$	
	Since $\lambda < \lambda_{lim}$, second order effects may be ignored and $M_{Ed} = M_{02} \ (\geq M_{min})$	
	Design of cross-section	
	Allowing 35 mm nominal cover, 8 mm links and 32 mm longitudinal bars, results in $d = 400 - (35 + 8 + 32/2) = 340$ mm say, $d/h = 340/400 = 0.85$. Reinforcement can be determined from the design chart in Table A3 as follows:	
	$\quad N_{Ed}/bhf_{ck} = (3707 \text{ or } 2365) \times 10^3/(400 \times 400 \times 32) = 0.73 \text{ or } 0.46$	
	$\quad M_{Ed}/bh^2f_{ck} = 213.3 \times 10^6/(400 \times 400^2 \times 32) = 0.104$	
	$\quad A_s f_{yk}/bhf_{ck} = 0.51$, which gives $A_s = 0.51 \times 400 \times 400 \times 32/500 = 5223$ mm^2	
9.5.2 (2)	Minimum amount of longitudinal reinforcement:	
	$\quad A_{s,min} = 0.1N_{Ed}/f_{yd} = 0.1 \times 3707 \times 10^3/(500/1.15) = 853$ mm^2 (4H20)	
	$\quad \geq 0.002A_c = 0.002 \times 400 \times 400 = 320$ mm^2	
9.5.2 (3)	Maximum amount of longitudinal reinforcement:	
	$\quad A_{s,max} = 0.04A_c = 0.04 \times 400 \times 400 = 6400$ mm^2 (8H32)	
	Similar calculations for the other storeys provide results as summarised below.	

Storey	N_{Ed} (kN) max/min	M_{Ed} (kN m)	$\dfrac{N_{Ed}}{bhf_{ck}}$	$\dfrac{M_{Ed}}{bh^2 f_{ck}}$	$\dfrac{A_s f_{yk}}{bhf_{ck}}$	A_s (mm^2)
4th floor–roof	574/442	191.9	0.11/0.08	0.094	0.20	2048
3rd–4th floor	1443/914	196.9	0.28/0.18	0.096	0.10	1024
2nd–3rd floor	2246/1397	201.5	0.44/0.27	0.098	0.20	2048
1st–2nd floor	3000/1881	205.8	0.59/0.37	0.101	0.36	3687
Grd–1st floor	3707/2365	213.3	0.73/0.46	0.104	0.51	5223

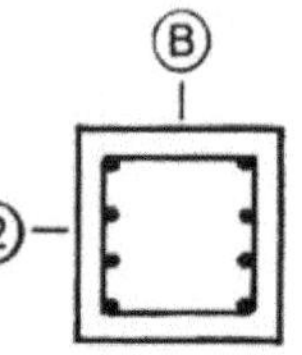

OUTPUT column (right side):

8H32 (Grd–1st floor)
8H25 (1st–2nd floor)
4H32 (2nd floor–roof)

Reference	CALCULATIONS
	A reasonable arrangement would be to provide 8H32 for the bottom storey, 8H25 for the next storey and 4H32 for the top three storeys.
	Tying requirements
BS EN 1990 A1.3.2	For the slab, accidental design load ($G_k + \psi_1 Q_k$)
	$\quad = 6.5 + 0.7 \times 4.0 = 9.3$ kN/m^2 (max), 6.5 kN/m^2 (min)

Example 1   **Calculation Sheet 50**

Reference	CALCULATIONS	OUTPUT
Table NA.A1.3	For the beam on line 2 (calculation sheet 42), accidental design load $= (9.3/14.1) \times 169.2 + 9.0/1.25 = 118.8$ kN/m (max), 85.2 kN/m (min) For the column, approximate accidental design load for load case 2 on beam plus edge loading (calculation sheets 43 and 48): $N_{Ad} = (118.8/178.2) \times 582.2 + 1.2 \times 9.6 \times (0.6 \times 9.3 + 6.0 + 5.0) = 579.2$ kN Minimum area of reinforcement required with $\sigma_s = 500$ MPa, $A_{s,min} = 579.2 \times 10^3/500 = 1159$ mm^2 (4H20 sufficient)	

INTERNAL COLUMN B2

The load from the sub-frame on line 2 is the total shear force at line B.

Values of axial load N (kN) and bending moment M (kN m)						
Loading	$1.25G_k + 1.5Q_k$				$1.5Q_k$	
Load case	1		2		1	2
Member	N	M	N	M	N	N
Roof beam	1025.2	0	981.7	14.0	(90.9)	
Column	15.3		15.3			
	1040.5	0	997.0	69.1		
4th floor beam	1446.6		1154.4		584.5	292.3
	2487.1	0	2151.4	69.1		
Column	15.3		15.3			
	2502.4	0	2166.7	69.1		
3rd floor beam	1446.6		1154.4		584.5	292.3
	3949.0	0	3321.1	69.1	1169.0	584.6
Column	15.3		15.3			
	3964.3	0	3336.4	69.1		
2nd floor beam	1446.6		1154.4		584.5	292.3
	5410.9	0	4490.8	69.1	1753.5	876.9
Column	15.3		15.3			
	5426.2	0	4506.1	69.1		
1st floor beam	1446.6		1154.4		584.5	292.3
	6872.8	0	5660.5	69.1	2338.0	1169.2
Column	17.8		17.8			
	6890.6	0	5678.3	69.1		
Grd. floor beam	1446.6		1154.4		584.5	292.3
	8337.2	0	6832.7	69.1	2922.5	1461.5
Column	17.8		17.8			
Foundation	8355.0		6850.5			

The maximum moment occurs when load case 2 is applied at the level considered. Maximum coexistent load occurs when load case 1 is applied at levels above, and minimum coexistent load occurs when $1.0G_k$ is applied at levels above. The latter arrangement can be critical for values of $N_{Ed} < 1160$ kN (see calculation sheet 48). The maximum load with a smaller coexistent moment results when load case 1 is applied at all levels.

For the basement storey with load case 1 at all levels:

$N_{Ed} = 8355 - 0.4 \times 2922.5 = 7186$ kN

6.1 (4)   Minimum total design moment, with $e_0 = h/30 = 400/30 \geq 20$ mm:

$M_{min} = N_{Ed} e_0 = 7186 \times 0.02 = 143.7$ kN m

For the basement storey with load case 2 at ground floor level:

$M_{top} = 69.1$ kN m, $M_{bot} = -0.5 M_{top} = -34.6$ kN m

$N_{Ed} = 1154.4 + 6890.6 - 0.4 \times (292.3 + 2338) = 6993$ kN (max)

$N_{Ed} = 1154.4 + [6890.6 - (2338 + 90.9)]/1.25 = 4724$ kN (min)

Example 1   Calculation Sheet 51

Reference	CALCULATIONS	OUTPUT
	Effective length and slenderness	
	As for column A2, $l_0 = 2.66$ m and $\lambda = 23.2$	
5.2 (9)	First order moment from imperfections (with load case 2 at ground floor level):	
	$M_i = Nl_0/400 = 6993 \times 2.66/400 = 46.5$ kN m	
	First order moments, including the effect of imperfections:	
	$M_{01} = -34.5 + 46.5 = 12.0$ kN m, $M_{02} = 69.1 + 46.5 = 115.6$ kN m	
5.8.3.1 (1)	Slenderness criterion: $\lambda_{lim} = 20(A \times B \times C)/\sqrt{n}$ where	
	$n = N/A_c f_{cd} = 6993/2901 = 2.41$ and	
	$A = 0.7$, $B = (1 + 2\omega)^{0.5}$, $C = 1.7 - 12.0/115.6 = 1.60$ where	
	$\omega = A_s f_{yd}/A_c f_{cd} = A_s \times 500/(1.15 \times 2901 \times 10^3) = A_s/6672$	
	Assuming 8H32, $\omega = 6434/6672 = 0.96$, $B = (1 + 2\omega)^{0.5} = 1.70$ and	
	$\lambda_{lim} = 20 \times 0.7 \times 1.7 \times 1.6/\sqrt{2.41} = 24.5$ $(>\lambda = 23.2)$	
	Since $\lambda < \lambda_{lim}$, second-order effects may be ignored and $M_{Ed} = M_{02}$ $(>M_{min})$	
	Design of cross-section	
	Although the nominal cover needed for durability is 25 mm, this will be increased to 35 mm to ensure a minimum cover to the H32 bars not less than the bar size.	
	Since $M_{min} > M_{02}$, the critical condition occurs with load case 1 at all levels. The reinforcement can be determined from the design chart in Table A3.	
	$N_{Ed}/bhf_{ck} = 7186 \times 10^3/(400 \times 400 \times 32) = 1.40$	
	$M_{Ed}/bh^2 f_{ck} = 143.7 \times 10^6/(400 \times 400^2 \times 32) = 0.070$	
	It is clear from the design chart that the size of the cross-section is too small. If this is increased to 500×500, the following values are obtained:	
	$N_{Ed}/bhf_{ck} = 7186 \times 10^3/(500 \times 500 \times 32) = 0.90$	
	$M_{Ed}/bh^2 f_{ck} = 143.7 \times 10^6/(500 \times 500^2 \times 32) = 0.036$	
	$A_s f_{yk}/bhf_{ck} = 0.50$, which gives $A_s = 0.5 \times 500 \times 500 \times 32/500 = 8000$ mm²	
	Maximum amount of longitudinal reinforcement:	
	$A_{s,max} = 0.04A_c = 0.04 \times 500 \times 500 = 10000$ mm² (12H32)	
	Similar calculations for the other storeys, with the size of the cross-section taken as 500×500 up to 1st floor level and 400×400 above this level, provide results as summarised below.	

Storey	N_{Ed} (kN) max/min	M_{Ed} (kN m)	$\dfrac{N_{Ed}}{bhf_{ck}}$	$\dfrac{M_{Ed}}{bh^2 f_{ck}}$	$\dfrac{A_s f_{yk}}{bhf_{ck}}$	A_s (mm²)
4th floor–roof	1041/760	75.1	0.20/0.15	0.037	0	320
3rd–4th floor	2503/1462	83.4	0.49/0.29	0.041	0.03	576
2nd–3rd floor	3848/2164	91.2	0.75/0.42	0.045	0.35	2048
1st–2nd floor	5076/2866	98.2	0.99/0.56	0.048	0.63	6452
Grd–1st floor	6190	123.8	0.78	0.031	0.34	5440
Basement	7186	143.7	0.90	0.036	0.50	8000

Reference	CALCULATIONS	OUTPUT
9.5.2 (2)	Minimum amount of longitudinal reinforcement (3rd–4th floor):	
	$A_{s,min} = 0.1N_{Ed}/f_{yd} = 0.1 \times 2503 \times 10^3/(500/1.15) = 576$ mm² (4H16)	
9.5.2 (3)	Maximum amount of longitudinal reinforcement ($0.04A_c$):	
	Above 1st floor: $A_{s,max} = 0.04 \times 400 \times 400 = 6400$ mm² (8H32)	12H32 (Bottom storey)
	Below 1st floor: $A_{s,max} = 0.04 \times 500 \times 500 = 10,000$ mm² (12H32)	12H25 (Grd–1st floor)
	Tying requirements	8H32 (1st–2nd floor)
BS EN 1990 A1.3.2	Accidental design load applied to column for load case 1 on beam on line 2:	4H32 (2nd–3rd floor) 4H25 (3rd–roof)
	$N_{Ad} = (118.8/178.2) \times 1446.6 = 964.4$ kN	

Example 1

Reference	CALCULATIONS	OUTPUT
Table NA.A1.3	Minimum area of reinforcement required with $\sigma_s = 500$ MPa, $A_{s,min} = 964.4 \times 10^3/500 = 1929$ mm^2 (4H25 sufficient) A suitable arrangement would be to provide 12H32 for the bottom storey, 12H25 for the next storey, 8H32 for the next storey, 4H32 for the next storey, and 4H25 for the top two storeys.	
BS EN 1990 A1.3.2 Table NA.A1.3	**EDGE COLUMN B1** The maximum design loads for the beam on line 1 are 78 kN/m at each floor (see calculation sheet 47) and 54 kN/m at the roof. These loads are 44% of those for the beam on line 2, and values pro-rata to those on column B2 will be taken. For the storey from ground to 1st floor, with load case 1 at all levels: $N_{Ed} = 0.44 \times (1025.2 + 1446.6 \times 4 - 0.3 \times 2338) + 15.3 \times 5 = 2765$ kN Bending moment at bottom of column, with load case 2 at ground floor level, $M_{02} = 0.44 \times 69.1 + 2765 \times 2.66/400 = 48.8$ kN m $M_{min} = N_{Ed} e_0 = 2765 \times 0.02 = 55.3$ kN m ($>M_{02}$) **Design of cross-section** $N_{Ed}/bhf_{ck} = 2765 \times 10^3/(400 \times 400 \times 32) = 0.54$ $M_{Ed}/bh^2 f_{ck} = 55.3 \times 10^6/(400 \times 400^2 \times 32) = 0.027$ $A_s f_{yk}/bhf_{ck} = 0.05$, which gives $A_s = 0.05 \times 400 \times 400 \times 32/500 = 512$ mm^2 $A_{s,min} = 0.1 N_{Ed}/f_{yd} = 0.1 \times 2765 \times 10^3/(500/1.15) = 636$ mm^2 (4H16) **Tying requirements** Accidental design load for beam on line 1 (see calculation sheet 47) is: $(0.4 \times 9.6 + 0.6) \times 9.3 + (1.2 \times 0.24 \times 25 + 5.0) = 53.5$ kN/m (max) Accidental design load applied to column for load case 1 on beam on line 1: $N_{Ad} = (53.5/178.2) \times 1446.6 = 434.3$ kN Minimum area of reinforcement required with $\sigma_s = 500$ MPa, $A_{s,min} = 434.3 \times 10^3/500 = 869$ mm^2 (4H20 sufficient) A reasonable arrangement would be to provide 4H20 at all levels.	 4H20 (Grd floor − roof)
	CORNER COLUMN A1 At each level, the load applied is the shear force from the sub-frame on line 1 plus the force due to the edge loading on line A (see calculation sheet 48). The dominant effect is biaxial bending, and the column will be checked for two cases as follows: (1) Bottom storey (ground to 1st floor), with load case 2 at all levels Line 1 $0.44 \times (573.8 + 838.1 \times 4 - 0.3 \times 994.4) + 15.3 \times 5 = 1672.8$ Line A $0.4 \times (205.4 + 255.9 \times 4)/1.2 = \underline{409.7}$ $N_{Ed} = \underline{2082.5}$ kN Line 1: $M_{Edy} = 0.44 \times 206.2 + 2082.5 \times 2.66/400 = 104.6$ kN m Line A: $M_{Edz} = 0.025 \times (255.9/1.2) \times 9.6 = 51.2$ kN m (2) Top storey, with $1.0 G_k$ at roof level and maximum load at 4th floor level $N_{Ed} = 0.44 \times (569.8 - 32.3)/1.25 + 0.4 \times 205.4/1.2 + 15.3 = 273$ kN $M_{0y} = 0.44 \times 206.2 + 273 \times 2.29/400 = 92.3$ kN m, $M_{0z} = 51.2$ kN m **Design of cross-section** From the chart in Table A3, the design resistance of the column for bending about either axis can be determined. Then, in the absence of a precise design for biaxial bending, a simplified criterion check for compliance may be made as follows: (1) Assuming 4H20, the following values are obtained: $A_s f_{yk}/bhf_{ck} = 1257 \times 500/(400 \times 400 \times 32) = 0.12$	

Example 1   Calculation Sheet 53

Reference	CALCULATIONS	OUTPUT
5.8.9 (4)	$N_{Ed}/bhf_{ck} = 2083 \times 10^3/(400 \times 400 \times 32) = 0.41$ $M_{Ed}/bh^2f_{ck} = 0.085, M_{Rdz} = M_{Rdy} = 0.085 \times 400 \times 400^2 \times 32 \times 10^{-6} = 174 \text{ kN m}$ $N_{Rd} = A_c f_{cd} + A_s f_{yd} = (400^2 \times 0.85 \times 32/1.5 + 1257 \times 500/1.15) \times 10^{-3} = 3448 \text{ kN}$ $N_{Ed}/N_{Rd} = 2083/3448 = 0.60.$ For values of $N_{Ed}/N_{Rd} \leq 0.7$, $a = 0.92 + 0.83(N_{Ed}/N_{Rd}) = 1.42$ $(M_{Edz}/M_{Rdz})^a + (M_{Edy}/M_{Rdy})^a = (51.2/174)^{1.42} + (104.6/174)^{1.42} = 0.66 \ (\leq 1.0)$ (2) Assuming 4H25, the following values are obtained: $A_s f_{yk}/bhf_{ck} = 1963 \times 500/(400 \times 400 \times 32) = 0.19$ $N_{Ed}/bhf_{ck} = 273 \times 10^3/(400 \times 400 \times 32) = 0.05$ $M_{Ed}/bh^2f_{ck} = 0.080, M_{Rdz} = M_{Rdy} = 0.080 \times 400 \times 400^2 \times 32 \times 10^{-6} = 163.8 \text{ kN m}$ $N_{Rd} = A_c f_{cd} + A_s f_{yd} = (400^2 \times 0.85 \times 32/1.5 + 1963 \times 500/1.15) \times 10^{-3} = 3755 \text{ kN}$ $N_{Ed}/N_{Rd} = 273/3755 = 0.073.$ For values of $N_{Ed}/N_{Rd} \leq 0.1$, $a = 1.0$ $(M_{Edz}/M_{Rdz})^a + (M_{Edy}/M_{Rdy})^a = (51.2/163.8)^{1.0} + (92.3/163.8)^{1.0} = 0.88 \ (\leq 1.0)$ A reasonable arrangement would be to provide 4H25 for the top storey, and 4H20 for the lower storeys.	 4H20 (Grd−4th floor) 4H25 (4th floor−roof)
	FIRE RESISTANCE The columns can be assumed to meet the requirements, since 300×300 columns were sufficient for the beam and slab construction (calculation sheets 19–20).	

Bar Marks	Commentary on Bar Arrangement (Drawing 13)
01	Bars (shape code 26) bearing on 75 mm kicker and cranked to fit alongside bars projecting from basement wall. Projection of starter bars $= 1.5 \times 35 \times 32 \times 5223/6434 + 75 = 1500$ mm say. Crank to begin 75 mm from end of starter bar. Length of crank $= 13\phi$ and overall offset dimension $= 2\phi$. Since 8H25 are sufficient at the next level, projection of bars above first floor $= 1.5 \times 35 \times 25 \times 3687/3927 + 75 = 1300$ mm say.
02, 06	Closed links (shape code 51), with 35 mm nominal cover, starting above kicker and stopping below integral beam at next floor level. See bar commentary in calculation sheet 20 for details of code requirements. The link spacing should not exceed 400 mm generally, or 240 mm for a distance above or below a beam equal to 400 mm generally, but 500 mm below first floor for column B2. The required areas of transverse bars in the lap zones are as follows: Column A2. Grd−1st floor: $1.5 \times 5223/6434 \times 804 = 979 \text{ mm}^2$ (13H10) Column B2. Fdn−Grd floor: $1.5 \times 8000/9651 \times 804 = 1000 \text{ mm}^2$ (13H10) Grd−1st floor: $1.5 \times 5440/5890 \times 491 = 681 \text{ mm}^2$ (9H10)
03	Bars (similar to bar mark 01) cranked to fit alongside bars projecting from foundation. Projection of starter bars $= 1.5 \times 35 \times 32 \times 8000/9651 + 75 = 1500$ mm say. Since 12H25 are sufficient at the next level, projection of bars above ground floor $= 1.5 \times 35 \times 25 \times 5440/5890 + 75 = 1300$ mm say.
04	Bars (shape code 35) cranked to fit alongside bars mark 03, and terminated with minimum end projection 100 mm below first floor level (due to reduction in column size above first floor).
05	Straight bars arranged in positions to suit reduced column size above first floor. Anchorage length of bar below first floor $= 35 \times 32 = 1120$ mm, but not less than 1300 mm to give lap similar to that between bars mark 03 and 04. Projection of bars above first floor $= 1.5 \times 35 \times 32 + 75 = 1750$ mm.
07	Closed links (shape code 51) in pairs, fixed to bars mark 06 and supporting bars mark 05. The combination of links (bars mark 06 and 07) provides sufficient transverse reinforcement in the lap zone.

Example 1: Reinforcement in Columns A2 and B2   Drawing 13

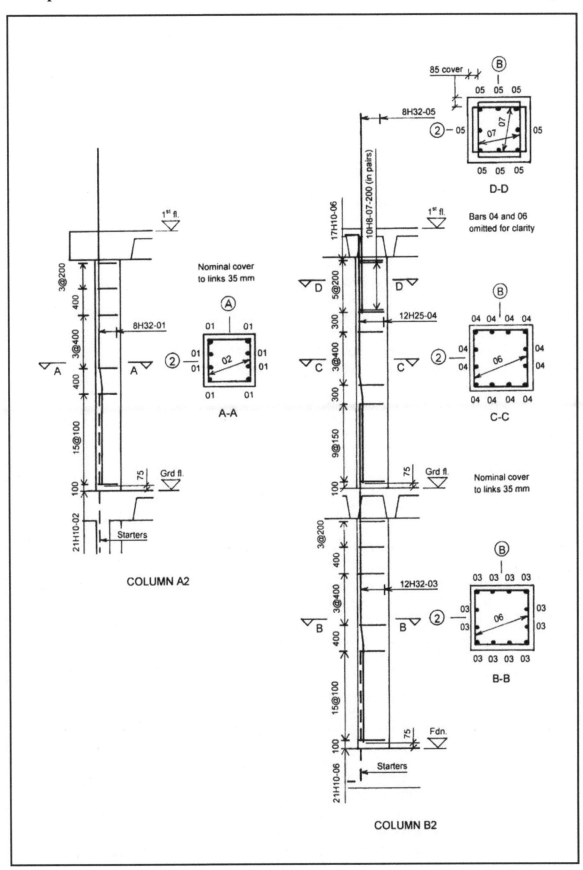

COLUMN A2

COLUMN B2

Example 1   **Calculation Sheet 54**

Reference	CALCULATIONS	OUTPUT
	STAIRS	
	The stair well is 2.2 m wide, with each flight 1 m wide and a 200 mm gap between flights. For optimum proportions, $2 \times$ rise + going = 600 mm. For the storeys above first floor, there are 10 steps in each flight, each with 175 mm rise and 250 mm going. For the storeys below first floor, there are 12 steps in each flight, each with 166.7 mm rise and 275 mm going. Waist thickness is 150 mm for flights above first floor, and 200 mm for flights below first floor. Landing thickness is 150 mm.	Stairs to offices
	Loading	
	For the landings, the characteristic loads are as follows:	
	Permanent load $\quad$ kN/m^2 $\quad$ Imposed load $\quad\quad$ kN/m^2 Self-weight of slab $0.150 \times 25 \quad = 3.75 \quad$ Offices $\quad\quad q_k = 2.5$ Finishes $\quad\quad\quad\quad\quad\quad = \underline{0.50} \quad$ Shopping areas $\quad q_k = 4.0$ $\quad\quad\quad\quad\quad\quad\quad\quad g_k = \underline{4.25}$	
	For the storeys above the first floor, the average concrete thickness of the flights over the plan area is $175/2 + 183 = 270$ mm. Additional load $= 0.120 \times 25 = 3.0$ kN/m^2.	
	Design ultimate load for landings $= 1.25 \times 4.25 + 1.5 \times 2.5 = 9.1$ kN/m^2	
	Additional design load for flights $= 1.25 \times 3.0 = 3.7$ kN/m^2	
	For the storeys below first floor, the average concrete thickness of the flights over the plan area is $167/2 + 234 = 318$ mm. Additional load $= 0.168 \times 25 = 4.2$ kN/m^2.	
	Design ultimate load for landings $= 1.25 \times 4.25 + 1.5 \times 4.0 = 11.3$ kN/m^2	
	Additional design load for flights $= 1.25 \times 4.2 = 5.3$ kN/m^2	
	Analysis	
	The stairs will be designed for the flights to span between landings, which are then designed to span transversely across the stair well. Effective span of flight is taken as clear horizontal distance plus, at each end, $0.5 \times$ width of landing ≤ 0.9 m. Thus, for the flights above first floor, effective span $= 10 \times 0.250 + 2 \times 0.9 = 4.3$ m.	
	Considering the flights to be simply supported, the bending moment at mid-span is	
	$M = 12.8 \times 1.25 \times (2.15 - 0.625) = 24.4$ kN m	
	The effective span of the landings between centres of walls is 2.4 m. Considering the load from the stair flights carried on a strip 1.8 m wide, the bending moment at mid-span and the shear force at d from face of support are:	
	$M = (9.1 + 12.8 \times 1.25/1.8) \times 2.4^2/8 = 18.0 \times 2.4^2/8 = 13.0$ kN m/m	
	$V = 18 \times 0.98 = 17.7$ kN/m	
	For the flights below first floor, effective span $= 12 \times 0.275 + 2 \times 0.9 = 5.1$ m.	
	$M = 16.6 \times 1.65 \times (2.55 - 0.825) = 47.3$ kN m	
	For the landings,	
	$M = (9.1 + 16.6 \times 1.65/1.8) \times 2.4^2/8 = 24.3 \times 2.4^2/8 = 17.5$ kN m/m	
	$V = 24.3 \times 0.93 = 22.6$ kN/m	
	Flexural design	
	Above the first floor, allowing for 12 mm bars, $d = 150 - (25 + 12/2) = 119$ mm	$d = 119$ mm
	According to the values of M/bd^2f_{ck}, where $b = 1000$ mm, appropriate values of z/d and A_s can be determined (Table A1), and suitable bars selected (Table A9)	
	For the flights:	
	$M/bd^2f_{ck} = 24.4 \times 10^6/(1000 \times 119^2 \times 32) = 0.054 \quad\quad z/d = 0.95$ (maximum) $A_s = M/(0.87f_{yk}z) = 24.4 \times 10^6/(0.87 \times 500 \times 0.95 \times 119)$ $\quad\quad = 496$ mm^2/m (5H12-225 gives 565 mm^2/m)	5H12 in flights (above first floor)
	For the landings, allowing for 12 mm bars, $d = 150 - (25 + 12 + 12/2) = 107$ mm	
	$A_s = 13.0 \times 10^6/(0.87 \times 500 \times 0.95 \times 107) = 294$ mm^2/m (H12-300)	H12-300 in landing
	Below first floor, allowing for 12 mm bars, $d = 200 - (25 + 12/2) = 169$ mm	$d = 169$ mm

Example 1   Calculation Sheet 55

Reference	CALCULATIONS	OUTPUT
	For the flights:	
	$M/bd^2f_{ck} = 47.3 \times 10^6/(1000 \times 169^2 \times 32) = 0.052$ $z/d = 0.95$ (maximum)	
	$A_s = 47.3 \times 10^6/(0.87 \times 500 \times 0.95 \times 169)$	
	$= 678\text{mm}^2/\text{m}$ (6H12-175 gives 679 mm^2/m)	6H12 in flights (below first floor)
	For the landings:	
	$A_s = (17.0/13.0) \times 294 = 385$ mm^2/m (H12-300 say)	H12-300 in landing
	Shear design	
	Maximum shear stress occurs in landings below the first floor:	
	$v = V/b_w d = 22.6 \times 10^3/(1000 \times 107) = 0.21$ MPa	
6.2.2 (1) Table NA.1	Minimum design shear strength, with $k = 2$ for $d \leq 200$ mm, is	
	$v_{min} = 0.035k^{3/2}f_{ck}^{1/2} = 0.035 \times 2^{3/2} \times 32^{1/2} = 0.56$ MPa	Shear satisfactory
	Deflection (see *Reynolds*, Table 4.21)	
7.4.1 (6)	Deflection requirements with regard to appearance and general utility of members apply in relation to the quasi-permanent load condition. The requirements may be met by limiting the span/effective depth ratio to specified values.	
	For the flights above first floor, actual span/effective depth $= 4300/119 = 36.1$	
	Service stress in reinforcement under quasi-permanent load is given by	
	$\sigma_s = (f_{yk}/\gamma_s)(A_{s,req}/A_{s,prov})[(g_k + \psi_2 q_k)/n]$	
	$= (500/1.15)(496/565)[(7.25 + 0.3 \times 2.5)/12.8] = 239$ MPa	
7.4.2 Table NA.5	From *Reynolds*, Table 4.21, limiting $l/d =$ basic ratio $\times \alpha_s \times \beta_s$ where:	
	For $100A_s/bd = 100 \times 496/(1000 \times 119) = 0.42 < 0.1f_{ck}^{0.5} = 0.1 \times 32^{0.5} = 0.56$,	
	$\alpha_s = 0.55 + 0.0075f_{ck}/(100A_s/bd) + 0.005f_{ck}^{0.5}[f_{ck}^{0.5}/(100A_s/bd) - 10]^{1.5}$	
	$= 0.55 + 0.0075 \times 32/0.42 + 0.005 \times 32^{0.5} \times (32^{0.5}/0.42 - 10)^{1.5} = 1.30$	
	$\beta_s = 310/\sigma_s = 310/239 = 1.30$	
	For a simply supported member, basic ratio $= 20$ and hence	
	Limiting $l/d = 20 \times \alpha_s \times \beta_s = 20 \times 1.3 \times 1.3 = 33.8$	
	The ratios apply to horizontal members and, for a stair flight, it is reasonable to increase the limiting value by 15%, as was done previously in BS 8110.	
	Increasing the value by 15% gives limiting $l/d = 38.9$ (> actual $l/d = 36.1$)	Check complies
	For the flights below first floor, actual span/effective depth $= 5100/169 = 30.2$	
	$\sigma_s = (500/1.15)(678/679)[(8.45 + 0.3 \times 4.0)/16.6] = 253$ MPa	
	For $100A_s/bd = 100 \times 678/(1000 \times 169) = 0.40$, $\alpha_s = 1.39$, $\beta_s = 310/253 = 1.22$	
	Limiting $l/d = 20 \times \alpha_s \times \beta_s = 20 \times 1.39 \times 1.22 = 33.9$ (> actual $l/d = 33.8$)	Check complies
	Cracking (see *Reynolds*, Table 4.23)	
7.3.2 (2)	Minimum area of reinforcement required in tension zone for crack control:	
	$A_{s,min} = k_c k f_{ct,eff} A_{ct}/\sigma_s$	
	Taking values of $k_c = 0.4$, $k = 1.0$, $f_{ct,eff} = f_{ctm} = 0.3f_{ck}^{(2/3)} = 3.0$ MPa (for general design purposes), $A_{ct} = bh/2$ (for plain concrete section) and $\sigma_s \leq f_{yk} = 500$ MPa	
	Above first floor: $A_{s,min} = 0.4 \times 1.0 \times 3.0 \times 1000 \times (150/2)/500 = 180$ mm^2/m	H10-400
	Flights below first floor: $A_{s,min} = 180 \times 200/150 = 240$ mm^2/m	H10-300
7.3.3 (1)	No other specific measures are necessary provided overall depth does not exceed 200 mm, and detailing requirements are observed.	Check complies
	Detailing requirements	
	Minimum area of longitudinal tension reinforcement:	
	$A_{s,min} = 0.26(f_{ctm}/f_{yk})bd = 0.26 \times (3.0/500)bd = 0.00156bd \geq 0.0013bd$	

Example 1   **Calculation Sheet 56**

Reference	CALCULATIONS	OUTPUT
	Above first floor: $A_{s,min} = 0.00156 \times 1000 \times 119 = 186$ mm^2/m	H10-400
	Flights below first floor: $A_{s,min} = 186 \times 169/119 = 264$ mm^2/m	H10-300
	Minimum area of secondary reinforcement (20% of principal reinforcement):	
	$A_{s,min} = 0.2 \times 496$ (or 678) $= 100$ (or 136 mm^2/m) Use H10-400.	H10-400
	Maximum spacing of principal reinforcement:	
	$3h = 450 \leq 400$ mm	Spacing satisfactory
	Maximum spacing of secondary reinforcement:	
	$3.5h = 525 \leq 450$ mm	Spacing satisfactory

Layout of stairs to offices

In this example, it is assumed that the finishes on the steps and the landings are of the same thickness, so that all the steps are of the same height. If the thickness of the finish on the landing is more than that on the steps, then the height of the top step should be reduced, and the height of the bottom step increased, to accommodate the difference in the thickness of the finishes. If the thickness of finish on the floor outside the stairwell is more than that on the floor landing, there should be a step in the top surface of the slab to accommodate the difference in the thickness of the finishes (see drawing 14).

The soffits of the stair flights are arranged to intersect with the landing on the same line. With 10 steps in each flight, the horizontal distance between the soffit intersection lines at the top and bottom of the flight is 10 times the going = 2500 mm. If the landings are made the same width at each end of the stairwell, then the distance from a soffit intersection line to the face of the end wall = $(7200 - 2500)/2 = 2350$ mm. In the figure below:

a = landing thickness $\times$ (going/rise) $= 150 \times 250/175 = 215$ mm say, $b =$ (going $- a$) $= 250 - 215 = 35$ mm.

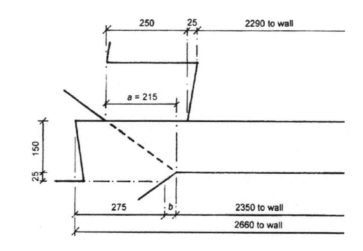

Bar Marks	Commentary on Bar Arrangement (Drawing 14)
01, 02, 03	Longitudinal main bars in flights, shape code 15, at a maximum spacing of 400 mm, provided in three lengths for ease of handling and fixing. Bars 01 and 03 to lap with bars projecting from walls at ends of stairwell. Length of lap between bars 01 and 02, $l_0 = 1.5 \times 35\phi = 650$ mm say. Bars 02 and 03 provided with an anchorage length beyond the point where they cross at top of flight, $l_b = 35\phi = 450$ mm say.
04, 05	Longitudinal bars, shape code 15, equal to at least 50% of span reinforcement, to cater for torsional restraint provided by landing and lap with bars projecting from walls at end of stairwell.
06, 08	Transverse bars in landings, at a maximum spacing of 400 mm, with top bars equal to at least 50% of bottom reinforcement, to lap with bars projecting from walls at sides of stairwell.
07	Secondary bars, at a maximum spacing of 450 mm, curtailed 25 mm from edges of flights.

Example 1: Reinforcement in Stairs to Offices

Drawing 14

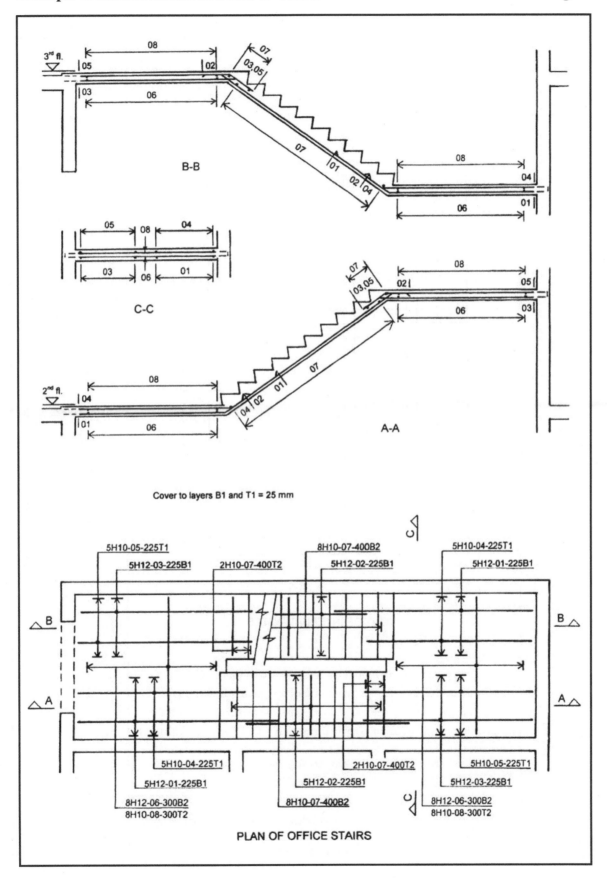

B-B

C-C

A-A

Cover to layers B1 and T1 = 25 mm

5H10-05-225T1

5H12-03-225B1

8H10-07-400B2

5H12-02-225B1

5H10-04-225T1

5H12-01-225B1

2H10-07-400T2

5H10-04-225T1

5H12-01-225B1

8H12-06-300B2
8H10-08-300T2

2H10-07-400T2

5H12-02-225B1

8H10-07-400B2

5H10-05-225T1

5H12-03-225B1

8H12-06-300B2
8H10-08-300T2

PLAN OF OFFICE STAIRS

Example 1

Reference	CALCULATIONS	OUTPUT
	INTERNAL WALLS	

The building is provided with a central core containing two lifts, a staircase, and an access well for services to all floors. The walls enclosing these areas provide lateral stability to the structure as a whole. Each wall is laterally supported by the floors and roof, and by adjacent perpendicular walls.

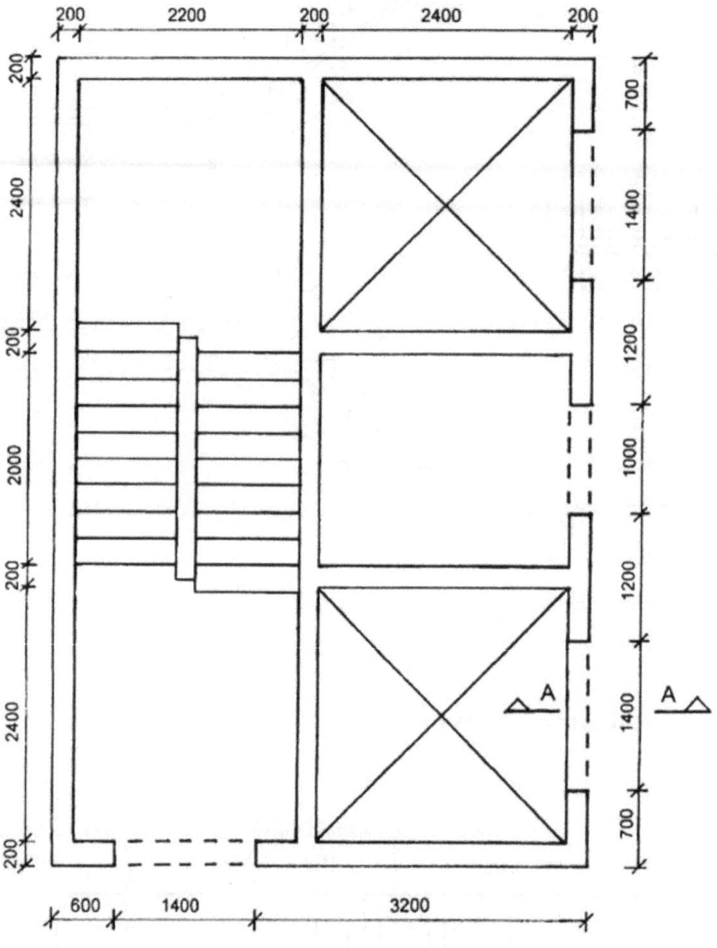

LAYOUT OF INTERNAL WALLS

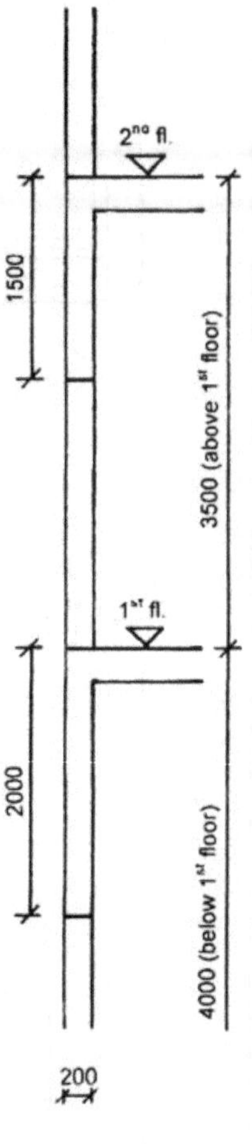

A-A

Wind loading

In this example, the assumed site location and ground terrain are as follows:

Site location: Thames Valley   Altitude of site above mean sea level: 50 m

Town terrain as defined in the UK National Annexe to Eurocode 1 Parts 1–4

BS EN 1991-1-4 4.2 NA.2.4 NA.2.5 Fig. NA.1

Height of building from ground to top of parapet is $h = 4 + 3.5 \times 4 + 1 = 19$ m

The fundamental value of the basic wind velocity, where A is altitude of site above mean sea level, and z is height of building above ground, is given by:

$$v_{b,0} = v_{b,map} \times c_{alt} = v_{b,map} \times \{1 + 0.001A \times (10/z)^{0.2}\}$$
$$= 21.5 \times \{1 + 0.001 \times 50 \times (10/19)^{0.2}\} = 22.5 \text{ m/s}$$

The basic wind velocity (for $c_{prob} = 1.0$) at height 10 m above ground, for country terrain, is given by:

BS EN 1991-1-4

$$v_b = v_{b,0} \times c_{dir} \times c_{season} = 22.5 \times 1.0 \times 1.0 = 22.5 \text{ m/s}$$

Example 1   **Calculation Sheet 58**

Reference	CALCULATIONS	OUTPUT
4.5 NA.2.18	The basic velocity pressure is given by $q_b = 0.5pv_b^2 = 0.5 \times 1.226\ v_b^2 = 0.613 \times 22.5^2 = 310$ Pa (0.31 kN/m^2)	
NA.2.17	The peak velocity pressure at height z above terrain, for sites in town terrain, when orography is not significant, is given by $q_p(z) = q_b \times c_e(z) \times c_{e,T}$	
BS EN 1991-1-4 A5 Figure A.5	For buildings in town terrain, where the distance to other buildings of height h_{ave} is no greater than $2h_{ave}$, $z = h - h_{dis}$ where h_{dis} is the lesser of $0.8h_{ave}$ or $0.6h$. In the absence of more accurate information, h_{ave} may be taken as 15 m. Hence, $z = h - h_{dis} = 19 - 0.6 \times 19 = 7.6$ m	
Figure NA.7 Figure NA.8	Assuming distances of 50 km upwind to shoreline, and 0.5 km inside town terrain: $c_e(z) = 2.2$, $c_{e,T} = 0.94$ Hence, $q_p(z) = 0.31 \times 2.2 \times 0.94 = 0.64$ kN/m^2 The wind force acting on the building may be determined as	
5.3 (2)	$F_w = c_s c_d \times c_f \times q_p(z) \times A_{ref}$ where $A_{ref} = bh$	
6.2 (1)	For framed buildings less than 100 m high with structural walls, and whose height is less than 4 times the in-wind depth, the value of $c_s c_d$ may be taken as 1.	
7.2.2 (1) Figure 7.4	A building whose height h is less than the in-wind depth b, should be considered as one part with the reference height $z_e = h$.	
NA.2.27 Table NA.4 7.2.2 (3)	For determining overall wind loads on the building, the net pressure coefficients given in the UK National Annex will be used. These coefficients may be multiplied by the factor for lack of correlation between the windward and leeward sides. For a wind force acting normal to the long face of the building: $b = 34.8$ m $h/d = 19/20.4 = 0.93$ giving $c_{pe,10} = 1.1$ and $c_f = 0.85\ c_{pe,10}$ $F_w = 1.0 \times 0.85 \times 1.1 \times 0.64 \times 34.8 \times 19 = 396$ kN For a wind force acting normal to the short face of the building: $b = 20.4$ m $h/d = 19/34.8 = 0.62$ giving $c_{pe,10} = 0.95$ and $c_f = 0.85\ c_{pe,10}$ $F_w = 1.0 \times 0.85 \times 0.95 \times 0.64 \times 20.4 \times 19 = 200$ kN **Analysis** The wall system will act as a single structure, provided the shear transference at the intersections is adequate. Although two walls have openings at each floor level, the lintels are so deep that the overall stiffness of the walls will not be greatly reduced. 	

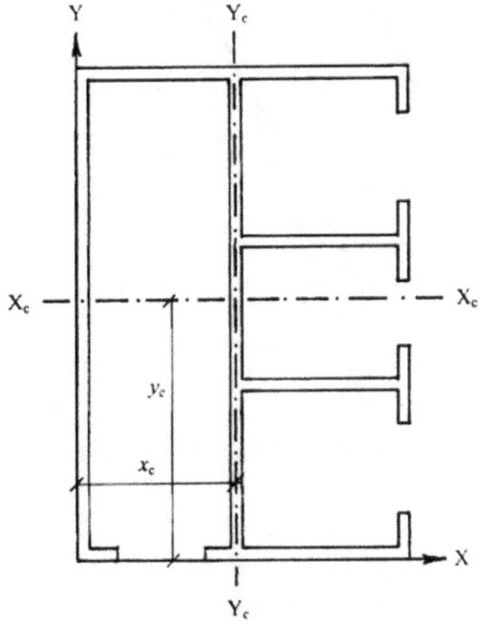

Example 1   Calculation Sheet 59

Reference	CALCULATIONS	OUTPUT
	The location of the centroidal axes, and values of the second moment of area about these axes, where t_w = wall thickness, can be determined as follows:	

$$\sum A_w = [2 \times 7.2 + (7.2 - 3.8) + (2 \times 5.2 - 1.4) + 2 \times 2.4]t_w = 31.6t_w \text{ m}^2$$

$$\sum A_w y = [(2 \times 7.2 + 3.4 + 2 \times 2.4) \times 3.8 + 5.2 \times 7.5 + 3.8 \times 0.1]t_w = 125.3t_w \text{ m}^3$$

$$y_c = 125.3/31.6 = 4.0 \text{ m}$$

$$I_{Xc} = [3 \times 7.2 \times (7.2^2/12 + 0.2^2) - 1.4 \times (2 \times 1.4^2/12 + 2.6^2 + 2.2^2)$$
$$- 1.0 \times (1.0^2/12 + 0.2^2) + 5.2 \times (0.2^2/12 + 3.5^2) + 3.8 \times (0.2^2/12 + 3.9^2)$$
$$= 198.9t_w \text{ m}^4$$

$$\sum A_w x = [2 \times 7.2 \times 1.3 + 3.4 \times 5.1 + 2 \times 5.2 \times 2.6 - 1.4 \times 1.3 + 4.8 \times 3.8]t_w$$
$$= 79.5t_w \text{ m}^3$$

$$x_c = 79.5/31.6 = 2.5 \text{ m}$$

$$I_{Yc} = [7.2 \times (2 \times 0.2^2/12 + 2.4^2) + 3.4 \times (0.2^2/12 + 2.6^2) + 10.4 \times (5.2^2/12 + 0.1^2)$$
$$- 1.4 \times (1.4^2/12 + 1.2^2) + 4.8 \times (2.4^2/12 + 1.3^2)]t_w = 96.2t_w \text{ m}^4$$

Assuming the wind force is distributed uniformly over the exposed surface height of the building ($h = 19$ m), total bending moment about the axis X_c at the top of the basement floor (4 m below ground level) is

$$M_x = 396 \times (19/2 + 4) = 5346 \text{ kN m}$$

Assuming a linear variation, the resulting intensity of vertical load in the walls:

$$w_k = \pm M_x (y - y_c)/I_{Xc} = \pm (5346/198.9) \times (y - y_c) = \pm 26.9(y - y_c) \text{ kN/m}$$

94.2 kN/m

24.2 kN/m

X_c ——— X_c

35.0 kN/m

104.9 kN/m

Vertical load intensity at base of walls resulting from overturning moment
(due to characteristic wind force) about axis X_c

The shear flow at a section distant $(y - y_c)$ from axis X_c is given by

$$v = F_w S/Ib \qquad \text{where } b \text{ is the total width of the walls cut by the section}$$

and S is the first moment of area of the walls to one side of the section

Example 1 **Calculation Sheet 60**

Reference	CALCULATIONS	OUTPUT

Maximum shear flow occurs at axis X_c, where $b = 3t_w$ and

$S = (3 \times 3.4^2/2 + 5.2 \times 3.5 + 2.4 \times 0.9 - 1.4 \times 2.2 - 0.3^2/2)t_w = 34.6t_w\ \text{m}^3$

$v = 396 \times 34.6t_w/(198.9t_w \times 3t_w) = 23.0/t_w\ \text{kN/m}^2$

Shear force and bending moment on lintel at ground floor level, where $h_s = 4.0$ m is the storey height and $l_b = 1.0$ m is the clear span of the lintel are:

$V_b = vh_st_w = 23 \times 4.0 = 92$ kN $\qquad M_b = Vl_b/2 = 92 \times 1.0/2 = 46$ kN m

Similarly, at the centre of 1.4 m span lintel, where $(y - y_c) = 2.2$ m:

$S = (2 \times 1.2 \times 2.9 + 0.5 \times 3.15 + 5.2 \times 3.5)t_w = 26.7t_w\ \text{m}^3$

$v = 396 \times 26.7t_w/(198.9t_w \times 3t_w) = 17.7/t_w\ \text{kN/m}^2$

$V_b = 17.7 \times 4.0 = 70.8$ kN $\qquad M_b = 70.8 \times 1.4/2 = 49.6$ kN m

Total bending moment about the axis Y_c at the top of the basement floor, and the resulting intensity of vertical load in the walls are:

$M_y = 200 \times (19/2 + 4) = 2700$ kN m

$w_k = \pm M_y (x - x_c)/I_{Yc} = (2700/96.2) \times (x - x_c) = \pm 28.1(x - x_c)$ kN/m

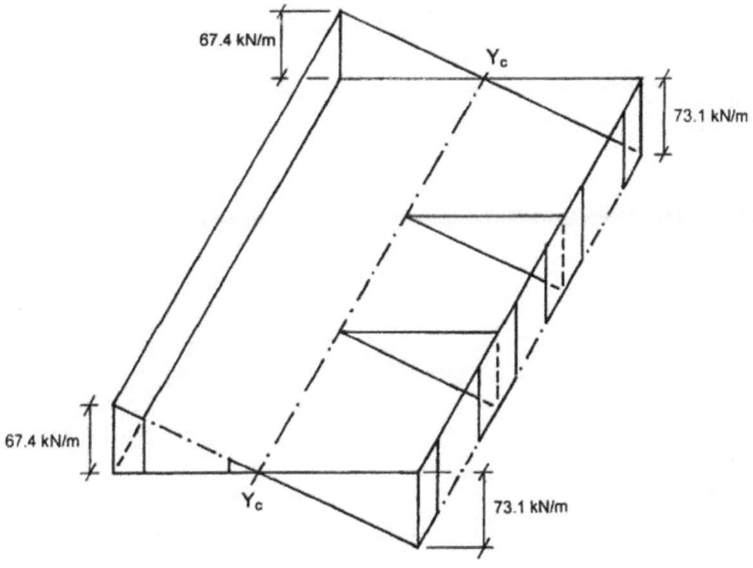

Vertical load intensity at base of walls resulting from overturning moment (due to characteristic wind force) about axis Y_c

At centre of 1.4 m span lintel, where $(x_c - x) = 1.3$ m:

$S = (7.2 \times 2.4 + 1.3 \times 1.85 + 0.6 \times 2.2)t_w = 21.0t_w\ \text{m}^3$

$v = 200 \times 21t_w/(96.2t_w \times 2t_w) = 21.8/t_w\ \text{kN/m}^2$

$V_b = 21.8 \times 4.0 = 87.2$ kN $\qquad M_b = 87.2 \times 1.4/2 = 61.0$ kN m

Dead and imposed loading

The loads applied to the walls from the floors and roof will vary according to the form of construction. For the beam and solid slab, or integral beam and ribbed slab, forms of construction, most of the load is concentrated at the positions where the beams are supported. Here, the local effects in each storey need to be considered, but it is reasonable to assume distributed loads at the base of the wall system.

Loads will be evaluated for the flat slab form of construction, in which the walls on gridlines C, D, 2 and 3 are considered to support an area of slab extending 3.6 m beyond the gridlines (3.4 m outside the walls). Load from the stairs and landings (see calculation sheet 54), and from the roof over this area, are considered to be supported equally by the wall on gridline C and the adjacent parallel wall.

Example 1   **Calculation Sheet 61**

Reference	CALCULATIONS	OUTPUT
	Area of flat slab assumed to be supported by internal walls	

WALLS ON GRIDLINES C AND D

Characteristic dead load at base of wall on gridline C:

		kN/m
Roof	$\{(7.6 + 3.4) \times 3.4/7.4 + 1.1)\} \times 7.5 =$	46.2
Floors	$5 \times (7.6 + 3.4) \times 3.4/7.4 \times 7.25 =$	183.2
Staircase area	$1.1 \times \{(5 \times 4.25 + 3 \times 3.0 \times 2.5/7.4 + 2 \times 4.2 \times 3.3/7.4)\} =$	30.8
Wall	$22.0 \times 0.2 \times 25 =$	110.0
Total		370.2

Characteristic imposed load, including reduction for number of floors:

		kN/m
Roof	$\{(7.6 + 3.4) \times 3.4/7.4 + 1.3)\} \times 0.6 =$	3.8
Floors	$0.6 \times [5 \times (7.6 + 3.4) \times 3.4/7.4 \times 4.0] = 0.6 \times 101.1 =$	60.7
Staircase area	$0.6 \times [1.1 \times (3 \times 2.5 + 2 \times 4.0)] = 0.6 \times 17.0 =$	10.2
Total		74.7

In-plane forces

For the wind force acting normal to the long face of the building, the maximum and minimum design ultimate vertical load intensities at the end of the wall on gridline C, with $\phi_0 = 0.7$ for imposed load and 0.5 for wind load, are as follows:

$$N_{Ed,max} = 1.25g_k + 1.5(0.7q_k + w_k) \text{ or } 1.25g_k + 1.5(q_k + 0.5w_k)$$

$$= 1.25 \times 370.2 + 1.5 \times (0.7 \times 74.7 + 104.9) = 698.5 \text{ kN/m}$$

$$N_{Ed,min} = 1.0g_k - 1.5w_k = 370.2 - 1.5 \times 104.9 = 212.8 \text{ kN/m (no tension)}$$

The load due to the tank room and lift motor room, which are located over the area between the wall on gridline D and the adjacent parallel wall, will be assumed to be the same as that for the staircase area. For the end portion of the wall on gridline D, allowing for the reaction from the lintel over the adjacent opening, the maximum design ultimate vertical load intensity at the end of the wall is

$$N_{Ed,max} = 2 \times (1.25 \times 370.2 + 1.5 \times 0.7 \times 74.7) + 1.5 \times 104.9 = 1240 \text{ kN/m}$$

Assuming F_w is resisted equally by the three walls parallel to the wind direction, the maximum design ultimate horizontal shear force in the plane of each wall, is

Example 1   **Calculation Sheet 62**

Reference	CALCULATIONS	OUTPUT
	$V_{Ed,max} = 1.5F_w/3 = 396/2 = 198$ kN	Transverse moments
	Transverse moments	
	From the analysis of the flat slab floor, in which the stiffness of the wall on gridline C was considered the same as that of two edge columns on gridline A (calculation sheets 25–26), the maximum moment applied to half the wall length is 226.8 kN m. Thus, at each floor, the distributed transverse moment in the wall, above and below the floor, is $0.5 \times 226.8/3.8 = 30$ kN m/m.	Floor connection
	From the analysis for the landings to the stairs (calculation sheet 54), assuming an end restraint moment equal to 50% of the span moment, the transverse moments applied to the wall at mid-storey height are 6.5 kN m/m (landings above first floor) and 8.8 kN m/m (landings below first floor). Since the effect of these moments is to reduce the moments at the floor connections, they may be ignored.	
	Resistance to bending and axial force	Landing connection
	For the wall on gridline D, since each vertical element is stiffened by a return wall, it is reasonable to ignore the effect of the transverse moments and design the wall for vertical load only. For the bottom storey, the maximum vertical load intensity at the end of the wall is $N_{Ed,max} = 1240$ kN/m.	
6.1 (4)	Minimum total design moment, with $e_0 = h/30 = 200/30 \geq 20$ mm, is	
	$M_{min} = N_{Ed}\,e_0 = 1240 \times 0.02 = 24.8$ kN m	
	$N_{Ed}/bhf_{ck} = 1240 \times 10^3/(1000 \times 200 \times 32) = 0.20$	
	$M_{Ed}/bh^2f_{ck} = 24.8 \times 10^6/(1000 \times 200^2 \times 32) = 0.02$	
	From the column design chart for $d/h = 0.8$ (Table A2), $A_s f_{yk}/bhf_{ck} = 0$.	
	For the wall on gridline C, the maximum vertical load intensity at the end of the wall, and the maximum transverse moments resulting from the connection to the ground floor slab are:	
	$N_{Ed,max} = 698.5$ kN/m $M_{top} = 30$ kN m/m $M_{bot} = -30/2 = -15$ kN m/m	
	Using the simplified method in Concise Eurocode 2, taking condition 1 (monolithic connection to members at least as deep as the overall thickness of the wall) at both top and bottom of the column,	
	$l_0 = 0.75l = 0.75 \times 3.76 = 2.82$ m (for storeys above 1st floor, $l_0 = 2.45$ m)	
5.2 (9)	First-order moment from imperfections (simplified procedure):	
	$M_i = Nl_0/400 = 698.5 \times 2.82/400 = 5.0$ kN m/m	
	First-order moments, including the effect of imperfections:	
	$M_{01} = -15 + 5 = -10$ kN m, $M_{02} = 30 + 5 = 35$ kN m	
5.8.3.2 (1)	Slenderness ratio $\lambda = l_0/(h/\sqrt{12}) = 2.82/(0.2/\sqrt{12}) = 48.9$	
5.8.3.1 (1)	Slenderness criterion, $\lambda_{lim} = 20(A \times B \times C)/\sqrt{n}$ where:	
	$n = N/A_c f_{cd} = N/(200 \times 0.85 \times 32/1.5) = 698.5/3626 = 0.20$	
	Taking $A = 0.7$, $B = 1.1$ and $C = 1.7 - M_{01}/M_{02} = 1.7 + 10/35 = 1.98$	
	$\lambda_{lim} = 20 \times 0.7 \times 1.1 \times 1.98/\sqrt{0.2} = 68.2$ ($> \lambda = 48.9$)	
	Since $\lambda < \lambda_{lim}$, second-order effects may be ignored and $M_{Ed} = M_{02}$ ($\geq M_{min}$)	
	A combination of maximum moment and minimum coexistent load can be critical for values of $N_{Ed} < N_{bal}$ where:	
5.8.8.3 (3)	$N_{bal} = 0.4A_c f_{cd} = 0.4 \times 1000 \times 200 \times 0.85 \times 32/1.5 \times 10^{-3} = 1450$ kN m	
	For the top storey of the wall on gridline C, the minimum vertical load intensity at the bottom of the wall, resulting from the characteristic dead load and uplift due to the design ultimate wind load acting on the short face of the building, is	
	$N_{Ed,min} = (46.2 + 3.5 \times 0.2 \times 25) - 67.4 \times (4.5 \times 2.25)/(19 \times 13.5) = 61.0$ kN/m.	
	The maximum transverse moment at the bottom of the wall, including the effect of imperfections is: $M_{02} = 30 + 61 \times 2.45/400 = 30.4$ kN m/m	

Example 1   **Calculation Sheet 63**

Reference	CALCULATIONS	OUTPUT
	$N_{Ed}/bhf_{ck} = 61.0 \times 10^3/(1000 \times 200 \times 32) = 0.010$	
	$M_{Ed}/bh^2f_{ck} = 30.4 \times 10^6/(1000 \times 200^2 \times 32) = 0.024$	
	From the column design chart for $d/h = 0.8$ (Table A2),	
	$A_s f_{yk}/bhf_{ck} = 0.06$, which gives $A_s = 0.06 \times 1000 \times 200 \times 32/500 = 768$ mm^2/m	
	Minimum amount of vertical reinforcement:	
9.6.2 (1)	$A_{s,vmin} = 0.002A_c = 0.002 \times 1000 \times 200 = 400$ mm^2/m	
	Similar calculations can be performed for the other storeys, and all the results are summarised below:	

Storey	N_{Ed} (kN min)	M_{Ed} (kN m)	$\dfrac{N_{Ed}}{bhf_{ck}}$	$\dfrac{M_{Ed}}{bh^2 f_{ck}}$	$\dfrac{A_s f_{yk}}{bhf_{ck}}$	A_s (mm^2)
4th floor–roof	61.0	30.4	0.010	0.024	0.06	768
3rd–4th floor	115.2	30.7	0.018	0.024	0.05	640
2nd–3rd floor	166.2	31.0	0.026	0.024	0.04	512
1st–2nd floor	213.9	31.3	0.033	0.025	0.03	min
Grd–1st floor	259.4	31.8	0.040	0.025	0.02	min
Basement	302.8	32.2	0.047	0.025	0.01	min

Reference	CALCULATIONS	OUTPUT
9.6.2 (2)	Maximum distance between two adjacent vertical bars = $3h \leq 400$ mm	
	Although H10-300 (EF) will suffice below 3rd floor level, providing H12-300 (EF) throughout will give extra rigidity.	Vertical bars H12-300 (EF)
	Shear design	
	Maximum shear stress at bottom of wall on gridline C, due to design ultimate wind load acting on long face of building:	
	$v_{Ed} = V_{Ed}/A_c = 198 \times 10^3/(7600 \times 200) = 0.13$ MPa	
	Normal stress due to characteristic dead load:	
	$\sigma_{cp} = N_{Ed}/A_c = 370.2 \times 10^3/(1000 \times 200) = 1.85$ MPa	
	Minimum design shear strength, with $k = 1.0$ (minimum) and $k_1 = 0.15$, is	
6.2.2 (1)	$v_{Rd,c} = 0.035k^{3/2}f_{ck}^{1/2} + k_1\sigma_{cp} = 0.035 \times 1.0 \times 32^{1/2} + 0.15 \times 1.85 = 0.47$ MPa	Horizontal shear OK
	Cracking	
	Minimum area of vertical reinforcement in tension zone to control cracking due to transverse moment (ignoring axial load) is given by	
7.3.2 (2)	$A_{s,min} = k_ckf_{ct,eff}A_{ct}/\sigma_s$ (where $k_c = 0.4$ for bending, $k = 1.0$ for $h \leq 300$ mm)	
	$= 0.4 \times 1.0 \times 3.0 \times 1000 \times 100/500 = 240$ mm^2/m (H12-400)	Vertical bars OK
7.3.3 (1)	No other specific measures are needed since the overall thickness does not exceed 200 mm, provided detailing requirements are observed.	
	A minimum area of horizontal reinforcement is required to control cracking due to restrained early thermal contraction. Assuming cracking occurs at age 3 days when $f_{cm}(t) = 24$ MPa say,	
	$f_{ct,eff} = f_{ctm}(t) = [f_{cm}(t)/(f_{ck} + 8)]f_{ctm} = [24/(32 + 8)] \times 3.0 = 1.8$ MPa	
7.3.2 (2)	$A_{s,min} = k_ckf_{ct,eff}A_{ct}/\sigma_s$ (where $k_c = 1.0$ for tension, $\sigma_s = f_{yk} = 500$ MPa)	Horizontal bars H10-200 (EF)
	$= 1.0 \times 1.0 \times 1.8 \times 1000 \times 100/500 = 360$ mm^2/m (EF) H10-200 (EF)	
7.3.4 PD 6687	For cracks from early thermal contraction, the crack width may be calculated as	
	$w_k = (0.8R\alpha\Delta T) \times s_{r,max}$ where	
	$s_{r,max} = 3.4c + 0.425k_1k_2(A_{c,eff}/A_s)\phi$	
	With $c = 25$ mm, $k_1 = 0.8$ (for high-bond bars), $k_2 = 1.0$ (for tension) and assuming $\phi = 10$ mm	
	$s_{r,max} = 3.4 \times 25 + 0.425 \times 0.8 \times 1.0 \times 1000 \times 100 \times 10/A_s = 85 + 340 \times 10^3/A_s$	
	Taking $R = 0.8$ (maximum), $\Delta T = 25°$C and $\alpha = 12 \times 10^{-6}$ per $°$C,	

Example 1   **Calculation Sheet 64**

Reference	CALCULATIONS	OUTPUT
	$w_k = 0.8 \times 0.8 \times 12 \times 10^{-6} \times 25 \times (85 + 340 \times 10^3/A_s)$	
	Hence, for $w_k = 0.3$ mm, $A_s = 230 \geq A_{s,min} = 360$ mm^2/m (EF)	
	With H10-200(EF), $w_k = 192 \times 10^{-6} \times (85 + 340 \times 10^3/393) = 0.18$ mm	Horizontal bars OK
	Floor-to-wall connection	
	The distributed moment in the slab (see calculation sheet 62) = 60 kN m/m	
	$M/bd^2f_{ck} = 60 \times 10^6/(1000 \times 206^2 \times 32) = 0.044$ $z/d = 0.95$ (max)	Bars at wall to floor
	$A_s = 60 \times 10^6/(0.87 \times 500 \times 0.95 \times 206) = 705$ mm^2/m (H12-150)	connection H12-150
	WALLS ON GRIDLINES 2 AND 3	
	Characteristic dead load at base of wall on gridline 3:	
	kN/m	
	Roof $(5.2 + 3.4) \times 3.4/5.0 \times 7.5 =$ 43.9	
	Floors $5 \times (5.2 + 3.4) \times 3.4/5.0 \times 7.25 = 212.0$	
	Wall $22.0 \times 0.2 \times 25 = \underline{110.0}$	
	Total 365.9	
	Characteristic imposed load, including reduction for number of floors:	
	kN/m	
	Roof $(5.2 + 3.4) \times 3.4/5.0 \times 0.6 =$ 3.5	
	Floors $0.6 \times [5 \times (5.2 + 3.4) \times 3.4/5.0 \times 4.0] = 0.6 \times 117.0 = \underline{70.2}$	
	Total 73.7	
	In-plane forces	
	For the wind force acting normal to the long face of the building, the maximum and minimum design ultimate vertical load intensities at the end of the wall on gridline 3, with $\phi_0 = 0.7$ for imposed load and 0.5 for wind load, are as follows:	
	$N_{Ed,max} = 1.25g_k + 1.5(0.7q_k + w_k)$ or $1.25g_k + 1.5(q_k + 0.5w_k)$	
	$= 1.25 \times 365.9 + 1.5 \times (0.7 \times 73.7 + 104.9) = 692.1$ kN/m	
	$N_{Ed,min} = 1.0g_k - 1.5w_k = 365.9 - 1.5 \times 104.9 = 208.5$ kN/m (no tension)	
	For the wind acting normal to the short face of the building, and assuming the force is resisted by the walls parallel to the wind direction in proportion to their length[3], the maximum design ultimate horizontal shear force in the plane of each long wall is approximately:	
	$V_{Ed,max} = 1.5 \times (4F_w/9) = 200/1.5 = 133$ kN	
	Transverse moments	
	From the analysis of the flat slab floor, it can be deduced that the moments applied to the walls will be similar but less than those for the walls on gridlines C and D.	
	Resistance to bending and axial force	
	Since the in-plane forces and the transverse moments are similar to those for the walls on gridlines C and D, the same reinforcement details will suffice.	
	Shear design	
	Maximum shear stress at bottom of wall on gridline 3, due to design ultimate wind load acting on short face of building:	
	$v_{Ed} = V_{Ed}/A_c = 133 \times 10^3/(5200 \times 200) = 0.13$ MPa	Reinforcement as
	This value is well within the minimum design shear strength as demonstrated in the calculations for the wall on gridline C.	shown for walls on gridlines C and D
BS EN 1992-1-2 5.4.2 Table 5.4	**FIRE RESISTANCE**	
	Required minimum dimensions for REI 90, with $\mu_{fi} = 0.7$ and wall exposed to fire on two sides, are as follows:	
	Wall thickness = 170 mm Axis distance (to centre of bars) = 25 mm	Sufficient for 1.5 h
	Since the cover required for durability is 25 mm, the axis distance is sufficient.	fire resistance

Example 1   **Calculation Sheet 65**

Reference	CALCULATIONS	OUTPUT
5.3.1 (3)	**LINTELS** Since the span is less than three times the overall depth, the lintel is considered as a deep beam. CIRIA Guide 2 *The Design of Deep Beams in Reinforced Concrete* gives recommendations on detailed analysis and design, although this publication was written for use with the then current British Standard CP 110. Clear span $l_o = 1.4$ m Effective span $l = 1.2l_o = 1.7$ m say Active height is taken as the lesser of h and l, giving $h_a = 1.7$ m The shear forces and bending moments due to the effects of the characteristic wind forces on the building are given in calculation sheet 60. For the wall on gridline D, the maximum coexisting design ultimate vertical load on the ground floor lintel is: $n = (1.25 \times 183.2 + 1.5 \times 0.7 \times 101.1)/5 + 1.25 \times 2.0 \times 0.2 \times 25 = 79.5$ kN/m The maximum design ultimate shear force on the lintel (including the effect of the wind acting on the short face of the building) is $V_{Ed,max} = 79.5 \times 1.4/2 + 1.5 \times 87.2 = 186.5$ kN The maximum bending moments due to the effects of the wind force are hogging at one end of the span and sagging at the other end of the span. Hence: The maximum hogging moment (including wind) at the end of the span is $M_{Ed,max} = 1.5 \times 61.0 + 79.5 \times 1.7^2/12 = 110.7$ kN m The maximum sagging moment (including wind) at the end of the span is $M_{Ed,max} = 1.5 \times 61.0 - (183.2/5 + 10) \times 1.7^2/12 = 80.3$ kN m **Flexural design** The required area of reinforcement is given by $A_s = M/(0.87f_{yk}z)$ where $z = 0.2l + 0.3h_a = 0.5 \times 1700 = 850$ mm For hogging: $A_s = 110.7 \times 10^6/(0.87 \times 500 \times 850) = 300$ mm^2 (4H10) Reinforcement should be contained in a band with boundaries at distances from the soffit of $0.2l = 0.34$ m and $0.8l = 1.36$ m, respectively. For sagging: $A_s = 80.3 \times 10^6/(0.87 \times 500 \times 850) = 218$ mm^2 (3H10) Reinforcement may be provided over a depth of $0.2h_a = 0.34$ m from the soffit. Minimum area of reinforcement in each face and each direction is given by:	
9.7 (1) Table NA.1 9.7.1 (2)	$A_{s,dbmin} = 0.2\% = 0.002 \times 1000 \times 200/2 = 200$ mm^2/m (H10-300) Maximum spacing of bars should not exceed twice the wall thickness ≤ 300 mm. **Shear design** The required inclination of the concrete strut (defined by $\cot\theta$), to obtain the least amount of shear reinforcement, can be shown to depend on the following factor: $v_w = V/[b_w z(1 - f_{ck}/250)f_{ck}]$ $\qquad = 186.5 \times 10^3/[200 \times 850 \times (1 - 32/250) \times 32] = 0.040$ From *Reynolds,* Table 4.18, for vertical links and values of $v_w < 0.138$, $\cot\theta = 2.5$ can be used. The area of links required is then given by $A_{sw}/s = V/f_{ywd}z\cot\theta$	Wall reinforcement is sufficient
6.2.3 (3)	$\qquad = 186.5 \times 10^3/(0.87 \times 500 \times 850 \times 2.5) = 0.20$ mm^2/mm From *Reynolds,* Table 4.20, H10-300 links gives 0.52 mm^2/mm	Wall reinforcement is sufficient

Example 1: Reinforcement in Internal Walls

Drawing 15

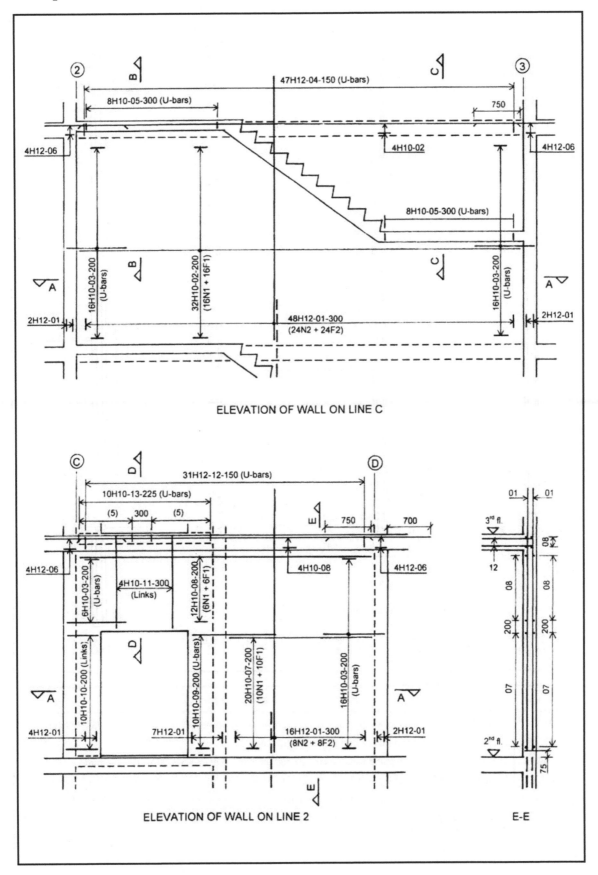

ELEVATION OF WALL ON LINE C

ELEVATION OF WALL ON LINE 2

E-E

Example 1: Cross-Sections of Internal Walls   Drawing 16

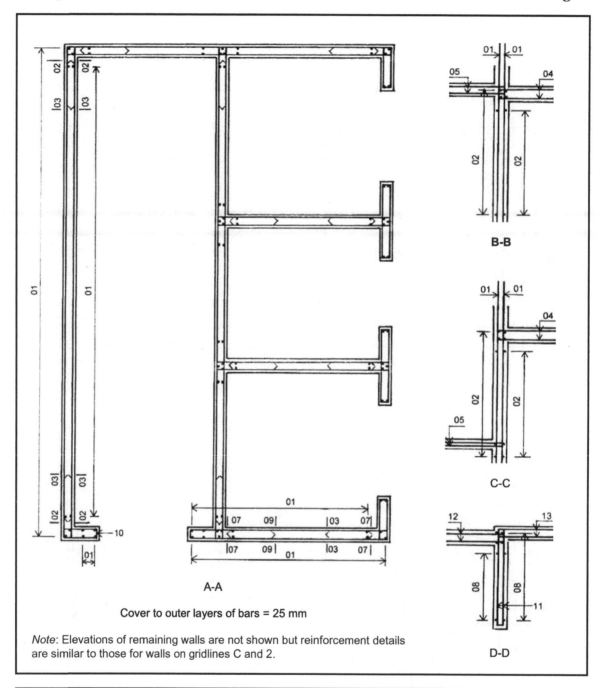

A-A

Cover to outer layers of bars = 25 mm

Note: Elevations of remaining walls are not shown but reinforcement details are similar to those for walls on gridlines C and 2.

B-B

C-C

D-D

Bar Marks	Commentary on Bar Arrangement (Drawings 15 and 16)
01	Bars bearing on 75 mm kicker. Projection of bars above next floor level = $1.5 \times 35 \times 12 + 75 = 700$ mm.
02, 07, 08	Bars curtailed 50 mm from face of return wall at each end.
03, 09	U-bars (shape code 21) lapping with bars 02, 07 and 08. Lap length = $1.5 \times 50 \times 10 = 750$ mm.
04, 12	U-bars (shape code 21) to lap with bars in floor slab. Projection of legs = $1.5 \times 35 \times 12 + 50 = 700$ mm.
05, 13	U-bars (shape code 21) to lap with bars in landings. Projection of legs = $1.5 \times 35 \times 10 + 50 = 600$ mm.
06	Straight bars lapping with bars 02 and 08, and projecting into floor slab. Lap length = 750 mm. Projection of bars into slab = 700 mm.
10, 11	Closed links (shape code 51).

4 Example 2: Foundations to Multi-Storey Building

Description

Different forms of foundation structure for the five-storey building designed in Example 1 are shown in drawings 1 and 2. In Example 1, three different forms of suspended floor construction were considered, resulting in closely spaced columns in the first case, and more widely spaced columns in the other cases. The result of a typical borehole examination, from which an estimate of the bearing capacity at different depths can be made, is shown below. The soil conditions indicate a low bearing capacity at the underside of the basement floor, but increasing significantly at greater depth. The water level in the ground is located below the basement floor, but the basement construction is required to be waterproof.

1. Where the suspended floors of the building are of beam and slab construction with the columns closely spaced, a continuous spread foundation of uniform thickness (raft) is shown in drawing 1. The construction of the basement is monolithic throughout with asphalt tanking applied to the concrete externally.

2. Where the suspended floors of the building are of flat slab construction with the columns more widely spaced, two forms of foundation construction are shown. A layout of isolated spread bases, with the remainder of the basement floor supported on a compressible sub-base, and tied to the isolated bases, is shown in drawing 2. External waterstops are provided at all joints, and an internal waterproof render is applied to the basement wall and floor.

3. An alternative arrangement of pile foundations with pile caps and beams and a suspended basement floor is shown in drawing 3. The basement is monolithic throughout with asphalt tanking applied to the concrete externally.

The basement will be constructed in open excavation, and the ground behind the wall will be reinstated by backfilling with a granular material. A graded drainage material will be provided behind the wall, with an adequate drainage system at the bottom. The fill to be retained is 3.8 m high above the top of the basement floor and the surcharge is 10 kN/m².

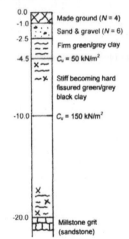

For the spread foundations (1 and 2), where the bearing stratum is firm clay with a representative value of $c_u = 50$ kN/m², presumed values are taken of 150 kN/m² for allowable bearing pressure, and 12 MN/m³ for modulus of subgrade reaction (see Table B1).

Assuming a 0.6 m thick base, with the underside at 4.5 m below original ground level, the difference between weight of soil excavated and (weight of concrete plus imposed load) added = $4.5 \times 16 - (0.6 \times 25 + 5) = 50$ kN/m² say.

Maximum allowable increase in loading intensity = $150 + 50 = 200$ kN/m².

Properties of the retained soil (uniform sub-angular gravel) are as follows:

Unit weight of soil: $\gamma = 20$ kN/m³   Angle of shearing resistance of soil: $\varphi' = 32°$

Coefficients of earth pressure (active and at rest):

$K_a = (1 - \sin \varphi')/(1 + \sin \varphi') = 0.30$ $K_o = (1 - \sin \varphi') = 0.47$

For the pile foundations (3), bored piles embedded in the stiff/hard fissured clay will be used, with the pile shaft lined through the upper sand and gravel layers. The load-carrying capacity of the piles will be provided by skin friction and end bearing in the clay taking a representative value of $c_u = 150$ kN/m². The size and depth of the piles, and the required numbers will be determined.

Typical borehole log

Schedule of Drawings and Calculations

Drawing	Components	Type of Construction	Calc. Sheets
1–3	General arrangement of basement with raft, isolated bases and pile foundations respectively		
	Basement with raft foundation		
4–5	Foundation/floor slab	Continuous raft	1–11
6–7	Wall	Propped cantilever wall continuous with raft foundation	12–13
	Basement with isolated spread foundations		
8–9	Foundations	Pad bases to internal columns and walls	14–18
10–11	Wall and floor slab	Propped cantilever wall on pad footing with tied floor slab	19–22
	Basement with pile foundations		
12–13	Foundations	Bored piles with pile caps and beams to columns and walls	23–26
14–17	Wall and floor slab	Propped cantilever wall continuous with suspended floor slab	27–34

Example 2: General Arrangement of Basement with Raft Foundation   **Drawing 1**

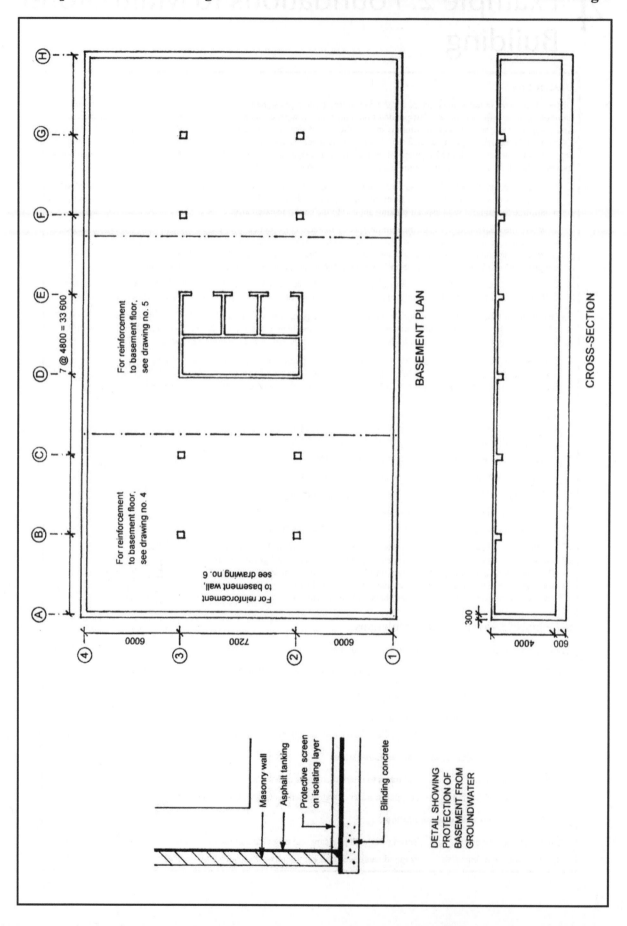

BASEMENT PLAN

For reinforcement to basement floor, see drawing no. 5

For reinforcement to basement floor, see drawing no. 4

For reinforcement to basement wall, see drawing no. 6

7 @ 4800 = 33 600

6000   7200   6000

CROSS-SECTION

300   4000   600

Masonry wall

Asphalt tanking

Protective screen on isolating layer

Blinding concrete

DETAIL SHOWING PROTECTION OF BASEMENT FROM GROUNDWATER

Example 2: General Arrangement of Basement with Isolated Bases   **Drawing 2**

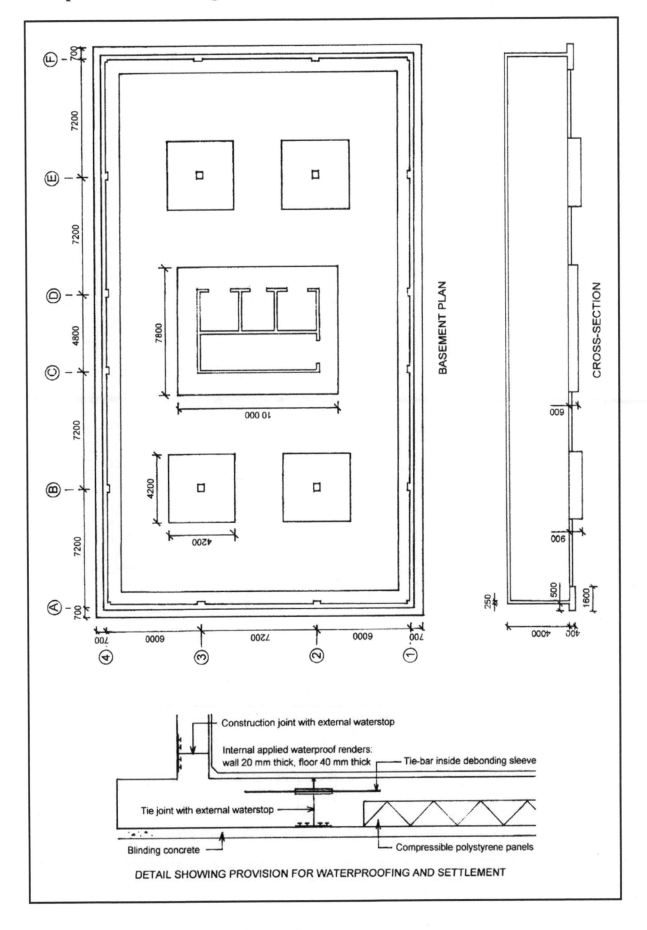

BASEMENT PLAN

CROSS-SECTION

Construction joint with external waterstop

Internal applied waterproof renders:
wall 20 mm thick, floor 40 mm thick — Tie-bar inside debonding sleeve

Tie joint with external waterstop

Blinding concrete — Compressible polystyrene panels

DETAIL SHOWING PROVISION FOR WATERPROOFING AND SETTLEMENT

Example 2: General Arrangement of Basement with Pile Foundations   **Drawing 3**

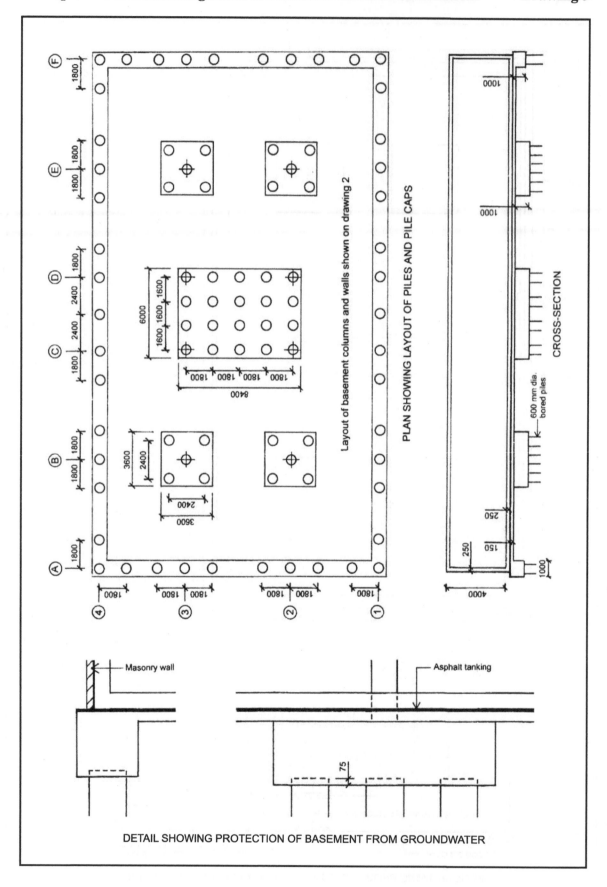

Layout of basement columns and walls shown on drawing 2

PLAN SHOWING LAYOUT OF PILES AND PILE CAPS

CROSS-SECTION

600 mm dia. bored piles

Masonry wall

Asphalt tanking

DETAIL SHOWING PROTECTION OF BASEMENT FROM GROUNDWATER

Example 2 **Calculation Sheet 1**

Reference	CALCULATIONS	OUTPUT
	CONTINUOUS RAFT FOUNDATION (Beam and slab superstructure)	

Loading

From the data tabulated on calculation sheets 14 and 16 of design example 1, and allowing for the appropriate reduction in the total imposed floor loads, the design loads for columns B1 and B2 are obtained as follows:

Column B1 (loads at top of basement wall)			
Load Case	Load	Dead (kN)	Imposed (kN)
1	Ultimate characteristic	$1274 - (409 + 12) = 853$ $853/1.25 = 683$	$0.6 \times 409 + 12 = 258$ $258/1.5 = 172$
2	Ultimate characteristic	$1332 - (465 + 14) = 853$ $853/1.25 = 683$	$0.6 \times 465 + 14 = 293$ $293/1.5 = 195$

Column B1	
Load (kN)	Moment (kN m)
1111 855	48.9 37.6
1146 878	56.4 41.0

Column B2 (loads at top of basement floor)			
Load Case	Load	Dead (kN)	Imposed (kN)
1	Ultimate characteristic	$2726 - (1113 + 34) = 1579$ $1579/1.25 = 1263$	$0.6 \times 1113 + 34 = 702$ $702/1.5 = 468$
2	Ultimate characteristic	$2080 - (486 + 15) = 1579$ $1579/1.25 = 1263$	$0.6 \times 486 + 15 = 307$ $307/1.5 = 205$

Column B2	
Load (kN)	Moment (kN m)
2281 1731	4.3 3.3
1886 1468	10.8 7.5

Similarly, the design loads for column A2 can be obtained as follows:

Column A2 (loads at top of basement wall)			
Load Case	Load	Dead (kN)	Imposed (kN)
1	Ultimate characteristic	$1355 - (404 + 12) = 939$ $939/1.25 = 751$	$0.6 \times 404 + 12 = 255$ $255/1.5 = 170$
2	Ultimate characteristic	$1121 - (177 + 5) = 939$ $939/1.25 = 751$	$0.6 \times 177 + 5 = 111$ $111/1.5 = 74$

Column A2	
Load (kN)	Moment (kN m)
1194 921	2.1 1.4
1050 825	5.2 4.0

The wall system forming the central core of the building is considered to support floor and roof areas that extend 2.4 m beyond gridlines D and E and 3.4 m beyond gridlines 2 and 3, as shown in the figure below.

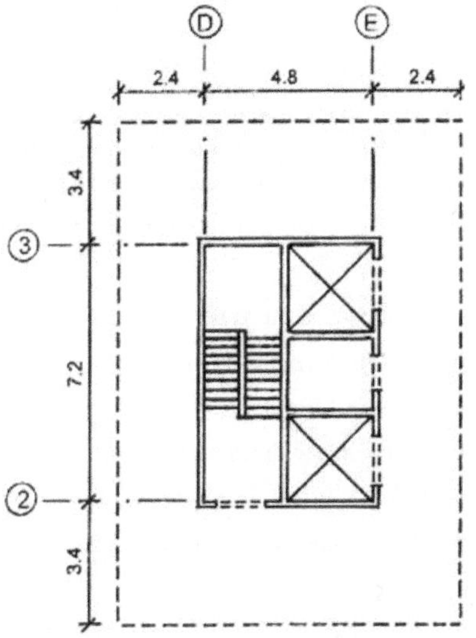

Area of floor assumed to be supported by internal walls

Example 2   Calculation Sheet 2

Reference	CALCULATIONS	OUTPUT
	The load due to the tank room and lift motor room, which are located over the area between the wall on gridline E and the adjacent parallel wall, will be assumed to be the same as that for the staircase area (see calculation sheet 61 of Example 1). For the analysis of the raft foundation, the total load on the wall system will be assumed to be distributed uniformly over the area enclosed by the walls.	

Total characteristic dead load at top of basement floor:

		kN
Roof	$9.6 \times 14.0 \times 5.25 =$	706
Floors	$5 \times (9.6 \times 14.0 - 5.2 \times 7.6) \times 5.0 =$	2372
Staircase areas ($\times 2$)	$2 \times 2 \times 7.2 \times 30.8 =$	887
Walls (ignoring openings)	$36.0 \times 22.0 \times 0.2 \times 25 =$	3960
Total		7925

Total characteristic imposed load:

		kN
Roof	$9.6 \times 14.0 \times 0.6 =$	81
Floors	$5 \times (9.6 \times 14.0 - 5.2 \times 7.6) \times 4.0 =$	1898
Staircase areas ($\times 2$)	$2 \times 2 \times 7.2 \times 10.2 =$	294
Total		2273

Overturning moment at top of basement floor, due to characteristic wind force in direction normal to long face of building: $M_x = 5346$ kN m

Overturning moment at top of basement floor, due to characteristic wind force in direction normal to short face of building: $M_y = 2700$ kN m

Internal Walls (Loads and Moments at Top of Basement Floor)			Wind Moment M_y (kN m)
Load and Combination	Vertical Load (kN)	Wind Moment M_x (kN m)	
Ultimate (1a)	$1.25 \times 7925 + 1.5 \times 0.7 \times 2273 = 12293$	$1.5 \times 5346 = 8020$	$1.5 \times 2700 = 4050$
(1b)	$1.0 \times 7925 = 7925$	$1.5 \times 5346 = 8020$	$1.5 \times 2700 = 4050$
(2)	$1.25 \times 7925 + 1.5 \times 2273 = 13316$	$0.5 \times 8020 = 4010$	$0.5 \times 4050 = 2025$
Service (1a)	$1.0 \times 7925 + 0.7 \times 2273 = 9516$	$1.0 \times 5346 = 5346$	$1.0 \times 2700 = 2700$
(1b)	$1.0 \times 7925 = 7925$	$1.0 \times 5346 = 5346$	$1.0 \times 2700 = 2700$
(2)	$7925 + 2273 = 10198$	$0.5 \times 5346 = 2673$	$0.5 \times 2700 = 1350$

Design loads due to weight of 300 mm thick basement wall are as follows:

Ultimate $1.25 \times 3.85 \times 0.3 \times 25 = 36.0$ kN/m.   Characteristic 28.8 kN/m

Raft thickness

The thickness will be taken as the minimum necessary to avoid the need for shear reinforcement at the internal columns. This can be estimated by considering the shear perimeter at distance $2d$ from the face of the column, and ignoring the shear reduction due to the bearing pressure inside the perimeter. Consider $h = 600$ mm.

Allowing 50 mm cover with 20 mm bars in each direction, mean effective depth

$d_{av} = 600 - (50 + 20) = 530$ mm.

The maximum shear stress should not exceed the value given by

6.2.2 (6) $v_{Rd,max} = 0.5 \nu f_{cd} = 0.5 \times 0.6(1 - f_{ck}/250) \times (\alpha_{cc} f_{ck}/1.5) = 0.2(1 - f_{ck}/250) f_{ck}$

$= 0.2 \times (1 - 32/250) \times 32 = 5.58$ MPa

For column B2, $N_{Ed} = 2281$ kN, and the shear stress at the column perimeter is

$v_{Ed} = N_{Ed}/ud = 2281 \times 10^3/(4 \times 300 \times 530) = 3.59$ MPa ($< v_{Rd,max}$)

The length of the control perimeter at distance $2d$ from the column face is

$u_1 = 4(c + \pi d) = 4 \times (300 + \pi \times 530) = 7860$ mm

Conservatively, taking $V_{Ed} = N_{Ed}$, the shear stress at the control perimeter is

$v_{Ed} = V_{Ed}/(u_1 d) = 2281 \times 10^3/(7860 \times 530) = 0.55$ MPa

Basic control perimeter for punching shear

Example 2 {style=left} **Calculation Sheet 3** {style=right}

Reference	CALCULATIONS	OUTPUT
6.2.2 (1)	The minimum shear strength, where $k = 1 + (200/d)^{1/2} = 1.61$, is given by $v_{min} = 0.035k^{3/2}f_{ck}^{1/2} = 0.035 \times 1.61^{3/2} \times 32^{1/2} = 0.40$ MPa Since V_{Ed} will be reduced due to the bearing pressure inside the control perimeter, and v_{Rd} depends on the tension reinforcement, $h = 600$ mm seems reasonable. **Analysis** The foundation will be analysed as a series of rectangular strips in two directions at right angles. The width of an internal strip is taken between the centrelines of adjacent panels. The width of an edge strip is taken from the centreline of the external wall to the centreline of the first panel. The loads on the edge strips are effectively distributed by the basement wall, and there is no need to analyse the strips for structural effects in the direction parallel to the wall. Each strip will be considered initially as a beam with free ends on an elastic soil. The moments required to restrain the resulting rotation at the ends of the beam will then be determined. The basement wall, considered initially as a beam propped at the top and fixed at the bottom, will be analysed to determine the moment at the edge of the foundation due to lateral earth pressure. The out-of-balance moments at the joints between the wall and the foundation will then be distributed according to the relative stiffness of the members, and the resulting effects determined. Effective span of basement wall between ground floor and basement floor is 4.0 m. Two soil conditions will be considered: at-rest pressure with surcharge, and active pressure without surcharge. Ultimate fixed-end moments at bottom of wall are: $M_{w,max} = 1.35 \times 0.47 \times (10 \times 4.0^2/8 + 20 \times 4.0^3/15) = 67.0$ kN m/m $M_{w,min} = 1.0 \times 0.30 \times 20 \times 4.0^3/15 = 25.6$ kN m/m *Note:* In the following calculations, the values of λL have been rounded down to 5.0 in one direction and 8.0 in the other direction, in order to use the data in Tables B2–B10. However, precise values of λL have been used to determine the flexural stiffness of the beams to avoid anomalies in the relative values. The flexural flexural stiffness, for equal and opposite unit rotations at each end of a beam on an elastic soil, is given by the relationship: $K_b = [k_s BL^3/4(\lambda L)^3] \times (\sinh \lambda L + \sin \lambda L)/(\cosh \lambda L - \cos \lambda L)$	Lateral soil pressure on basement wall Characteristic values of fixed-end moment at bottom of wall $M_{w,max} = 49.6$ kN m/m $M_{w,min} = 25.6$ kN m/m
	BEAM ON LINE B This is considered as a 4.8 m wide beam loaded with two sets of symmetrically placed loads, $0.5F_1$ at columns B1 and B4, and $0.5F_2$ at columns B2 and B3. With a = distance from ends of beam to point of application of loads, $a/L = 0$ for F_1, and $6.0/19.2 = 0.3$ say for F_2. Bending moments at the bottom of B1 and B4 will be combined with the moments at the bottom of the basement wall. Bending moments at the bottom of B2 and B3 are very small and will be ignored. Loads on beam	

Sum of maximum loads at ends of beam (including weight of basement wall):

Load	Load Case 1	Load Case 2
Ultimate characteristic	$F_1 = 2 \times (1111 + 36.0 \times 4.8) = 2568$ kN $F_1 = 2 \times (855 + 28.8 \times 4.8) = 1986$ kN	$F_1 = 2638$ kN $F_1 = 2032$ kN

Sum of maximum loads at columns B2 and B3:

Load	Load Case 1	Load Case 2
Ultimate characteristic	$F_2 = 2 \times 2281 = 4562$ kN $F_2 = 2 \times 1731 = 3462$ kN	$F_2 = 3772$ kN $F_2 = 2936$ kN

Example 2   **Calculation Sheet 4**

Reference	CALCULATIONS	OUTPUT

The end slopes for beams on elastic foundations can be determined from the data in Table B2, where $\lambda L = (3k_s L^4/Eh^3)^{1/4}$. With $L = 19.2$ m between the centrelines of columns B1 and B4, $E_c = 33$ GN/m^2 for C32/40 concrete, and $h = 0.6$ m:

$$\lambda L = [3 \times 12 \times 10^3 \times 19.2^4/(33 \times 10^6 \times 0.6^3)]^{1/4} = 5.12 \text{ (say 5.0)}$$

From the table for two symmetrically placed loads,

$$\theta_A = -\theta_B = (-25.66 F_1 + 5.642 F_2)/(k_s B L^2)$$

Thus, the fixed-end moments required to offset the end slopes are

$$M_A = -M_B = -(k_s B L^3/504.7) \times \theta_A = 0.0508 F_1 L - 0.0112 F_2 L$$

Flexural stiffness values of the beam (with $\lambda L = 5.12$) and wall are

$$K_b = k_s B L^3/537 = 12 \times 10^3 \times 1.0 \times 19.2^3/537 = 0.158 \times 10^6 \text{ kN m/m}$$

$$K_w = E_c B h_w^3/4L_w = 33 \times 10^6 \times 1.0 \times 0.3^3/(4 \times 4.0) = 0.056 \times 10^6 \text{ kN m/m}$$

Moment distribution factors: $D_b = 0.158/(0.158 + 0.056) = 0.738$, $D_w = 0.262$

Case 1. Characteristic loads with load case 2 on the building and minimum soil pressure on the wall. Fixed-end moments for beam and wall are:

Beam: $M_A = -M_B = (0.0508 \times 2032 - 0.0112 \times 2936) \times 19.2 = 1350.5$ kN m

Wall: $M_A = -M_B = -25.6 \times 4.8 + 0.5 \times 41.0 = -102.5$ kN m

Final moments at ends of beam, after releasing fixed-end moments, are

$$M_{A0} = 1350.5 - 0.738 \times (1350.5 - 102.5) = 429 \text{ kN m} \qquad M_{B0} = -429 \text{ kN m}$$

From Tables B3, B5 and B6, maximum bearing pressure (at $x/L = 0$) is

$$q = -51.31(M_0/BL^2) + 5.085(F_1/BL) - 0.070(F_2/BL)$$

$$= -51.31 \times 429/(4.8 \times 19.2^2) + (5.085 \times 2032 - 0.070 \times 2936)/(4.8 \times 19.2)$$

$$= -12.4 + 109.9 = 98 \text{ kN/m}^2 \ (<200 \text{ kN/m}^2)$$

Case 2. Ultimate design loads with load case 1 for the building and maximum soil pressure on the wall. Fixed-end moments for beam and wall are:

Beam: $M_A = -M_B = (0.0508 \times 2568 - 0.0112 \times 4562) \times 19.2 = 1524$ kN m

Wall: $M_A = -M_B = -67.0 \times 4.8 + 0.5 \times 56.4 = -294$ kN m

Final moments at ends of beam, after releasing fixed-end moments, are

$$M_{A0} = 1524 - 0.738 \times (1524 - 294) = 616 \text{ kN m} \qquad M_{B0} = -616 \text{ kN m}$$

From Tables B3, B5 and B6, $M_x = c_0 M_0 + (c_1 F_1 + c_2 F_2)L$ where

x/L	0	0.1	0.2	0.3	0.4	0.5
c_0	1.0	0.817	0.489	0.203	0.026	−0.032
c_1	0	−0.029	−0.030	−0.022	−0.013	−0.010
c_2	0	0.001	0.006	0.022	0.001	−0.006
M_x kN m	616	−839	−653	968	−538	−1038

Maximum shear force (at $x/L = 0$) is

$$V = -0.5 F_1 = -0.5 \times 2638 = -1319 \text{ kN}$$

Case 3. Ultimate design loads with minimum load on the building (constructed up to ground floor only) and maximum soil pressure on the wall.

$$F_1 = 2 \times (108.7 + 28.8 \times 4.8) = 494 \text{ kN} \qquad F_2 = 2 \times 209.0 = 418 \text{ kN}$$

Fixed-end moments for beam and wall are

Beam: $M_A = -M_B = (0.0508 \times 494 - 0.0112 \times 418) \times 19.2 = 392$ kN m

Wall: $M_A = -M_B = -67.0 \times 4.8 = -322$ kN m

Moments at bottom of wall after releasing fixed-end moments, are

$$M_{A0} = -322 - 0.262 \times (392 - 322) = -340 \text{ kN m} \qquad M_{B0} = 340 \text{ kN m}$$

Example 2

Calculation Sheet 5

Reference	CALCULATIONS	OUTPUT
	BEAM ON LINES D AND E COMBINED	

This is considered as a 9.6 m wide beam loaded with a set of symmetrically placed loads, $0.5F_1$ at columns D1/E1 and D4/E4, respectively, plus a centrally placed distributed load F_2 on the area enclosed by the internal walls. The values of F_1 will be conservatively taken as twice those calculated for columns B1 and B4. If $2c$ is the length of the distributed load F_2, $c/L = 3.8/19.2 = 0.2$ say (i.e., $2c = 7.68$ m).

The distribution of vertical loading resulting from the overturning moments due to the wind loading can be represented by equal and opposite forces F_3, F_4 and F_5 at distances $a/L = 0.3$, 0.4 and 0.5, respectively, from the ends of the beam. Values of the forces are $F_3 = (5/8)M/7.68$, $F_3 = (3/4)M/7.68$ and $F_5 = (1/8)M/7.68$, where M is the overturning moment at the top of the basement floor. Forces F_5 cancel each other and may be ignored.

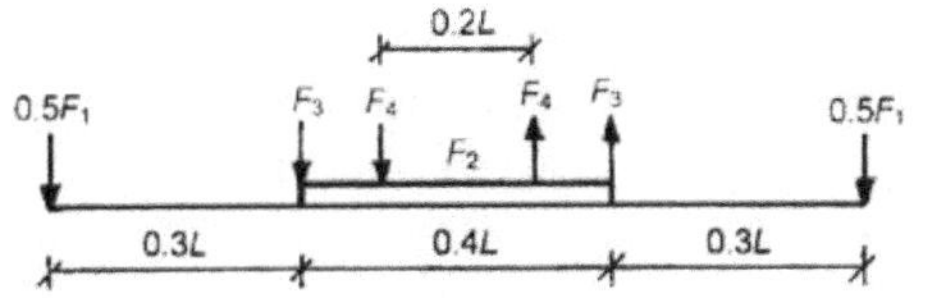

Loads on beam

Sum of maximum loads at $a/L = 0$ (including weight of basement wall):

Load	Load Case 1 (kN)	Load Case 2 (kN)
Ultimate characteristic	$F_1 = 2 \times 2568 = 5136$ $F_1 = 2 \times 1986 = 3972$	$F_1 = 2 \times 2638 = 5276$ $F_1 = 2 \times 2032 = 4064$

Maximum loads from internal walls, F_2 with $c/L = 0.2$, F_3 with $a/L = 0.3$ and F_4 with $a/L = 0.3$ are:

Load	Dead and Imposed (kN)	Wind (kN)
Ultimate service	$F_2 = 12293$ $F_2 = 9516$	$F_3 = \pm (5/8) \times 8020/7.68 = 653$ $F_4 = \pm (3/4) \times 8020/7.68 = 783$ $F_3 = \pm (5/8) \times 5346/7.68 = 435$ $F_4 = \pm (3/4) \times 5346/7.68 = 522$

From Table B2, the end slopes for a beam with free ends are:

$$\theta_A = [-25.66F_1 + 5.887F_2 \pm (10.36 - 0.922)F_3 \pm (9.015 - 2.894)F_4] / (k_s BL^2)$$

$$\theta_B = [+25.66F_1 - 5.887F_2 \pm (10.36 - 0.922)F_3 \pm (9.015 - 2.894)F_4] / (k_s BL^2)$$

Thus, the fixed-end moments required to offset the end slopes, are

$$M_A = -(k_s BL^3/504.7) \times \theta_A = +0.0508F_1L - 0.0117F_2L \pm 0.0187F_3L \pm 0.0121F_4L$$

$$M_B = -(k_s BL^3/504.7) \times \theta_B = -0.0508F_1L + 0.0117F_2L \pm 0.0187F_3L \pm 0.0121F_4L$$

Case 1. Characteristic/service loads with maximum values for load on building and soil pressure on wall. Fixed-end moments for beam and wall are:

Beam: $M_A = (+ 0.0508 \times 4064 - 0.0117 \times 9516 \pm 0.0187 \times 435$
$\pm 0.0121 \times 522) \times 19.2 = 2104$ or 1549 kN m

$M_B = (-0.0508 \times 4064 + 0.0117 \times 9516 \pm 0.0187 \times 435$
$\pm 0.0121 \times 522) \times 19.2 = -1549$ or $- 2104$ kN m

Wall: $M_A = -M_B = 2 \times (-49.6 \times 4.8 + 0.5 \times 41.0) = -435$ kN m

Final moments at ends of beam, after releasing fixed-end moments, are

$M_{A0} = 2104 - 0.738 \times (2104 - 435) = +872$ kN m (or $+727$)

$M_{B0} = -1549 + 0.738 \times (1549 - 435) = -727$ kN m (or -872)

From Tables B3 to B5 and B10, maximum bearing pressure (at $x/L = 0.5$) is

Example 2 **Calculation Sheet 6**

Reference	CALCULATIONS	OUTPUT

$q = 5.814[(M_{A0} - M_{B0})/BL^2] - 0.671(F_1/BL) + 2.103(F_2/BL) + 0 \times [(F_3 + F_4)/BL]$

$= 5.814 \times (872 + 727)/(9.6 \times 19.2^2) +$
$\qquad\qquad (-0.671 \times 4064 + 2.103 \times 9516)/(9.6 \times 19.2)$

$= 2.6 + 93.8 = 97 \text{ kN/m}^2 \ (<200 \text{ kN/m}^2)$

Case 2. Ultimate design loads with maximum load on building and maximum soil pressure on wall. Fixed-end moments for beam and wall are:

Beam: $M_A = (+0.0508 \times 5276 - 0.0117 \times 12293 \pm 0.0187 \times 653$
$\pm 0.0121 \times 783) \times 19.2 = 2800 \text{ or } 1968 \text{ kN m}$

$M_B = (-0.0508 \times 5276 + 0.0117 \times 12293 \pm 0.0187 \times 653$
$\pm 0.0121 \times 783) \times 19.2 = -1968 \text{ or } -2800 \text{ kN m}$

Wall: $M_A = -M_B = -2 \times 294 = -588 \text{ kN m}$

Final moments at ends of beam, after releasing fixed-end moments, are

$M_{A0} = 2800 - 0.738 \times (2800 - 588) = +1168 \text{ kN m (or +950)}$

$M_{B0} = -1968 + 0.738 \times (1968 - 588) = -950 \text{ kN m (or -1168)}$

From Tables B3 to B5 and B10,

$M_x = c_{A0}M_{A0} - c_{B0}M_{B0} + (c_1F_1 + c_2F_2 \pm c_3F_3 \pm c_4F_4)L$

x/L	0	0.1	0.2	0.3	0.4	0.5
c_{A0}	1.0	0.823	0.508	0.238	0.067	-0.016
c_{B0}	0	-0.006	-0.020	-0.035	-0.040	-0.016
c_1	0	-0.029	-0.030	-0.022	-0.013	-0.010
c_2	0	-0.001	-0.002	0.006	0.014	0.016
c_3	0	0.004	0.022	0.061	0.024	0
c_4	0	0	0.007	0.024	0.056	0
M_x kN m	1168	-2168	-2556	558	3170	2730
or	950	-2449	-3433	-1753	862	2730

Maximum shear force (at $x/L = 0$) is

$V = -0.5F_1 = -0.5 \times 5276 = -2638 \text{ kN}$

Case 3. Ultimate design loads with minimum load on building (constructed up to ground floor only) and maximum soil pressure on wall.

$F_1 = 2 \times 494 = 988 \text{ kN}, \qquad F_2 = 1390 \text{ kN}$

Fixed-end moments for beam and wall are:

Beam: $M_A = -M_B = (0.0508 \times 988 - 0.0117 \times 1390) \times 19.2 = 652 \text{ kN m}$

Wall: $M_A = -M_B = -67.0 \times 9.6 = -643 \text{ kN m}$

Moments at bottom of wall after releasing fixed-end moments, are:

$M_{A0} = -643 - 0.262 \times (652 - 643) = -645 \text{ kN m} \qquad M_{B0} = 645 \text{ kN m}$

BEAM ON LINES 2 AND 3 COMBINED

This is considered as a 13.2 m wide beam loaded with three sets of symmetrically placed concentrated loads, $0.5F_1$ at columns A2/A3 and H2/H3, $0.5F_2$ at columns B2/B3 and G2/G3, and $0.5F_3$ at columns C2/C3 and F2/F3, plus a centrally placed distributed load F_4 on the area enclosed by the internal walls. For the concentrated loads, $a/L = 0$ for F_1, $4.8/33.6 = 0.15$ say for F_2, and $9.6/33.8 = 0.30$ say for F_3. For the distributed load, $c/L = 2.6/33.6 = 0.1$ say (i.e., $2c = 6.72$ m)

The distribution of vertical loading resulting from the overturning moments due to the wind loading can be represented by equal and opposite forces F_5, F_6 and F_7 at distances $a/L = 0.3, 0.4$ and 0.5, respectively from the ends of the beam. Values of the forces are $F_5 = (5/8)M/6.72$, $F_6 = (3/4)M/6.72$ and $F_7 = (1/8)M/6.72$, where M is the overturning moment at the top of the basement floor. Forces F_7 cancel each other and may be ignored.

Example 2   **Calculation Sheet 7**

Reference	CALCULATIONS	OUTPUT
	Loads on beam	

Sum of maximum loads at ends of beam (including weight of basement wall):

Load	Load Case 1 (kN)	Load Case 2 (kN)
Ultimate characteristic	$F_1 = 4 \times (1194 + 36.0 \times 6.6) = 5726$ $F_1 = 4 \times (921 + 28.8 \times 6.6) = 4444$	$F_1 = 5150$ $F_1 = 4060$

Sum of maximum loads at columns B2/B3 and G2/G3 (with $a/L = 0.15$), and also at columns C2/C3 and F2/F3 (with $a/L = 0.3$):

Load	Load Case 1 (kN)	Load Case 2 (kN)
Ultimate characteristic	$F_2 = F_3 = 4 \times 2281 = 9124$ $F_2 = F_3 = 4 \times 1731 = 6924$	$F_2 = F_3 = 7544$ $F_2 = F_3 = 5872$

Maximum loads from internal walls, F_4 with $c/L = 0.1$, F_5 with $a/L = 0.4$ and F_6 with $a/L = 0.45$:

Load	Dead and Imposed (kN)	Wind (kN)
Ultimate service	$F_4 = 12293$ $F_4 = 9516$	$F_5 = \pm (5/8) \times 4050/6.72 = 377$ $F_6 = \pm (3/4) \times 4050/6.72 = 452$ $F_5 = \pm (5/8) \times 2700/6.72 = 251$ $F_6 = \pm (3/4) \times 2700/6.72 = 302$

With $L = 33.6$ m between the centrelines of columns A2/A3 and H2/H3,

$$\lambda L = (3k_s L^4/Eh^3)^{1/4} = [3 \times 12 \times 10^3 \times 33.6^4/(33 \times 10^6 \times 0.6^3)]^{1/4} = 8.96 \text{ (say 8.0)}.$$

From Table B2, the end slopes for a beam with free ends are:

$$\theta_A = [-63.92F_1 + 10.96F_2 + 7.865F_3 + 0.464F_4 \pm (4.905 + 1.147)F_5 \\ \pm (1.589 + 1.015)F_6]/(k_s BL^2)$$

$$\theta_B = [+63.92F_1 - 10.96F_2 - 7.865F_3 - 0.464F_4 \pm (4.905 + 1.147)F_5 \\ \pm (1.589 + 1.015)F_6]/(k_s BL^2)$$

Thus, the fixed-end moments required to offset the end slopes, are

$$M_A = -(k_s BL^3/2047) \times \theta_A = +0.0312F_1 L - 0.0054F_2 L - 0.0038F_3 L - 0.0002F_4 L \\ \pm 0.0030F_5 L \pm 0.0013F_6 L$$

$$M_B = -(k_s BL^3/2047) \times \theta_B = -0.0312F_1 L + 0.0054F_2 L + 0.0038F_3 L + 0.0002F_4 L \\ \pm 0.0030F_5 L \pm 0.0013F_6 L$$

Flexural stiffness values of the beam (with $\lambda L = 8.96$) and wall are

$$K_b = k_s BL^3/2877 = 12 \times 10^3 \times 1.0 \times 33.6^3/2877 = 0.158 \times 10^6 \text{ kN m/m}$$

$$K_w = E_c Bh_w^3/4L_w = 33 \times 10^6 \times 1.0 \times 0.3^3/(4 \times 4.0) = 0.056 \times 10^6 \text{ kN m/m}$$

Moment distribution factors: $D_b = 0.158/(0.158 + 0.056) = 0.738$, $D_w = 0.262$

Case 1. Characteristic/service loads with maximum load on building and minimum soil pressure on wall. Fixed-end moments for beam and wall are:

Beam: $M_A = (+0.0312 \times 4444 - 0.0054 \times 6924 - 0.0038 \times 6924 - 0.0002 \times 9516$
 $\pm 0.0030 \times 251 \pm 0.0013 \times 302) \times 33.6 = 2493$ or 2416 kN m

$M_B = (-0.0312 \times 4444 + 0.0054 \times 6924 + 0.0038 \times 6924 + 0.0002 \times 9516$
 $\pm 0.0030 \times 251 \pm 0.0013 \times 302) \times 33.6 = -2416$ or -2493 kN m

Wall: $M_A = -M_B = -25.6 \times 13.2 = -338$ kN m

Example 2   **Calculation Sheet 8**

Reference	CALCULATIONS	OUTPUT
	Final moments at ends of beam, after releasing fixed-end moments, are:	

Final moments at ends of beam, after releasing fixed-end moments, are:

$M_{A0} = 2493 - 0.738 \times (2493 - 338) = +903$ kN m (or +882)

$M_{B0} = -2416 + 0.738 \times (2416 - 338) = -882$ kN m (or -903)

From Tables B3 to B6 and B10, maximum bearing pressure (at $x/L = 0.5$) is

$q = -0.243[(M_A0 - M_B0)/BL^2] - 0.192(F_1/BL) - 0.160(F_2/BL) + 0.789(F_3/BL)$
$\quad + 3.438(F_4/BL) + 0 \times [(F_5 + F_6)/BL]$

$= -0.243 \times (903 + 882)/(13.2 \times 33.6^2)$
$\quad + (-0.192 \times 4444 - 0.160 \times 6924 + 0.789 \times 6924 + 3.438 \times 9516)/(13.2 \times 33.6)$

$= 82$ kN/m^2 (<200 kN/m^2)

Case 2. Ultimate design loads with maximum load on building and maximum soil pressure on wall. Fixed-end moments for beam and wall are:

Beam: $M_A = (+0.0312 \times 5726 - 0.0054 \times 9124 - 0.0038 \times 9124 - 0.0002 \times 12293$
$\quad\quad \pm 0.0030 \times 377 \pm 0.0013 \times 452) \times 33.6 = 3158$ or 3042 kN m

$\quad M_B = (-0.0312 \times 5726 + 0.0054 \times 9124 + 0.0038 \times 9124 + 0.0002 \times 12293$
$\quad\quad \pm 0.0030 \times 377 \pm 0.0013 \times 452) \times 33.6 = -3042$ or -3158 kN m

Wall: $M_A = -M_B = -67.0 \times 13.2 = -885$ kN m

Final moments at ends of beam, after releasing fixed-end moments, are:

$M_{A0} = 3158 - 0.738 \times (3158 - 885) = 1480$ kN m (or +1450)

$M_{B0} = -3042 + 0.738 \times (3042 - 885) = -1450$ kN m (or -1480)

From Tables B3 to B10,

$M_x = c_{A0}M_{A0} - c_{B0}M_{B0} + (c_1F_1 + c_2F_2 + c_3F_3 + c_4F_4 \pm c_5F_5 \pm c_6F_6)L$

x/L	0	0.1	0.15	0.2	0.3	0.4	0.45	0.5
c_{A0}	1.0	0.635	0.390	0.196	-0.006	-0.043	-0.037	-0.026
c_{B0}	0	0.001	0.001	0.002	0.001	-0.008	-0.016	-0.026
c_1	0	-0.020	-0.018	-0.013	-0.004	0.001	0.002	0.002
c_2	0	0.006	0.016	0.006	-0.003	-0.003	-0.003	-0.003
c_3	0	-0.001	-0.001	0.001	0.015	-0.002	-0.006	-0.007
c_4	0	-0.001	-0.003	-0.004	-0.004	0.003	0.008	0.013
c_5	0	-0.003	-0.004	-0.005	0.004	0.038	0.067	0
c_6	0	-0.001	-0.002	-0.003	0	0.017	0.032	0
M_x kN m	1480	-1840	394	-1823	1300	563	2187	2613
or	1450	-1753	544	-1611	1199	-915	-482	2613

Note: All values for c_6, and other values for $x/L = 0.15$ and 0.45 are not included in Tables B3 to B10.

Case 3. Ultimate design loads with minimum load on building (constructed up to ground floor only) and maximum soil pressure on wall.

$F_1 = 4 \times (9.6 + 28.8) \times 6.6 = 1014$ kN, $F_2 = F_3 = 418$ kN, $F_4 = 1390$ kN

Fixed-end moments for beam and wall are:

Beam: $M_A = -M_B = (0.0312 \times 1014 - 0.0054 \times 418 - 0.0038 \times 418$
$\quad\quad - 0.002 \times 1390) \times 33.6 = 840$ kN m

Wall: $M_A = -M_B = -67.0 \times 13.2 = -885$ kN m

Moments at bottom of wall after releasing fixed-end moments, are

$M_{A0} = -885 + 0.262 \times (885 - 840) = -873$ kN m $M_{B0} = 873$ kN m

DESIGN OF BASEMENT FLOOR

The bending moments determined for each beam will be spread uniformly across the beam width, except at the internal columns where 75% of the total moment will be allocated to the middle half of the beam width (similar to a flat slab design). The design is based on concrete strength class C32/40 and reinforcement grade 500.

Example 2 — Calculation Sheet 9

Reference	CALCULATIONS	OUTPUT

Durability

4.4.1.3 / Table NA.1 — Cover recommended for concrete cast against blinding is as follows:

$c_{min} = 40$ mm $\Delta c_{dev} = 10$ mm $c_{nom} = 40 + 10 = 50$ mm (bottom)

BS 8500 — For the top surface, exposure class XC1 applies and $c_{nom} = 25$ mm is sufficient.

BEAM ON LINE B

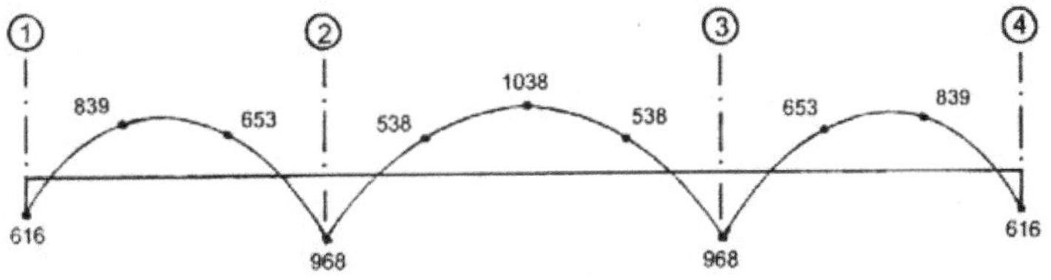

Bending moment diagram for 4.8 m wide beam

Flexural design

Allowing for 50 mm cover and 20 mm bars, $d = 600 - (50 + 20/2) = 540$ mm

Minimum area of longitudinal tension reinforcement:

9.2.1.1 — $A_{s,min} = 0.26(f_{ctm}/f_{yk})b_t d = 0.26 \times (3.0/500) \times 1000 \times 540 = 843$ mm^2/m

Maximum spacing of bars:

9.3.1.1 (3) — $s_{max} = 3h \leq 400$ mm (but ≤ 250 mm in areas with concentrated loads)

According to the calculated values of M/bd^2f_{ck}, appropriate values of z/d and A_s can be determined (Table A1) and suitable reinforcement selected (Table A9).

At column B2 ($x/L = 0.3$), moment for middle half of beam is

$M_{Ed} = 0.75 \times 968 = 726$ kN m where $b = 2400$ mm $A_{s,min} = 2024$ mm^2

$M/bd^2f_{ck} = 726 \times 10^6/(2400 \times 540^2 \times 32) = 0.033$ $z/d = 0.95$ (maximum)

$A_s = 726 \times 10^6/(0.87 \times 500 \times 0.95 \times 540) = 3254 \geq 2024$ mm^2 (12H20-200)

Similarly, moment for each outer quarter of beam is

$M_{Ed} = 0.125 \times 968 = 121$ kN m where $b = 1200$ mm $A_{s,min} = 1012$ mm^2

$A_s = 121 \times 10^6/(0.87 \times 500 \times 0.95 \times 540) = 543 \geq 1012$ mm^2 (4H20-300)

Values at other positions are shown in the following table, where:

$b = 4.8$ m and $A_{s,min} = 843 \times 4.8 = 4047$ mm^2 (16H20-300 say)

x/L	M_{Ed} (kN m)	M/bd^2f_{ck}	z/d	A_s (mm^2)	Bars
0	616	0.014	0.95	2761	16H20(B)
0.1	–839	0.019	0.95	3760	16H20(T)
0.2	–653	0.015	0.95	2927	16H20(T)
0.4	–538	0.012	0.95	2411	16H20(T)
0.5	–1038	0.023	0.95	4652	16H20(T)

OUTPUT: Provide H20-300 (top and bottom) throughout with additional bars at internal columns (see shear design)

Punching shear at columns B2 and B3:

For a check at distance $2d$ from the column face (see calculation sheet 2):

$d_{av} = 530$ mm, $u_1 = 7860$ mm and $v_{min} = 0.4$ MPa (where $k = 1.61$).

Area inside the control perimeter is given by

$A_1 = c^2 + 8cd + 4\pi d^2 = 0.3^2 + 8 \times 0.3 \times 0.53 + 4 \times \pi \times 0.53^2 = 4.89$ m^2

From Tables B3, B5 and B6, average bearing pressure at $x/L = 0.3$ is

$q_{av} = 11.28(M_0/BL^2) - 0.070(F_1/BL) + 1.397(F_2/BL)$

$= 11.28 \times 616/(4.8 \times 19.2^2) + (-0.070 \times 2568 + 1.397 \times 4562)/(4.8 \times 19.2)$

$= 3.9 + 67.2 = 71$ kN/m^2

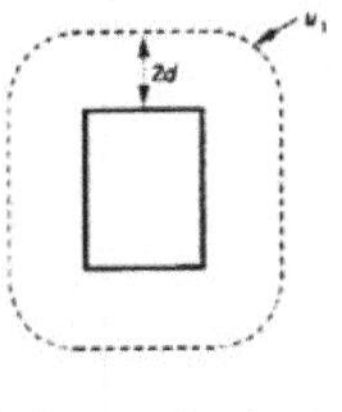

Basic control perimeter for punching shear

Example 2   **Calculation Sheet 10**

Reference	CALCULATIONS	OUTPUT
	For the middle half of the beam, increasing q_{av} by 50%, gives $q = 106.5$ kN/m Net applied shear force is $V_{Ed} = N_{Ed} - A_1 \times q = 2281 - 4.89 \times 106.5 = 1760$ kN Average shear stress at the control perimeter is $v_{Ed} = V_{Ed}/(u_1 d) = 1760 \times 10^3/(7860 \times 530) = 0.43$ MPa $(>v_{min})$ The design shear strength at the basic control perimeter is given by $v_{Rd,c} = 0.12k(100A_{sl}fck/b_w d)^{1/3} \geq v_{min}$ Thus, the required reinforcement area for $v_{Rd,c} = v_{Ed}$ is given by $A_{sl} = (v_{Ed}/0.12k)^3 \times (b_w d/100f_{ck})$ $= [0.43/[0.12 \times 1.61)]^3 \times (1000 \times 530)/(100 \times 32) = 1826$ mm^2/m (H20-150) Further checks should be made for control perimeters at distances less than $2d$ from the column face to find the critical perimeter. A check at distance $a = d$ from the column face will usually be sufficient. In this case, $u_1 = 4c + 2\pi d = 4 \times 300 + 2 \times \pi \times 530 = 4530$ mm $A_1 = c^2 + 4cd + \pi d^2 = 0.3^2 + 4 \times 0.3 \times 0.53 + \pi \times 0.53^2 = 1.61$ m^2 $V_{Ed} = 2281 - 1.61 \times 106.5 = 2110$ kN $v_{Ed} = 2110 \times 10^3/(4530 \times 530) = 0.88$ MPa The design shear strength for values of $a < 2d$ is given by	Provide H20-150(B) in each direction at internal columns for strip of width $4d + c = 4 \times 0.53 + 0.3 = 2.4$ m
6.4.4 (1) Table NA.1		
6.4.4 (2)	$v_{Rd,c} = 0.12k(100A_{sl}fck/b_w d)^{1/3} \times 2d/a \geq v_{min} \times 2d/a$ $= 0.12 \times 1.61 \times [100 \times 2094 \times 32/(1000 \times 530)]^{1/3} \times 2 = 0.90$ MPa $(>v_{Ed})$ **Shear at walls on lines 1 and 4** At distance d from the face of the wall, $x = 150 + 540 = 690$ mm. From Table B5, with $\lambda L = 5.0$ and $x/L = 0.69/19.2 = 0.035$, $V/F = -0.366$ by interpolation. If the value is determined accurately by calculation, $V/F = -0.338$. From calculation sheet 3, for a 4.8 m wide beam, $F_1 = 2638$ kN. Hence, at critical section, maximum shear stress is $v_{Ed} = V_{Ed}/(bd) = 0.366 \times 2638 \times 10^3/(4800 \times 540) = 0.38$ MPa $(<v_{min})$	 Shear satisfactory

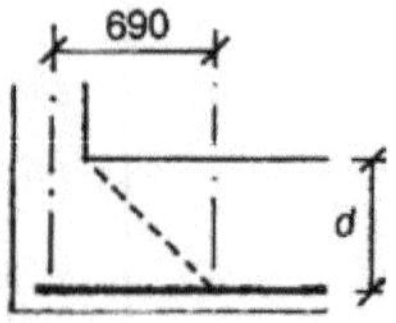

BEAM ON LINES D AND E COMBINED

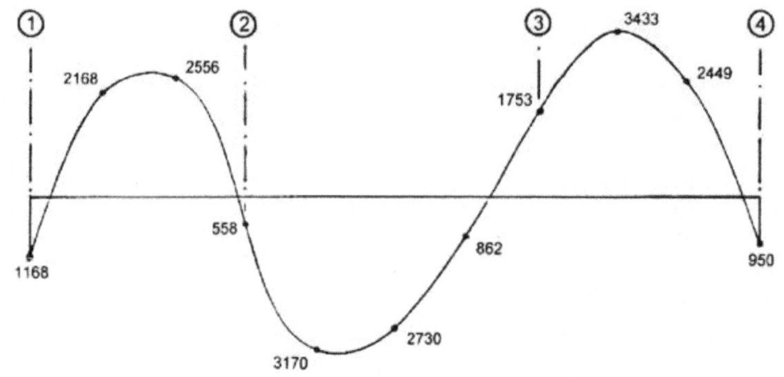

Bending moment diagram for 9.6 m wide beam

The diagram shown applies where the wind moment is anti-clockwise. For a clockwise wind moment, the diagram should be to the opposite hand. In the following table, there are two sets of values for M_{Ed} according to the direction of the wind, but the required reinforcement is shown for the more critical value of M_{Ed} only, where:

$b = 9.6$ m and $A_{s,min} = 843 \times 9.6 = 8093$ mm^2 (32H20-300 say)

Example 2   **Calculation Sheet 11**

Reference	CALCULATIONS	OUTPUT

x/L	M_{Ed} (kN m)	$M/bd^2 f_{ck}$	z/d	A_s (mm^2)	Bars
0	1168/950	0.013	0.95	5234	32H20(B)
0.1	−2168/−2449	0.028	0.95	10975	40H20(T)
0.2	−2556/−3433	0.039	0.95	15384	48H20(T)
0.3	566/−1753	0.020	0.95	7856	32H20(T)
0.4	3170/862	0.036	0.95	14206	48H20(B)
0.5	2730	0.031	0.95	12234	40H20(B)

BEAM ON LINES 2 AND 3 COMBINED

Allowing for 20 mm bars in each direction, $d = 540 - 20 = 520$ mm

Minimum area of longitudinal tension reinforcement:

9.2.1.1

$A_{s,min} = 0.26(f_{ctm}/f_{yk})b_t d = 0.26 \times (3.0/500) \times 1000 \times 520 = 812$ mm^2/m

With $b = 13.2$ m, $A_{s,min} = 812 \times 13.2 = 10720$ mm^2 (44H20-300 say)

The maximum moment occurs at $x/L = 0.5$, where $M_{Ed} = 2613$ kN m

$A_s = 2613 \times 10^6/(0.87 \times 500 \times 0.95 \times 520) = 12160 \leq 10720$ mm^2

Clearly, minimum reinforcement, H20-300, will be sufficient for the full width of the raft in this direction.

Cracking due to loading

Minimum area of reinforcement required in tension zone for crack control:

7.3.2 (2)

$A_{s,min} = k_c k f_{ct,eff} A_{ct}/\sigma_s$ where

$k_c = 0.4$ for bending, $k = 0.8$ for $h = 600$ mm, $f_{ct,eff} = f_{ctm} = 0.3 f_{ck}^{(2/3)} = 3.0$ MPa for general design purposes, $A_{ct} = bh/2$ and $\sigma_s \leq f_{yk} = 500$ MPa.

$A_{s,min} = 0.4 \times 0.8 \times 3.0 \times 1000 \times 300/500 = 576$ mm^2/m (H20-300 provided)

7.3.3 (2)

BS EN 1990 Table NA.A1.1

The crack width criterion can be satisfied by limiting either the bar size or the bar spacing, according to the stress in the reinforcement under the quasi-permanent service load, given by $N_2 = $ (dead load $+ \psi_2 \times$ imposed load) where $\psi_2 = 0.3$.

For column B2, the design ultimate and quasi-permanent service loads are

$N_{Ed} = 2281$ kN, $N_2 = 1263 + 0.3 \times 468 = 1404$ kN

For the middle half of the beam on line B, $A_{s,req} = 3254$ mm^2 and 16H20-150 will be provided. The stress in the reinforcement under quasi-permanent service load is given approximately pro rata to the stress under the ultimate design load as

$\sigma_s = N_2/N_{Ed} \times 0.87 f_{yk} \times A_{s,req}/A_{s,prov} = 0.62 \times 435 \times 3254/5026 = 175$ MPa

Table 7.2
Table 7.3

From *Reynolds*, Table 4.24, for $w_k = 0.3$ mm and $\sigma_s = 175$ MPa, the maximum values for compliance, by interpolation, are either bar size $\phi_s^* = 29$ mm or bar spacing $= 280$ mm. Clearly, both of these criteria are satisfied by using H20-200.

At sections where minimum reinforcement is sufficient, H20-300 is provided. The most critical condition is at $x/L = 0.5$ for the beam on line B, where

$\sigma_s = 0.62 \times 435 \times 4652/5026 = 250$ MPa

Hence, the maximum criterion is either $\phi_s^* = 15$ mm or bar spacing $= 240$ mm. The maximum bar size, with 25 mm top cover, is given by

$\phi_s = \phi_s^*(f_{ct,eff}/2.9)[k_c h_{ct}/2(h - d)] = 15 \times (3.0/2.9) \times 0.4 \times 300/(2 \times 35) = 26$ mm

Cracking due to restrained early thermal contraction

Assuming this occurs at about $t = 3$ days, when $f_{cm}(t) = 24$ MPa say:

3.1.2 (9)

$f_{ct,eff} = f_{ctm}(t) = [f_{cm}(t)/f_{cm}] f_{ctm} = [24/(f_{ck} + 8)] \times 3.0 = 1.8$ MPa

With $k_c = 1.0$ for tension and $k = 0.8$ for $h = 600$ mm,

7.3.2 (2)

$A_{s,min} = k_c k f_{ct,eff} A_{ct}/f_{yk} = 1.0 \times 0.8 \times 1.8 \times 1000 \times 300/500 = 864$ mm^2/m (EF)

For cracks resulting from early thermal contraction, crack width is calculated as

Example 2   Calculation Sheet 12

Reference	CALCULATIONS	OUTPUT
7.3.4 PD 6687 2.16	$w_k = (0.8R\alpha\Delta T) \times s_{r,max}$ where $s_{r,max} = 3.4c + 0.425k_1k_2(A_{c,eff}/A_s)\phi$ With $c = 50$ mm, $k_1 = 0.8$ for high bond bars, $k_2 = 1.0$ for tension, $h_{c,ef}$ as the lesser of $2.5(h - d)$ and $h/2$, and H20-300 as minimum reinforcement: $s_{r,max} = 3.4 \times 50 + 0.425 \times 0.8 \times 1.0 \times (2.5 \times 60 \times 1000/1047) \times 20 = 1144$ mm Taking $R = 0.8$ for infill bays, $\Delta T = 34°C$ for 350 kg/m³ Portland cement concrete and 650 mm section thickness (*Reynolds*, Table 2.18), and $\alpha = 12 \times 10^{-6}$ per °C: $w_k = 0.8 \times 0.8 \times 12 \times 10^{-6} \times 34 \times 1144 = 0.30$ mm	
	Detailing requirements	
8.4.2 Figure 8.2	The bond conditions are described as 'good' for the bottom bars and 'poor' for the top bars. For simplicity, the design anchorage length will be taken as the basic required anchorage length. From *Reynolds*, Table 4.30, with $f_{ck} = 32$ MPa: $l_{b,rqd} = 35\phi$ (good bond), $l_{b,rqd} = 50\phi$ (poor bond) If all the bars are lapped at the same position, the design lap length is 1.5 times the design anchorage length, giving values: $l_0 = 1.5 \times 35 \times 20 = 1050$ mm (bottom) $l_0 = 1.5 \times 50 \times 20 = 1500$ mm (top)	
	DESIGN OF BASEMENT WALL	
	Durability	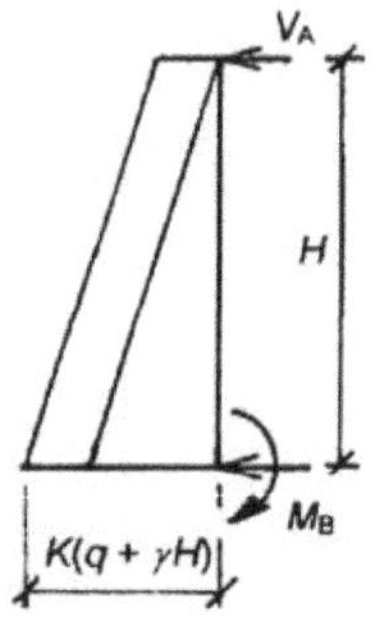
BS 8500	Since the external surface of the wall is protected by a continuous barrier system, it is reasonable to consider exposure class XC1 for both surfaces of the wall. $c_{min} = 15$ mm $\Delta c_{dev} = 10$ mm $c_{nom} = 15 + 10 = 25$ mm	
	Flexural design	
	Allowing for 25 mm cover, 16 mm diameter bars (EW) and horizontal bars in the outer layers, for the vertical bars, $d = 300 - (25 + 16 + 16/2) = 250$ mm. Minimum area of vertical tension reinforcement:	
9.2.1.1	$A_{s,min} = 0.26 \times (3.0/500) \times 1000 \times 250 = 390$ mm²/m (H12-300) Minimum area of reinforcement required in tension zone for crack control:	
7.3.2 (2)	$A_{s,min} = k_c k f_{ct,eff} A_{ct}/\sigma_s$ where $k_c = 0.4$ for bending, $k = 1.0$ for $h = 300$ mm, $f_{ct,eff} = f_{ctm} = 0.3f_{ck}^{(2/3)} = 3.0$ MPa for general design purposes, $A_{ct} = bh/2$ and $\sigma_s \leq f_{yk} = 500$ MPa. $A_{s,min} = 0.4 \times 1.0 \times 3.0 \times 1000 \times 150/500 = 360$ mm²/m (H12-300) Moment of resistance provided by minimum reinforcement is given by $M = A_s(0.87f_{yk})z = 377 \times 0.87 \times 500 \times 0.95 \times 250 \times 10^{-6} = 39$ kN m/m Maximum design ultimate moment at junction of wall and floor occurs for beam on line B (case 2), where: $M_B = 616/4.8 = 128$ kN m/m. Hence, from Table A1: $M/bd^2f_{ck} = 128 \times 10^6/(1000 \times 250^2 \times 32) = 0.064$ $z/d = 0.940$ $A_s = 128 \times 10^6/(0.87 \times 500 \times 0.94 \times 250) = 1252$ mm²/m (H16-150) Shear force at top of wall (at positions away from influence of column moment) is $V_A = 1.35 \times 0.47 \times (10 \times 4.0/2 + 20 \times 4.0^2/6) - 128/4.0 = 14.5$ kN/m Bending moment at distance a from top of wall is given by $M = V_A \times a - 1.35 \times 0.47 \times (10a^2/2 + 20a^3/6)$	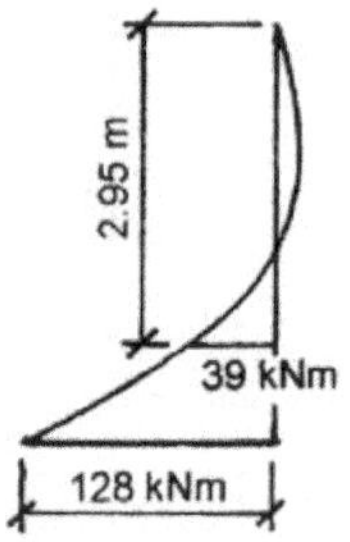 Case 2 for beam on line B 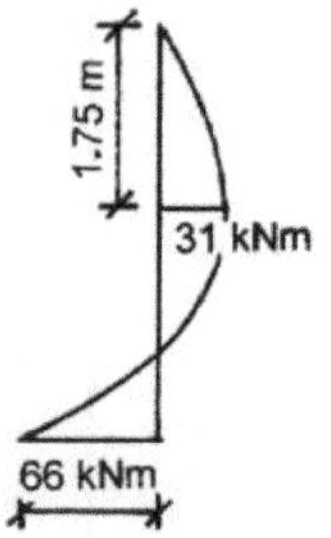
9.2.1.3 (2)	Solving this equation, with $M = -39$ kN m/m, gives $a = 2.95$ m. Here, H12-300 is sufficient, but bars to be curtailed should continue for a distance $a_l = d = 250$ mm. Projection of bars from bottom of wall $= (4.0 - 2.95) + 0.25 = 1.3$ m Minimum design ultimate moment at junction of wall and floor, occurs for beam on lines 2 and 3 (case 3), where: $M_B = 873/13.2 = 66$ kN m/m.	 Case 3 for beam on lines 2 and 3

Example 2   **Calculation Sheet 13**

Reference	CALCULATIONS	OUTPUT
	Corresponding shear force at top of wall is	
	$\quad V_A = 1.35 \times 0.47 \times (10 \times 4.0/2 + 20 \times 4.0^2/6) - 66.0/4.0 = 30.0$ kN/m	
	If a_o is distance from top of wall to point of zero shear, then	Provide H12-300 (EF)
	$\quad V_A - 1.35 \times 0.47 \times (10a_o + 20a_o^2/2) = 0$ which gives $a_o = 1.75$ m	as minimum vertical
	Hence, maximum sagging moment (at $a_o = 1.75$ m) is	reinforcement with
	$\quad M = V_A \times a_o - 1.35 \times 0.47 \times (10a_o^2/2 + 20a_o^3/6) = 31$ kN m	H16-150 on outer face
	$\quad A_s = 31 \times 10^6/(0.87 \times 500 \times 0.95 \times 250) = 300 \geq 375$ mm^2/m (H12-300)	of wall for height of
		1.3 m above base
	Shear design	
	Maximum shear force at bottom of wall occurs for beam on line B (case 2), where	
	$\quad V_{Ed} = 1.35 \times 0.47 \times (10 \times 4.0/2 + 20 \times 4.0^2/3) + 128/4.0 = 112.4$ kN/m	
	$\quad v_{Ed} = V_{Ed}/(bd) = 112.4 \times 10^3/(1000 \times 250) = 0.45$ MPa	
	Minimum design shear strength, where $k = 1 + (200/d)^{1/2} = 1.89$, is	
6.2.2 (1)	$\quad v_{min} = 0.035k^{3/2}f_{ck}^{1/2} = 0.035 \times 1.89^{3/2} \times 32^{1/2} = 0.51$ MPa ($>v_{Ed}$)	
	Cracking due to restrained early thermal contraction	
	Minimum area of horizontal reinforcement, with $f_{ct,eff} = 1.8$ MPa for cracking at age of 3 days, $k_c = 1.0$ for tension and $k = 1.0$ for $h = 300$ mm, is given by:	
7.3.2 (2)	$\quad A_{s,min} = k_c k f_{ct,eff} A_{ct}/f_{yk} = 1.0 \times 1.0 \times 1.8 \times 1000 \times 150/500 = 540$ mm^2/m (EF)	
7.3.4	With $c = 25$ mm, $k_1 = 0.8$ for high bond bars, $k_2 = 1.0$ for tension, $h_{c,ef}$ as the lesser of $2.5(h - d)$ and $h/2$, and H16-300 (EF) as minimum reinforcement:	
	$\quad s_{r,max} = 3.4c + 0.425k_1k_2(A_{c,eff}/A_s)\,\phi$	
	$\quad\quad = 3.4 \times 25 + 0.425 \times 0.8 \times 1.0 \times (2.5 \times 33 \times 1000/670) \times 16 = 755$ mm	
PD 6687 2.16	With $R = 0.8$ for wall on a thick base, $\Delta T = 25°C$ for 350 kg/m^3 Portland cement concrete and 300 mm thick wall (*Reynolds*, Table 2.18), and $\alpha = 12 \times 10^{-6}$ per $°C$:	Provide H16-300 (EF) minimum horizontal reinforcement
	$\quad w_k = (0.8R\alpha\Delta T) \times s_{r,max} = 0.8 \times 0.8 \times 12 \times 10^{-6} \times 25 \times 755 = 0.15$ mm	
	Wall as deep beam	
	The wall also acts as a deep beam in distributing the concentrated column loads to produce a uniform loading at the bottom of the wall. CIRIA Guide 2 *The design of deep beams in reinforced concrete* gives recommendations on detailed analysis and design, although this publication was written for use with the then current British standard CP110. The effective span is taken as lesser of distance between centres of columns or 1.2 times clear span, and the active height is taken as lesser of actual height or effective span.	
	For the wall on gridline A, the design ultimate column loads (case 1) and resulting bending moments are as shown in the figure below:	

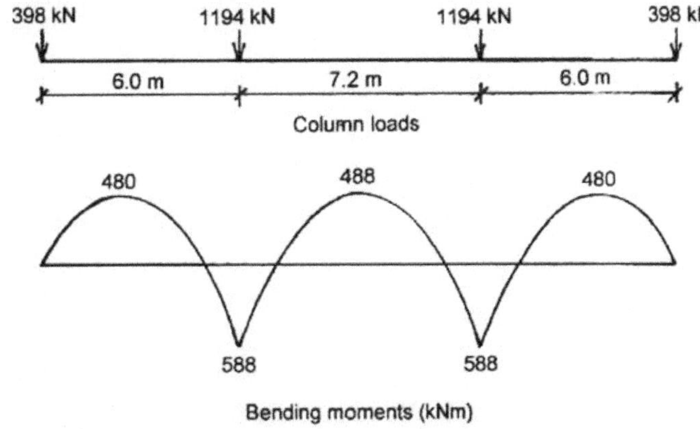

Column loads

Bending moments (kNm)

Example 2   **Calculation Sheet 14**

Reference	CALCULATIONS	OUTPUT
	The required areas of tension reinforcement are given by $A_s = M/(0.87 f_{yk} z)$ where, for multi-span beams with $l/h < 2.5$, $z = 0.2l + 0.3h_a = 0.2 \times 7200 + 0.3 \times 4000 = 2640$ mm (middle span) For sagging: $A_s = 488 \times 10^6/(0.87 \times 500 \times 2640) = 425$ mm^2 (3H16) Reinforcement may be spread over a depth of $0.2h_a = 0.8$ m from top of beam For hogging: $A_s = 588 \times 10^6/(0.87 \times 500 \times 2640) = 512$ mm^2 (3H16) Reinforcement should be contained within a depth of $0.8h_a = 3.2$ m from bottom of beam. Clearly, provision of H16-300 (EF) as minimum horizontal reinforcement is more than sufficient.	
6.2.1 (8)	Since the loads and the reactions are applied to opposite edges of the beam, and the load is uniformly distributed over the whole span, the design shear force may be checked at distance d from the face of the column. In addition, the maximum shear at the column should not exceed $V_{Ed,max}$. For the internal columns: $V_{Ed,max} = 0.5 \times 1194 = 597$ kN. Hence, taking $d = 0.9h_a = 3600$ mm, $V_{Ed,max}/b_w d = 597 \times 10^3/(300 \times 3600) = 0.56$ MPa $\leq v_{Rd,max} = 5.58$ MPa (calculation sheet 2) Taking $d = 3600$ mm, the shear force at distance d from the face of the column is negligible. In any case, with H12-300 (EF) as the minimum vertical reinforcement, the design shear resistance can be shown to be more than sufficient. Considering the bars on one face only, since the bars on the other face are necessary to resist the bending moment due to the lateral earth loading, and taking $z = 0.8h_a = 3200$ mm, and $\cot\theta = 2.5$ say:	
6.2.3 (3)	$V_{Rd,s} = (A_{sw}/s) z f_{ywd} \cot\theta$ $= 0.377 \times 3200 \times (0.87 \times 500) \times 2.5 \times 10^{-3} = 1312$ kN ($>V_{Ed,max}$) From *Reynolds*, Table 4.18, $\cot\theta = 2.5$ may be used for values of $v_w \leq 0.138$, where $v_w = V/[b_w z (1 - f_{ck}/250) f_{ck}]$ $= 1312 \times 10^3/[300 \times 3200 \times (1 - 32/250) \times 32] = 0.049$ (<0.138)	Wall reinforcement sufficient for deep beam requirements

Example 2: Reinforcement in Raft Foundation (1)   **Drawing 4**

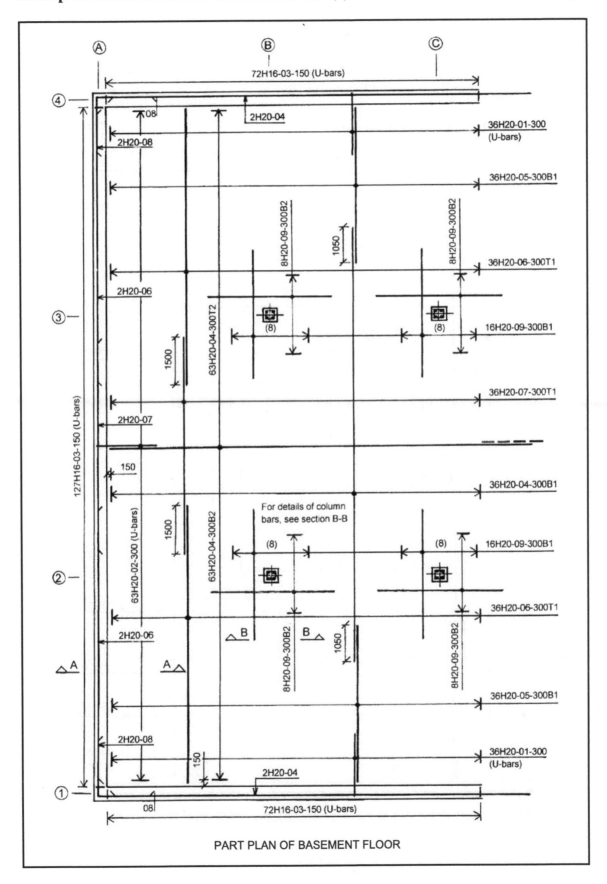

PART PLAN OF BASEMENT FLOOR

Example 2: Reinforcement in Raft Foundation (2)   **Drawing 5**

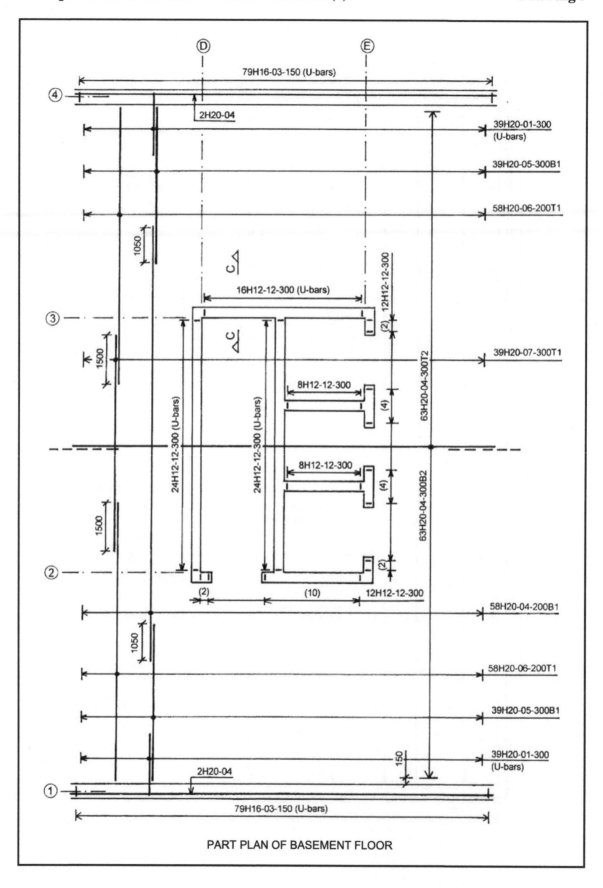

PART PLAN OF BASEMENT FLOOR

Example 2: Reinforcement in Basement Walls

Drawing 6

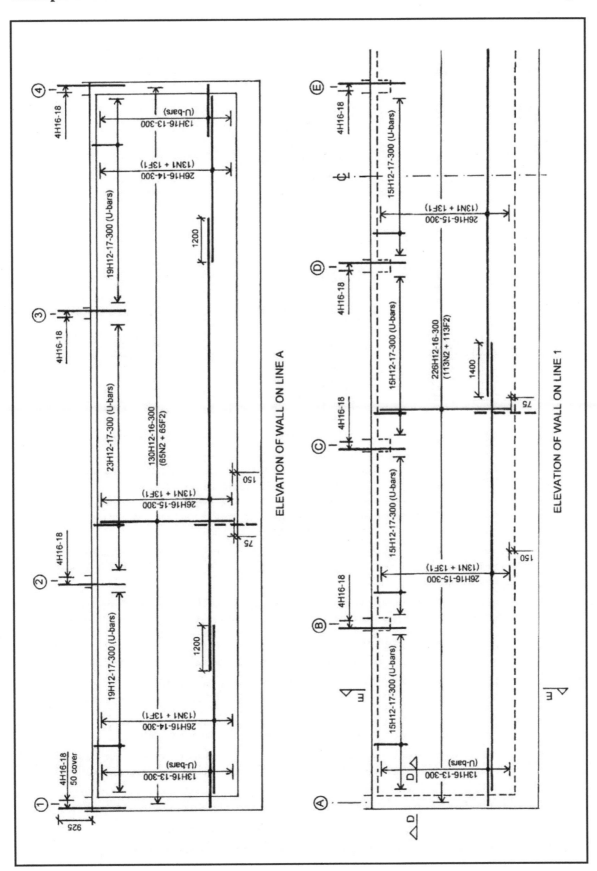

ELEVATION OF WALL ON LINE A

ELEVATION OF WALL ON LINE 1

Example 2: Cross-Sections for Raft Foundation and Basement Walls   Drawing 7

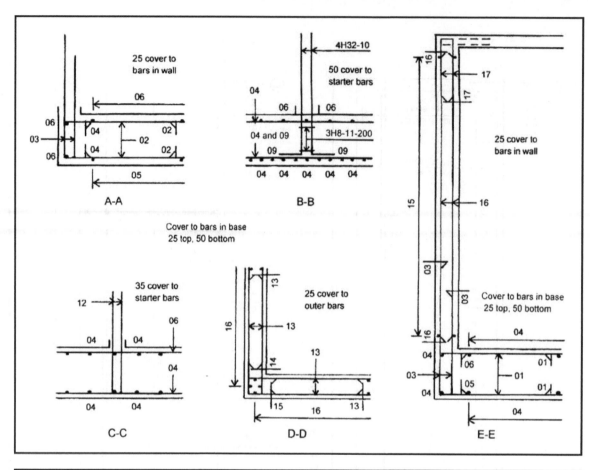

Bar Marks	Commentary on Bar Arrangement (Drawings 4–7)
01, 02	Bars (shape code 21) lapping with bars 04, 05 and 06. Lap length = 1500 mm, based on requirement for top leg (see calculation sheet 12). Cover = 50 mm (bottom), and 25 mm (top and ends).
03	Bars (shape code 21) to lap with vertical bars in basement wall. Projection above basement floor to provide a lap length above kicker = $1.5 \times 35 \times 12 + 75 = 725$ mm say. For the outer leg (see calculation sheet 12), the required projection = 1200 mm.
04, 05, 06 07, 14, 15	Straight bars (maximum length 12 m) curtailed 150 mm from wall face. Bars 04 to 07 with laps of 1050 mm (bottom) and 1500 mm (top). Bars 14 and 15 with minimum lap length = $1.5 \times 50 \times 16 = 1200$ mm.
08	Bars (shape code 11) lapping with bars 04 and 06.
09	Straight bars (additional to minimum reinforcement). Length = $6d + c = 6 \times 530 + 300 = 3500$ mm.
10	Column starter bars (shape code 11) standing on mat formed by bars 04. Projection above basement floor to provide lap length above 75 mm kicker = $1.5 \times 35 \times 32 + 75 = 1800$ mm. Cover to bars in column = 50 mm to enable 35 mm cover to links.
11	Closed links (shape code 51) to hold column starter bars in place during construction.
12	Starter bars (shape code 21) for internal walls. Projection above basement floor to provide a lap length above 75 mm kicker = $1.5 \times 35 \times 12 + 75 = 725$ mm say. Cover to bars in wall = 35 mm to enable 25 mm cover to horizontal bars.
13	Bars (shape code 21) lapping with bars 14 and 15. Lap length = 1200 mm.
16	Straight bars bearing on kicker and curtailed 50 mm below ground floor slab.
17	Bars (shape code 21) lapping with bars 16. Lap length = $1.5 \times 35 \times 12 = 650$ mm say.

Example 2

Calculation Sheet 15

Reference	CALCULATIONS	OUTPUT
	ISOLATED PAD FOUNDATIONS (Flat slab superstructure)	

Loading

From the data tabulated on calculation sheets 32 and 34 of design example 1, and allowing for the appropriate reduction in the total imposed floor loads, the design loads for columns B1 and B2 are obtained as follows:

Column B1 (Loads at Top of Basement Wall)			
Load Case	Load	Dead (kN)	Imposed (kN)
2	Ultimate characteristic	$2768 - (739 + 22) = 2007$ $2007/1.25 = 1606$	$0.6 \times 739 + 22 = 466$ $466/1.5 = 311$

Column B1	
Load (kN)	Moment (kN m)
2473	82.1
1917	53.8

Column B2 (Loads at Top of Basement Floor)			
Load Case	Load	Dead (kN)	Imposed (kN)
1	Ultimate characteristic	$5341 - (1839 + 56) = 3446$ $3446/1.25 = 2757$	$0.6 \times 1839 + 56 = 1159$ $1159/1.5 = 773$

Column B2	
Load (kN)	Moment (kN m)
4605	5.2
3530	4.0

For column B2, maximum moment occurs with load case 3 at ground floor level, when the ultimate values are $M = 26.2$ kN m and $N = 4488$ kN (max).

Similarly, the design loads for column A2 can be obtained as follows:

Column A2 (Loads at Top of Basement Wall)			
Load Case	Load	Dead (kN)	Imposed (kN)
2	Ultimate characteristic	$2672 - (862 + 26) = 1784$ $1784/1.25 = 1427$	$0.6 \times 862 + 26 = 543$ $543/1.5 = 362$

Column A2	
Load (kN)	Moment (kN m)
2327	113.4
1789	84.3

The wall system forming the central core of the building is considered to support floor and roof areas that extend 3.6 m beyond the gridlines (i.e., 3.4 m outside the walls), as shown in the figure below.

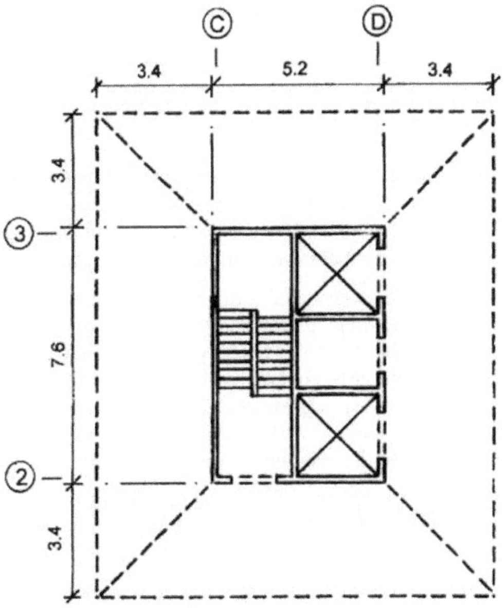

Area of flat slab assumed to be supported by internal walls

Example 2   **Calculation Sheet 16**

Reference	CALCULATIONS	OUTPUT
	The load due to the tank room and lift motor room, which are located over the area between the wall on gridline D and the adjacent parallel wall, will be assumed to be the same as that for the staircase area (see calculation sheet 61 of Example 1). For the analysis of the spread foundation, the total load on the wall system will be assumed to be distributed uniformly over the area enclosed by the walls.	

Total characteristic dead load at top of basement floor:

		kN
Roof	$12.0 \times 14.4 \times 7.5 =$	1296
Floors	$5 \times (12.0 \times 14.4 - 5.2 \times 7.6) \times 7.25 =$	4832
Staircase areas ($\times 2$)	$2 \times 2 \times 7.2 \times 30.8 =$	887
Walls (ignoring openings)	$36.0 \times 22.0 \times 0.2 \times 25 =$	<u>3960</u>
Total		10975

Total characteristic imposed load, including reduction for number of floors:

		kN
Roof	$12.0 \times 14.4 \times 0.6 =$	104
Floors	$0.6 \times [5 \times (12.0 \times 14.4 - 5.2 \times 7.6) \times 4.0] =$	1600
Staircase areas ($\times 2$)	$2 \times 2 \times 7.2 \times 10.2 =$	<u>294</u>
Total		1998

Overturning moment at top of basement floor, due to characteristic wind force in direction normal to long face of building: $M_x = 5346$ kN m

Overturning moment at top of basement floor, due to characteristic wind force in direction normal to short face of building: $M_y = 2700$ kN m

Internal Walls (Loads and Moments at Top of Basement Floor)			Wind Moment M_y (kN m)
Load and Combination	Vertical Load (kN)	Wind Moment M_x (kN m)	Wind Moment M_y (kN m)
Ultimate (1a)	$1.25 \times 10975 + 1.5 \times 0.7 \times 1998 = 15817$	$1.5 \times 5346 = 8020$	$1.5 \times 2700 = 4050$
(1b)	$1.0 \times 10975 = 10975$	$1.5 \times 5346 = 8020$	$1.5 \times 2700 = 4050$
(2)	$1.25 \times 10975 + 1.5 \times 1998 = 16716$	$0.5 \times 8020 = 4010$	$0.5 \times 4050 = 2025$
Service (1a)	$1.0 \times 10975 + 0.7 \times 1998 = 12374$	$1.0 \times 5346 = 5346$	$1.0 \times 2700 = 2700$
(1b)	$1.0 \times 10975 = 10975$	$1.0 \times 5346 = 5346$	$1.0 \times 2700 = 2700$
(2)	$1.0 \times (10975 + 1998) = 12973$	$0.5 \times 5346 = 2673$	$0.5 \times 2700 = 1350$

Design loads due to weight of 250 mm thick basement wall are as follows:

Ultimate $1.25 \times 3.76 \times 0.25 \times 25 = 29.4$ kN/m.   Characteristic 23.5 kN/m

Analysis

Isolated bases will be provided to the columns, the external walls, and the internal wall system, respectively, with the remainder of the basement floor supported on a compressible sub-base and tied to the bases. For the bases to the columns and the internal wall system, a linear distribution of bearing pressure will be assumed.

The base to the external walls will be considered initially as a beam with free ends on an elastic soil. The moment required to restrain the resulting rotation at the external end of the beam will then be determined. The wall, considered initially as a beam propped at the top and fixed at the bottom, will be analysed to determine the moment at the base due to lateral earth pressure. The out-of-balance moments at the joint between the wall and the base will then be distributed according to the relative stiffness of the members, and the resulting effects determined.

Effective span of basement wall between ground floor and basement floor is 4.0 m. Two soil conditions will be considered: at-rest pressure with surcharge, and active pressure without surcharge. Ultimate fixed-end moments at bottom of wall are

$M_{w,max} = 1.35 \times 0.47 \times (10 \times 4.0^2/8 + 20 \times 4.0^3/15) = 67.0$ kN m/m

$M_{w,min} = 1.0 \times 0.30 \times 20 \times 4.0^3/15 = 25.6$ kN m/m

OUTPUT:

Lateral soil pressure on basement wall

Characteristic values of fixed-end moment at bottom of wall

$M_{w,max} = 49.6$ kN m/m

$M_{w,min} = 25.6$ kN m/m

Example 2   **Calculation Sheet 17**

Reference	CALCULATIONS	OUTPUT
	BASE TO COLUMN B2	Base size
	The maximum characteristic load is 3530 kN and the associated bending moment is negligible. Hence, assuming an allowable increase in the base loading intensity of 200 kN/m², the required base area = 3530/200 = 17.65 m² (4.2 m square).	4.2×4.2 m
	The maximum design ultimate load is 4605 kN, and the resulting uniform bearing pressure $q = 4605/4.2^2 = 261$ kN/m².	
	The design bending moment at the face of the column is	
	$M_{Ed} = 261 \times 4.2 \times 1.9^2/2 = 1979$ kN m	
	Flexural design	
	Assuming a 900 mm deep base, and allowing for 50 mm cover with 20 mm bars in each direction, mean effective depth: $d_{av} = 900 - (50 + 20) = 830$ mm	Thickness 900 mm
9.2.1.1	Minimum area of longitudinal tension reinforcement: $A_{s,min} = 0.26(f_{ctm}/f_{yk})b_t d = 0.26 \times (3.0/500) \times 1000 \times 830 = 1295$ mm²/m	
9.3.1.1 (3)	Maximum spacing of bars: $s_{max} = 3h \leq 400$ mm (but ≤ 250 mm in areas with concentrated loads)	
	According to the calculated value of $M/bd^2 f_{ck}$, appropriate values of z/d and A_s can be determined (Table A1) and suitable reinforcement selected (Table A9).	
	$M_{Ed}/bd^2 f_{ck} = 1979 \times 10^6/(4200 \times 830^2 \times 32) = 0.022,$ $z/d = 0.95$ (max)	Provide H20-200B
	$A_s = 1979 \times 10^6/(0.87 \times 500 \times 0.95 \times 830) = 5770$ mm² (21H20-200)	in both directions
	Shear design	
	The maximum shear stress should not exceed the value given by	
6.2.2 (6)	$v_{Rd,max} = 0.5 \nu f_{cd} = 0.5 \times 0.6(1 - f_{ck}/250) \times (\alpha_{cc}f_{ck}/1.5) = 0.2(1 - f_{ck}/250)f_{ck}$	
	$= 0.2 \times (1 - 32/250) \times 32 = 5.58$ MPa	
	The shear stress at the column perimeter is	
	$v_{Ed} = N_{Ed}/ud = 4605 \times 10^3/(4 \times 400 \times 830) = 3.47$ MPa ($<v_{Rd,max}$)	
	Punching shear checks should be made for control perimeters at distances $a \leq 2d$ to determine the critical condition, where:	
	Length of control perimeter and circumscribed area are, respectively:	
	$u_1 = 4c + 2\pi a,$ $A_1 = c^2 + 4ca + \pi a^2$	
	Net applied shear force and average shear stress are, respectively:	
	$V_{Ed} = N_{Ed} - A_1 \times q,$ $v_{Ed} = V_{Ed}/(u_1 d)$	
6.4.4 (2)	The minimum design shear strength for values of $a \leq 2d$ is given by	
	$v_{Rd,c} = v_{min} \times 2d/a$ where $v_{min} = 0.035k^{3/2}f_{ck}^{1/2}$ and	Basic control perimeter
	$k = 1 + (200/d)^{1/2} = 1.49,$ $v_{min} = 0.035 \times 1.49^{3/2} \times 32^{1/2} = 0.36$ MPa	for punching shear
	Values for different control perimeters are shown in the following table:	

a (mm)	u_1 (mm)	A_1 (m²)	V_{Ed} (kN)	v_{Ed} (MPa)	$v_{Rd,c}$ (MPa)
600	5370	2.25	4018	0.90	0.99
$d = 830$	6815	3.65	3652	0.65	0.72
1000	7883	4.90	3326	0.51	0.60

Shear satisfactory

Reference	CALCULATIONS	OUTPUT
	Cracking due to flexure	
	Minimum area of reinforcement required in tension zone for crack control:	
7.3.2 (2)	$A_{s,min} = k_c k f_{ct,eff} A_{ct}/\sigma_s$ where	
	$k_c = 0.4$ for bending, $k = 0.65$ for $h \geq 800$ mm, $f_{ct,eff} = f_{ctm} = 0.3 f_{ck}^{(2/3)} = 3.0$ MPa for general design purposes, $A_{ct} = bh/2$ and $\sigma_s \leq f_{yk} = 500$ MPa.	
	$A_{s,min} = 0.4 \times 0.65 \times 3.0 \times 1000 \times 450/500 = 702$ mm²/m (H20-300)	

Example 2

Reference	CALCULATIONS	OUTPUT
	BASE TO INTERNAL WALLS Under service conditions, the maximum bending moments and co-existent vertical load are as follows: $M_x = 5346$ kN m, $M_y = 2700$ kN m, $N = 12{,}374$ kN. For a 10 m long base, with wind acting in a direction normal to the long face of the building, the required base width is $\quad B = (N/L + 6M_x/L^2)/200 = (12374/10 + 6 \times 5346/10^2)/200 = 7.8$ m say. For wind acting in a direction normal to the short face of the building, maximum bearing pressure is $\quad q = (N/L + 6M_y/L^2)/B = (12374/7.8 + 6 \times 2700/7.8^2)/10 = 185$ kN/m² (<200) The maximum design ultimate bending moments and co-existent vertical load are as follows: $M_x = 8020$ kN m, $M_y = 4050$ kN m, $N = 15817$ kN. The maximum and minimum values at the ends of the resulting linear distributions of bearing pressure are as follows: For wind acting in a direction normal to the long face of the building: $\quad q = (15817/10 \pm 6 \times 8020/10^2)/7.8 = 265$ and 141 kN/m² Bearing pressures (kN/m²) Maximum design bending moment at outer face of walls on lines 2 and 3: $\quad M_{Ed} = -(265/2 - 15/6) \times 7.8 \times 1.2^2 = -1460$ kN m For wind acting in a direction normal to the short face of the building: $\quad q = (15817/7.8 \pm 6 \times 4050/7.8^2)/10 = 243$ and 163 kN/m² Bearing pressures (kN/m²) Maximum design bending moment at outer face of walls on lines C and D: $\quad M_{Ed} = -(243/2 - 13/6) \times 10 \times 1.3^2 = -2017$ kN m **Flexural design** Consider a 600 mm deep base, with 50 mm cover and 20 mm bars. For bars in the longitudinal direction, $d = 600 - (50 + 20/2) = 540$ mm. Minimum area of longitudinal tension reinforcement:	Base size 10.0 m × 7.8 m Thickness 600 mm
9.2.1.1	$\quad A_{s,min} = 0.26(f_{ctm}/f_{yk})b_t d = 0.26 \times (3.0/500) \times 1000 \times 540 = 843$ mm²/m Maximum spacing of bars:	
9.3.1.1 (3)	$\quad s_{max} = 3h \leq 400$ mm (but ≤ 250 mm in areas with concentrated loads) $\quad M_{Ed}/bd^2 f_{ck} = 1460 \times 10^6/(7800 \times 540^2 \times 32) = 0.020 \qquad z/d = 0.95$ (max) $\quad A_s = 1460 \times 10^6/(0.87 \times 500 \times 0.95 \times 540) = 6543$ mm² (26H20-300) For bars in the transverse direction, $d = 600 - (50 + 20 + 20/2) = 520$ mm. $\quad M_{Ed}/bd^2 f_{ck} = 2017 \times 10^6/(10000 \times 520^2 \times 32) = 0.024 \qquad z/d = 0.95$ (max) $\quad A_s = 2017 \times 10^6/(0.87 \times 500 \times 0.95 \times 520) = 9386$ mm² (33H20-300) Since the sagging moments in the regions between the walls are small, it is clear that minimum reinforcement H20-300 (top and bottom) will suffice.	 Provide H20-300 (top and bottom) throughout

Example 2 **Calculation Sheet 19**

Reference	CALCULATIONS	OUTPUT
	Shear design	
	Maximum design shear force at distance d from outer face of walls on lines C and D (i.e., 0.78 m from edge of base) is: $V_{Ed} = (243 - 8/2) \times 0.78 = 187$ kN/m	
	$v_{Ed} = V_{Ed}/(bd) = 187 \times 10^3/(1000 \times 520) = 0.36$ MPa ($<v_{min}$)	
	Cracking	
	From the calculations for the raft foundation (calculation sheet 11), it is clear that the provision of H20-300 as minimum reinforcement satisfies the criteria.	
	BASE TO EXTERNAL WALL	
	Consider the wall on line 1, and assume that the concentrated load and bending moment at column B1 are distributed uniformly over a length of 7.2 m. If the base is positioned so that the bending moment at the centre of the base due to vertical load is equal to the fixed-end moment at the bottom of the wall due to lateral soil pressure, a uniform distribution of bearing pressure results. Thus, there is no tilting of the base, and the assumption of full fixity at the bottom of the wall is correct. Clearly, this condition cannot always be achieved, and the effect of base tilting on the bending moment at the bottom of the wall often needs to be considered.	
	In such cases, the following iterative procedure can be used, where F is the vertical load, M_0 is the fixed-end moment at the base of the wall, EI is the flexural rigidity of the wall, k_s is the modulus of subgrade reaction of the bearing stratum and θ is the slope of the base. The dimensions of the wall and base are shown in the figure.	
	(1) Assume a value M_1 for the bending moment at the bottom of the wall, such that $M_0 > M_1 > N(a - b/2)$. Calculate $c = (a - M_1/N)$.	
	(2) If $c \leq b/3$, calculate $\theta = N/4.5c^2k_s$. If $c > b/3$, calculate $\theta = 6N(1 - 2c/b)/b^2k_s$.	
	(3) Calculate $(M_0 - M_2)$, where $M_2 = 3EI\theta/H^2$.	
	(4) If $M_1 = (M_0 - M_2)$, M_1 is the correct value for the moment at the bottom of the wall. Otherwise, assume a new value for M_1 and repeat procedure until the correct value is found.	
	Case 1. Characteristic loads with maximum vertical load on building and either (a) maximum, or (b) minimum, soil pressure on wall. Load and fixed-end moments at bottom of wall are as follows:	
	Total load: $N = 23.5 + 1917/7.2 = 290$ kN/m	
	Minimum width of base $= 290/200 = 1.45$ m	
	(a) Fixed-end moment: $M_0 = -49.6 + 0.5 \times 53.8/7.2 = -46$ kN m/m	
	Eccentricity of vertical load for uniform bearing $= 46/290 = 0.16$ m	
	(b) Fixed-end moment: $M_0 = -25.6 + 0.5 \times 53.8/7.2 = -22$ kN m/m	
	Eccentricity of vertical load for uniform bearing $= 22/290 = 0.08$ m	
	Since the column load is so dominant, the centre of the total load is approximately at the column centre. For a 1.6 m wide base, with outer edge 0.7 m from centre of column, $a = 1.6 - 0.7 = 0.9$ m. Modulus of subgrade reaction $k_s = 12$ MN/m^3.	
	(a) If $M_1 = 35$ kN m/m, $c = 0.9 - 35/290 = 0.780$ m ($>b/3 = 0.53$ m)	
	$\theta = 6 \times 290 \times (1 - 2 \times 0.780/1.6)/(1.6^2 \times 12 \times 10^3) = 1.42 \times 10^{-3}$	
	$M_2 = 3 \times 33 \times 10^6 \times (0.25^3/12) \times 1.42 \times 10^{-3}/4.0^2 = 11.4$ kN m/m	
	$M_0 - M_2 = 46 - 11.4 = 34.6$ kN m/m ($= M_1$ approx)	
	Maximum bearing pressure $= 290/1.6 + 6 \times (35 - 290 \times 0.1)/1.6^2 = 196$ kN/m^2	Case 1 (a)
	(b) If $M_1 = 27$ kN m/m, $c = 0.9 - 27/290 = 0.807$ m ($>b/3 = 0.53$ m)	
	$\theta = 6 \times 290 \times (1 - 2 \times 0.807/1.6)/(1.6^2 \times 12 \times 10^3) = -0.50 \times 10^{-3}$	
	$M_2 = 3 \times 33 \times 10^6 \times (0.25^3/12) \times (-0.50) \times 10^{-3}/4.0^2 = -4.0$ kN m/m	
	$M_0 - M_2 = 22 + 4.0 = 26.0$ kN m/m ($= M_1$ approx)	
	Maximum bearing pressure $= 290/1.6 + 6 \times (290 \times 0.1 - 26)/1.6^2 = 188$ kN/m^2	Case 1 (b)

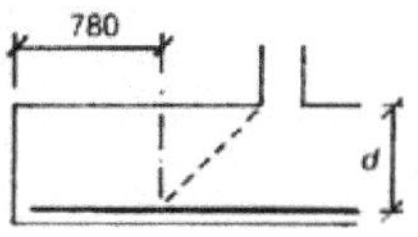

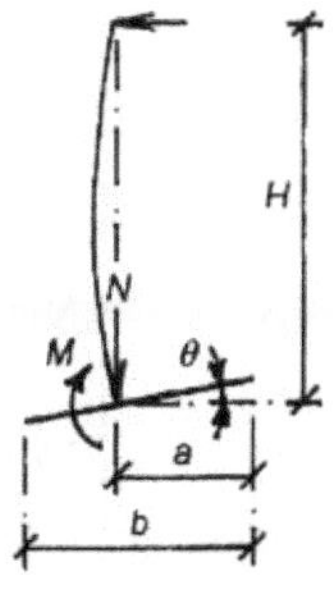

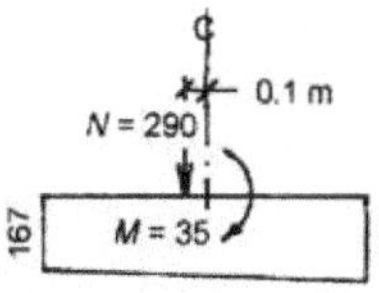

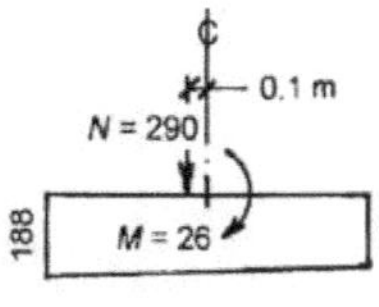

Example 2   **Calculation Sheet 20**

Reference	CALCULATIONS	OUTPUT
	Case 2. Ultimate design loads with maximum soil pressure on wall and either (a) maximum, or (b) minimum (i.e. constructed up to ground floor only), load on the building. Loads and fixed-end moments at base of wall are as follows:	
	(a) Total load: $N = 29.4 + 2473/7.2 = 373$ kN/m	
	Fixed-end moment: $M_0 = -67.0 + 0.5 \times 82.1/7.2 = -61$ kN m/m	
	If $M_1 = 45.5$ kN m/m, $c = 0.9 - 45.5/373 = 0.778$ m $(>b/3 = 0.53$ m)	
	$\theta = 6 \times 373 \times (1 - 2 \times 0.778/1.6)/(1.6^2 \times 12 \times 10^3) = 2.00 \times 10^{-3}$	
	$M_2 = 3 \times 33 \times 10^6 \times (0.25^3/12) \times 2.00 \times 10^{-3}/4.0^2 = 16$ kN m/m	
	$M_0 - M_2 = 61 - 16 = 45$ kN m/m $(= M_1$ approx)	
	Bearing pressures at inner and outer edges of base are:	
	$q = 373/1.6 \pm 6 \times (46 - 373 \times 0.1)/1.6^2 = 253$ kN/m² and 213 kN/m²	Case 2 (a)
	(b) $N = 23.5 + 261/7.2 = 60$ kN/m $M_0 = -67.0$ kN m/m	
	If $M_1 = 26$ kN m/m, $c = 0.9 - 26/60 = 0.467$ m $(< b/3 = 0.53$ m)	
	$\theta = 60/(4.5 \times 0.467^2 \times 12 \times 10^3) = 5.09 \times 10^{-3}$	
	$M_2 = 3 \times 33 \times 10^6 \times (0.25^3/12) \times 5.09 \times 10^{-3}/4.0^2 = 41$ kN m/m	
	$M_0 - M_2 = 67 - 41 = 26$ kN m/m $(= M_1)$	Case 2 (b)
	Flexural design	
	Consider a 400 mm deep base, with 50 mm cover and 12 mm bars. For bars in the transverse direction, $d = 400 - (50 + 12/2) = 340$ mm say.	
	Minimum area of transverse reinforcement:	
9.2.1.1	$A_{s,min} = 0.26(f_{ctm}/f_{yk})b_t d = 0.26 \times (3.0/500) \times 1000 \times 340 = 531$ mm²/m	
	Maximum spacing of bars:	
9.3.1.1 (3)	$s_{max} = 3h \leq 400$ mm (but ≤ 250 mm in areas with concentrated loads)	
	Maximum moment occurs for case 2 (a), where value at inside face of wall is	
	$M_{Ed} = 253 \times 0.85^2/2 - 25 \times 0.85^3/3 = 86.3$ kN m/m	
	$M_{Ed}/bd^2 f_{ck} = 86.3 \times 10^6/(1000 \times 340^2 \times 32) = 0.024$ $z/d = 0.95$ (max)	Provide H16-300B in both directions
	$A_s = 86.3 \times 10^6/(0.87 \times 500 \times 0.95 \times 340) = 615$ mm²/m (H16-300)	
	Shear design	
	Maximum shear force at distance d from inside face of the wall (i.e. 0.51 m from inner edge of base) is	
	$V_{Ed} = 253 \times 0.51 - 25 \times 0.51^2/2 = 126$ kN/m	
	Maximum shear stress is	
	$v_{Ed} = V_{Ed}/(bd) = 126 \times 10^3/(1000 \times 340) = 0.37$ MPa	
	The minimum shear strength, where $k = 1 + (200/d)^{1/2} = 1.76$, is given by	
6.2.2 (1)	$v_{min} = 0.035k^{3/2}f_{ck}^{1/2} = 0.035 \times 1.76^{3/2} \times 32^{1/2} = 0.46$ MPa $(>v_{Ed})$	Shear satisfactory
	DESIGN OF BASEMENT WALL	
	Durability	
BS 8500	For the external surface, exposure class XC2 applies and $c_{nom} = 35$ mm is needed.	$c_{nom} = 35$ mm
	Flexural design	
	Allowing for 25 mm cover, 16 mm diameter bars (EW) and horizontal bars in the outer layers, for the vertical bars, $d = 250 - (35 + 16 + 16/2) = 190$ mm.	
	Minimum area of vertical tension reinforcement:	
9.2.1.1	$A_{s,min} = 0.26 \times (3.0/500) \times 1000 \times 190 = 297$ mm²/m (H12-300)	

Example 2

Calculation Sheet 21

Reference	CALCULATIONS	OUTPUT
7.3.2 (2)	Minimum area of reinforcement required in tension zone for crack control: $A_{s,min} = k_c k f_{ct,eff} A_{ct}/\sigma_s$ where $k_c = 0.4$ for bending, $k = 1.0$ for $h \leq 300$ mm, $f_{ct,eff} = f_{ctm} = 0.3 f_{ck}^{(2/3)} = 3.0$ MPa for general design purposes, $A_{ct} = bh/2$ and $\sigma_s \leq f_{yk} = 500$ MPa. $A_{s,min} = 0.4 \times 1.0 \times 3.0 \times 1000 \times 125/500 = 300$ mm^2/m (H12-300) Maximum design ultimate moment at bottom of wall occurs for case 2 (a), where $M_{Ed} = 45.5$ kN m/m. Hence, from Table A4: $M/bd^2 f_{ck} = 45.5 \times 10^6/(1000 \times 190^2 \times 32) = 0.040$ $z/d = 0.95$ (max) $A_s = 45.5 \times 10^6/(0.87 \times 500 \times 0.95 \times 190) = 580$ mm^2/m (H16-300) Minimum design ultimate moment at bottom of wall occurs for case 2 (b), where $M_{Ed} = 26$ kN m/m. Corresponding shear force at top of wall is $V_{Ed} = 1.35 \times 0.47 \times (10 \times 4.0/2 + 20 \times 4.0^2/6) - 26/4.0 = 40$ kN/m If a_o is distance from top of wall to point of zero shear, then $V_{Ed} - 1.35 \times 0.47 \times (10 a_o + 20 a_o^2/2) = 0$ which gives $a_o = 2.1$ m Hence, maximum sagging moment (at $a_o = 2.1$ m) is $M_{Ed} = V_{Ed} \times a_o - 1.35 \times 0.47 \times (10 a_o^2/2 + 20 a_o^3/6) = 51$ kN m $A_s = 51 \times 10^6/(0.87 \times 500 \times 0.95 \times 190) = 650$ mm^2/m (H16-300) **Shear design** Maximum shear force at bottom of wall occurs for case 2 (a), where: $V_{Ed} = 1.35 \times 0.47 \times (10 \times 4.0/2 + 20 \times 4.0^2/3) + 45.5/4.0 = 91.8$ kN/m $v_{Ed} = V_{Ed}/(bd) = 91.8 \times 10^3/(1000 \times 190) = 0.49$ MPa Minimum design shear strength, where $k = 2.0$ for $d \leq 200$ mm, is	45.5 kNm Case 2 (a) 2.1 m 51 kNm 26 kNm Case 2 (b) Provide H16-300 (EF) vertical reinforcement
6.2.2 (1)	$v_{min} = 0.035 k^{3/2} f_{ck}^{1/2} = 0.035 \times 2.0^{3/2} \times 32^{1/2} = 0.56$ MPa ($> v_{Ed}$) **Cracking due to restrained early thermal contraction** Minimum area of horizontal reinforcement, with $f_{ct,eff} = 1.8$ MPa for cracking at age of 3 days, $k_c = 1.0$ for tension and $k = 1.0$ for $h \leq 300$ mm, is given by	
7.3.2 (2) 7.3.4	$A_{s,min} = k_c k f_{ct,eff} A_{ct}/f_{yk} = 1.0 \times 1.0 \times 1.8 \times 1000 \times 125/500 = 450$ mm^2/m (EF) With $c = 35$ mm, $k_1 = 0.8$ for high bond bars, $k_2 = 1.0$ for tension, $h_{c,ef}$ as the lesser of $2.5(h - d)$ and $h/2$, and H12-200 (EF) as minimum reinforcement: $s_{r,max} = 3.4c + 0.425 k_1 k_2 (A_{c,eff}/A_s) \phi$ $= 3.4 \times 35 + 0.425 \times 0.8 \times 1.0 \times (2.5 \times 41 \times 1000/565) \times 12 = 860$ mm	Provide H12-200 (EF) minimum horizontal reinforcement
PD 6687 2.16	With $R = 0.8$ for wall on a thick base, $\Delta T = 25°$C for 350 kg/m^3 Portland cement concrete and 250 mm thick wall (*Reynolds*, Table 2.18), and $\alpha = 12 \times 10^{-6}$ per °C: $w_k = (0.8 R \alpha \Delta T) \times s_{r,max} = 0.8 \times 0.8 \times 12 \times 10^{-6} \times 25 \times 860 = 0.17$ mm **Wall as deep beam** (see calculation sheets 13 and 14) For the wall on gridline A, the design ultimate column loads and resulting bending moments are as shown in the figure below:	

776 kN   2327 kN   2327 kN   776 kN

6.0 m   7.2 m   6.0 m

Column loads

Example 2   **Calculation Sheet 22**

Reference	CALCULATIONS	OUTPUT
	936 951 936 1146 1146 Bending moments (kN m) The required areas of longitudinal tension reinforcement, with $z = 2640$ mm, are For sagging: $A_s = 951 \times 10^6/(0.87 \times 500 \times 2640) = 828$ mm^2 (8H12) For hogging: $A_s = 1146 \times 10^6/(0.87 \times 500 \times 2640) = 998$ mm^2 (9H12) Providing H12-200 (EF) as minimum horizontal reinforcement will be sufficient. The maximum shear force is at column B1, where $V_{Ed,max} = 0.5 \times 2473 = 1237$ kN. $V_{Ed,max}/b_w d = 1237 \times 10^3/(250 \times 3600) = 1.38$ MPa ($<v_{Rd,max} = 5.58$ MPa) At distance d from the face of the columns, the shear force is negligible. Moreover, with H12-300 as minimum vertical bars on one face of the wall, $z = 3200$ mm and $\cot\theta = 2.5$ (see calculation sheet 14): $V_{Rd,s} = (A_{sw}/s)zf_{ywd}\cot\theta$ 6.2.3 (3) $= 0.377 \times 3200 \times (0.87 \times 500) \times 2.5 \times 10^{-3} = 1312$ kN ($>V_{Ed,max}$)	
7.3.2 (2)	**DESIGN OF BASEMENT FLOOR** The 200 mm thick slab, which is tied throughout, is effectively continuous and will be reinforced to control cracking due to early thermal contraction. Minimum area of horizontal reinforcement, with $k_c = 1.0$ for tension and $k = 1.0$ for $h \le 300$ mm, is given by $A_{s,min} = k_ckf_{ct,eff}A_{ct} = 1.0 \times 1.0 \times 1.9 \times 1000 \times 100/500 = 380$ mm^2/m (EF) This can be provided by A393 fabric (EF), with the sheet widths chosen to suit the widths of the slab panels, and a longitudinal lap length $= 1.5 \times 35 \times 10 = 525$ mm. Tie bars between the bases and the slab panels provide slightly less area than the minimum reinforcement. The bars are able to yield inside the de-bonding sleeves, and the joints are provided with external waterstops and top surface sealants.	

Bar Marks	Commentary on Bar Arrangement (Drawings 8–9)
01, 08, 10	Straight bars with 50 mm cover against blinding (bottom and ends).
02, 09, 11	Straight bars with 50 mm cover at ends, to provide support to bars 03 and 04.
03	Tie bars de-bonded either side of joint between base and adjacent floor slab. Projection length for anchorage and de-bonding $= 35 \times 16 + 150 = 700$ mm say. Bars approximately at mid-depth of floor slab, but displaced 10 mm in each direction (up one way and down the other way) to avoid bars clashing at corners of bases.
04, 05, 12	Bars (shape code 21), with covers in each direction differing by 20 mm to suit layering of bars.
06	Column starter bars (shape code 11) standing on mat formed by bars 01. Projection of bars above basement floor to provide a lap length above 75 mm kicker $= 1.5 \times 35 \times 32 + 75 = 1800$ mm. Cover to bars in column $= 50$ mm to enable 35 mm cover to links.
07	Closed links (shape code 51) to hold column starter bars in place during construction.
13	Starter bars (shape code 21) for internal walls. Projection above basement floor to provide a lap length above 75 mm kicker $= 1.5 \times 35 \times 12 + 75 = 725$ mm say. Cover to bars in wall $= 35$ mm to enable 25 mm cover to horizontal bars.

Example 2: Reinforcement in Base to Internal Columns Drawing 8

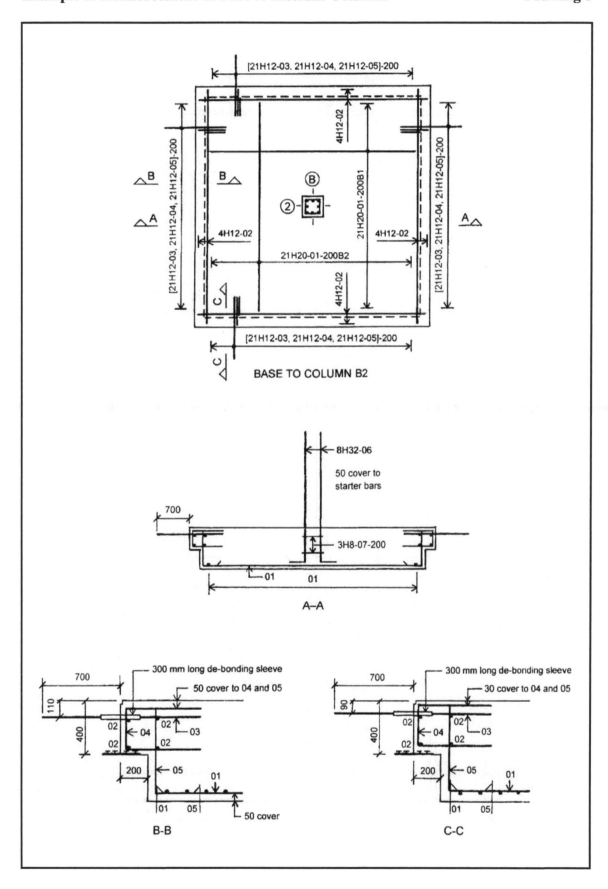

Example 2: Reinforcement in Base to Internal Walls

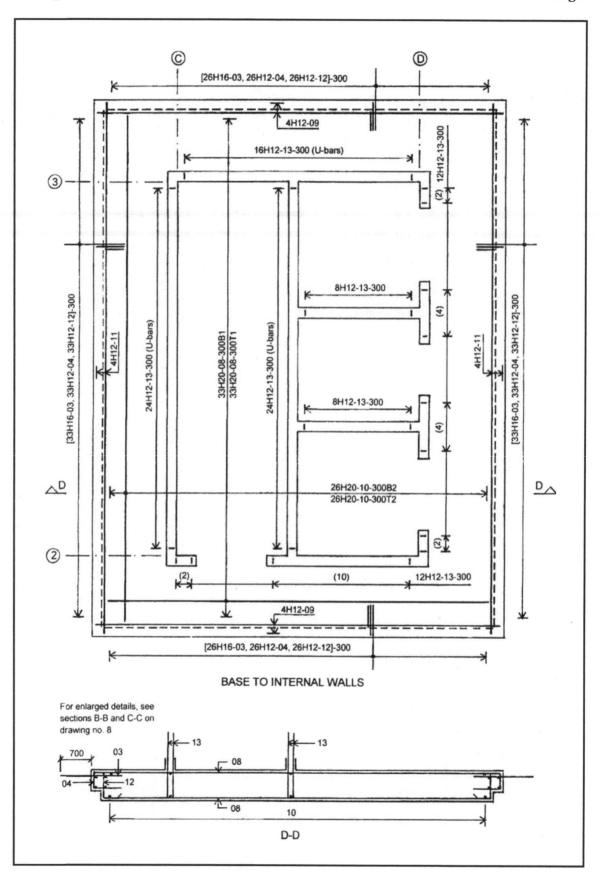

BASE TO INTERNAL WALLS

For enlarged details, see
sections B-B and C-C on
drawing no. 8

D-D

Example 2: Reinforcement in Basement Wall and Footing

Drawing 10

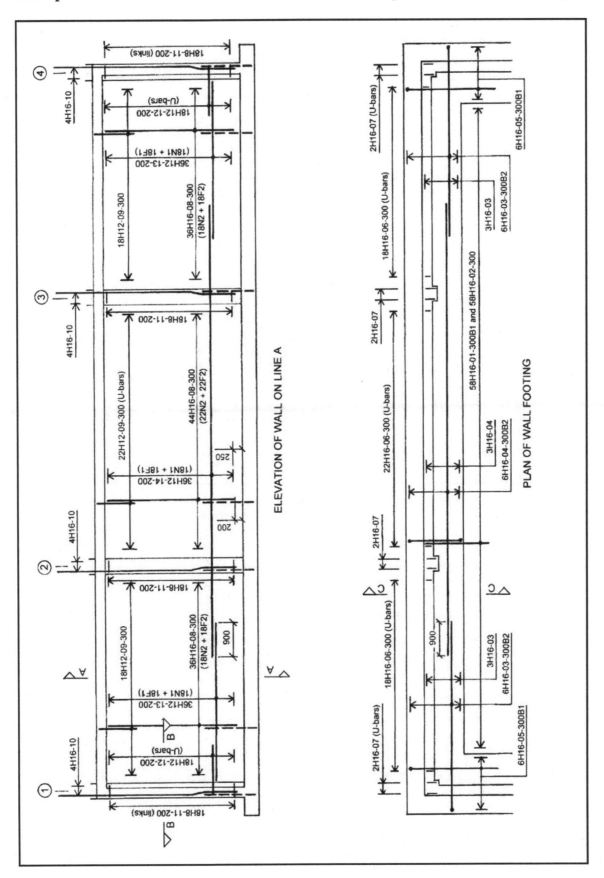

ELEVATION OF WALL ON LINE A

PLAN OF WALL FOOTING

Example 2: Cross-Sections for Basement Wall and Footing   Drawing 11

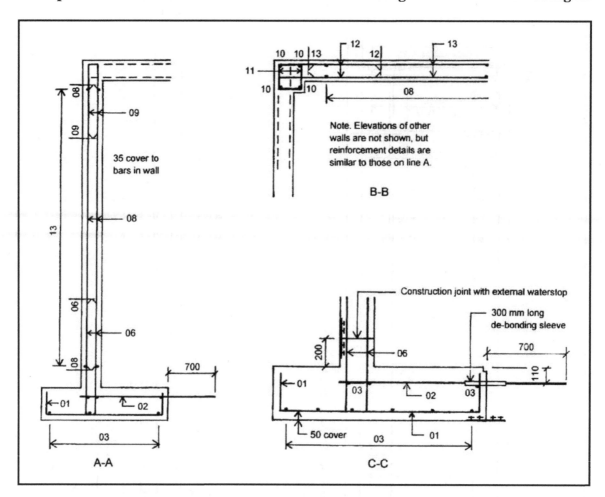

Bar Marks	Commentary on Bar Arrangement (Drawings 10–11)
01	Bars (shape code 21) with 50 mm cover against blinding (bottom and ends).
02	Tie bars de-bonded either side of joint between base and adjacent floor slab. Projection length for anchorage and de-bonding = $35 \times 16 + 150 = 700$ mm say. Bars approximately at mid-depth of floor slab, but displaced 10 mm down for walls on lines A and F, and 10 mm up for walls on lines 1 and 4.
03, 05	Bars (shape code 11) with 50 mm cover against blinding (bottom and ends).
04	Straight bars (12 m long) with lap length = $1.5 \times 35 \times 16 = 900$ mm say.
06, 07	Starter bars (shape code 21) for wall and columns. Projection above basement floor to provide a lap length above 200 mm kicker = $1.5 \times 35 \times 16 + 200 = 1050$ mm say.
08	Straight bars bearing on 200 mm kicker, and curtailed 50 mm below ground floor slab.
09	Bars (shape code 21) with lap length = $1.5 \times 35 \times 12 = 650$ mm say.
10	Bars (shape code 26) bearing on 200 mm kicker and cranked to fit alongside bars projecting from footing. Projection of bars above ground floor level = $1.5 \times 35 \times 16 + 75 = 925$ mm.
11	Closed links (shape code 51), with 35 mm nominal cover, starting above kicker and stopping below ground floor slab.
12	Bars (shape code 21) with lap length = $1.5 \times 50 \times 12 = 900$ mm.
13, 14	Straight bars (12 m long maximum), starting above kicker and stopping below ground floor slab. Lap length = $1.5 \times 50 \times 12 = 900$ mm.

Example 2   Calculation Sheet 23

Reference	CALCULATIONS	OUTPUT
	PILE FOUNDATIONS (Flat slab superstructure)	

Loading

Critical design values determined on calculation sheets 15-16, for the columns and internal wall assembly, are as follows:

Limit State	Ultimate		Characteristic/Serviceability	
Member	Load (kN)	Moment (kN m)	Load (kN)	Moment (kN m)
Column B1	2473	82.1	1917	53.8
Column B2	4605	5.2	3530	4.0
Column A2	2327	113.4	1789	84.3
Internal walls	15,817	$M_x = 8020$	12 374	$M_x = 5346$
ditto	15,817	$M_y = 4050$	12 374	$M_y = 2700$

Design loads due to weight of the 250 mm thick basement wall are as follows:

Ultimate: 29.4 kN/m   Characteristic: 23.5 kN/m

Fixed-end moments at the bottom of the wall due to maximum or minimum lateral soil pressures are:

Ultimate: 67.0 or 25.6 kN m/m   Characteristic: 49.6 or 25.6 kN m/m

Characteristic loads for the 250 mm thick basement floor slab are as follows:

Slab and finishes: $0.250 \times 25 + 1.25 = 7.5$ kN/m^2 Imposed: 4.0 kN/m^2

Total design loads for basement floor slab are as follows:

Ultimate: $1.25 \times 7.5 + 1.5 \times 4.0 = 15.4$ kN/m^2 Characteristic: 11.5 kN/m^2

Design of piles

A preliminary examination of the foundation loads shows that the load at external columns B1 and A2 is approximately 60% of the load at internal column B2. An arrangement of 3 piles at the external columns and 5 piles at the internal columns will be considered. Initial calculations show that 600 mm diameter piles embedded about 14 m into the underlying clay stratum ($c_u = 150$ kN/m^2) would be suitable.

BS-EN 1997-1

The characteristic bearing resistance of a pile is given by $R_{ck} = R_{bk} + R_{sk}$ where:

$R_{bk} = q_{bk} A_b$ is the base resistance, and $R_{sk} = q_{sk} A_s$ is the shaft resistance.

For a 14 m length with $q_{bk} = 9c_u$ and $q_{sk} = 0.5c_u$:

$R_{ck} = 9 \times 150 \times (3.14 \times 0.6^2/4) + 0.5 \times 150 \times (3.14 \times 0.6) \times 14$

$= 382 + 1978 = 2360$ kN

In practice the value of q_{sk} will vary with depth and the shaft resistance provided by each stratum would be summed over the embedment length. Similarly, the value of q_{bk} at the bottom of the pile would be used to assess the base resistance.

The design ultimate bearing resistance is given by $R_{cd} = R_{bk}/\gamma_b + R_{sk}/\gamma_s$. Thus, with $\gamma_b = 1.6$ (for bored piles) and $\gamma_s = 1.3$: $R_{cd} = 382/1.6 + 1978/1.3 = 1760$ kN

Note: This value applies when the γ_F values on actions are $\gamma_G = 1.0$ and $\gamma_Q = 1.3$.

Fundamentally, the settlement of pile foundations under serviceability and ULS conditions should be assessed, but it is common practice to restrict the characteristic/service pile load to 40% of the characteristic bearing resistance. For the purpose of the example, the characteristic/service load for a 14 m embedment length will be restricted to $2360/2.5 = 944$ kN.

9.8.5 (3) Table 9.6N

Minimum area of longitudinal reinforcement in cast-in-place bored pile:

$A_{s,bpmin} = 0.005A_c = 0.005 \times (3.14 \times 600^2/4) = 1413$ mm^2 (6H20)

With 75 mm cover to 10 mm links, spacing of longitudinal bars around periphery of pile = $(600 - 190) \times 3.14/6 = 215$ mm (clear distance between bars ≤ 200 mm).

Example 2   **Calculation Sheet 24**

Reference	CALCULATIONS	OUTPUT
	BASE TO COLUMN B2	Pile cap 3.6 m square and 1.0 m deep with a 5 pile arrangement as shown below.
	Assume a group of 5 × 600 mm diameter bored piles, with one centre pile and four perimeter piles spaced at 2.4 m centres, and a 3.6 m square × 1.0 m deep pile cap.	
	Maximum characteristic load at bottom of column is 3530 kN and the associated bending moment is negligible. Additional characteristic load from the basement floor slab and pile cap (plus 150 mm concrete overlay) is as follows:	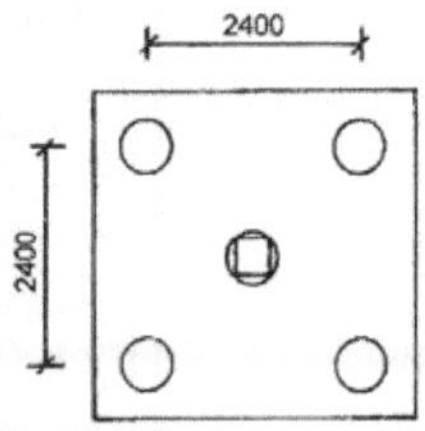
	$F_{k,add} = 1.2 \times 7.2 \times 7.2 \times 11.5 + 3.6^2 \times 1.15 \times 25 = 1088$ kN	
	Total characteristic load = 3530 + 1088 = 4618 kN (924 kN per pile).	
	The maximum design ultimate load at the bottom of the column $N_{ud} = 4605$ kN and the corresponding additional load is as follows:	
	$F_{u,add} = 1.2 \times 7.2 \times 7.2 \times 15.4 + 1.25 \times 3.6^2 \times 1.15 \times 25 = 1424$ kN	
	Total design ultimate load = 4605 + 1424 = 6029 kN (1206 kN per pile).	
	Design of pile cap	
	The forces in the pile cap resulting from the design ultimate column load may be determined by truss analogy. Since one of the piles is directly below the column, only 80% of the column load contributes to the forces in the truss. This is of a triangulated form with a node at the centre of the loaded area and four lower nodes located at the intersections of the centrelines of the perimeter piles with the tensile reinforcement. If d is the effective depth of the reinforcement, and l is the spacing of the perimeter piles, the tensile force along each side of the square formed by the piles is given by $F_t = 0.8N_{ud} \times l/8d$.	
	With $d = 900$ mm (i.e., bars 100 mm from bottom of pile cap) and $l = 2100$ mm:	
	$F_t = 0.1 \times 4605 \times 2400/900 = 1228$ kN	
	Area of reinforcement required between each pair of perimeter piles:	Provide 6H25-200 at bottom between each pair of perimeter piles
	$A_s = 1228 \times 10^3/(0.87 \times 500) = 2823$ mm^2 (6H25-200)	
	The bars should be contained within a band extending not more than $1.5d$ each side of the pile centreline, and provided with a tension anchorage beyond the centres of the piles. For bent bars with half the gap between adjacent bars $\geq 3\phi$, and 'good'	
8.4.4 (1) Table 8.2	bond conditions, $l_{bd} = \alpha_1 l_{b,rqd} = 0.7 \times 35 \times 25 = 625$ mm say.	
	With 75 mm cover, distance from end of bar to centre of pile = 600 − 75 = 525 mm. Thus, the addition of a minimum radius bend will provide sufficient anchorage.	
9.2.1.1 (1)	Minimum total area of longitudinal reinforcement in each direction:	
	$A_{s,min} = 0.26 \times (3.0/500) \times 3600 \times 900 = 5055$ mm^2	
9.8.1 (3)	Since, in each direction, the area of reinforcement provided in two ties (12H25) is sufficient to meet the total minimum requirements, there is no need to provide any further reinforcement in the bottom of the pile cap. Also, the sides and top surface may be unreinforced when there is no risk of tension developing in these areas.	
	Consider the critical section for shear to be on a line inside the piles at a distance of 20% of the pile diameter from the inner face. Distance of this section from the face of the column is $a_v = 0.5 \times (2400 - 400) - 0.3 \times 600 = 820$ mm.	
6.2.2 (6)	The critical shear occurs on a vertical section extending across the full width of the pile cap. The contribution of the column load to the shear force may be reduced by applying a factor $\beta = a_v/2d$ for $0.5d \leq a_v \leq 2d$. Here, $\beta = 820/(2 \times 900) = 0.46$.	
	Thus, applying this factor to 40% of the column load (load on two piles):	
	$v_{Ed} = V_{Ed}/bd = 0.46 \times 0.4 \times 4605 \times 10^3/(4000 \times 900) = 0.24$ MPa	
6.2.2 (1)	The minimum shear strength, where $k = 1 + (200/d)^{1/2} = 1.47$, is given by	
	$v_{min} = 0.035k^{3/2}f_{ck}^{1/2} = 0.035 \times 1.47^{3/2} \times 32^{1/2} = 0.35$ MPa $(>v_{Ed})$	
	Shear stress due to 80% of column load calculated at perimeter of column:	
	$v_{Ed,max} = V_{Ed,max}/ud = 0.8 \times 4605 \times 10^3/(4 \times 400 \times 900) = 2.56$ MPa	
6.2.2 (6)	$v_{Rd,max} = 0.2(1 - f_{ck}/250)f_{ck} = 0.2 \times (1 - 32/250) \times 32 = 5.58$ MPa $(>v_{Ed,max})$	

Example 2 — Calculation Sheet 25

Reference	CALCULATIONS	OUTPUT
	BASE TO EXTERNAL WALLS	
	Assume a 1.0 m wide × 1.0 m deep pile beam with 3 × 600 mm diameter bored piles, spaced at 1.8 m centres, located under each external column.	Pile beam 1.0 m wide and 1.0 m deep with 3 piles at 1.8 m centres located under each external column.
	The maximum characteristic load at the top of the wall for column B1 is 1917 kN. The associated bending moment will be incorporated in the design of the basement.	
	The additional characteristic load from the basement wall, floor slab and pile beam (plus concrete overlay) is as follows:	
	$F_{k,add} = 1.2 \times 7.2 \times (23.5 + 2.4 \times 11.5 + 1.0 \times 1.15 \times 25) = 690$ kN	
	Total characteristic load = 1917 + 690 = 2607 kN (869 kN per pile).	
	The maximum design ultimate load at the bottom of the column is 2473 kN and the corresponding additional load is as follows:	
	$F_{u,add} = 1.2 \times 7.2 \times (29.4 + 2.4 \times 15.4 + 1.25 \times 28.75) = 884$ kN	
	Total design ultimate load = 2473 + 884 = 33,357 kN (1119 kN per pile).	
	Design of pile beam	
	Load from the column and basement wall will be transferred to the piles mainly by deep beam action in the wall. For the pile beam:	
9.2.1.1 (1)	Minimum area of longitudinal reinforcement:	
	$A_{s,min} = 0.26 \times (3.0/500) \times 1000 \times 900 = 1404$ mm^2 (3H25)	
	Minimum requirements for vertical links are given by	
9.2.2 (5)	$A_{sw}/s = (0.08\sqrt{f_{ck}})\, b_w / f_{yk} = (0.08\sqrt{32}) \times 1000/500 = 0.91$ mm^2/mm	Provide 3H25-300 at top and bottom with H16-400 links
9.2.2 (6)	$s \leq 0.75d = 0.75 \times 900 = 675$ mm (H16-400 say)	
	BASE TO INTERNAL WALLS	
	Assume a group of 20 × 600 mm diameter bored piles, arranged in a rectangular grid as shown in the figure below, with an 8.4 m × 6.0 m × 1.0 m deep pile cap.	Pile cap 8.4 m × 6.0 m and 1.0 m deep with a 20 pile arrangement as shown in figure.
	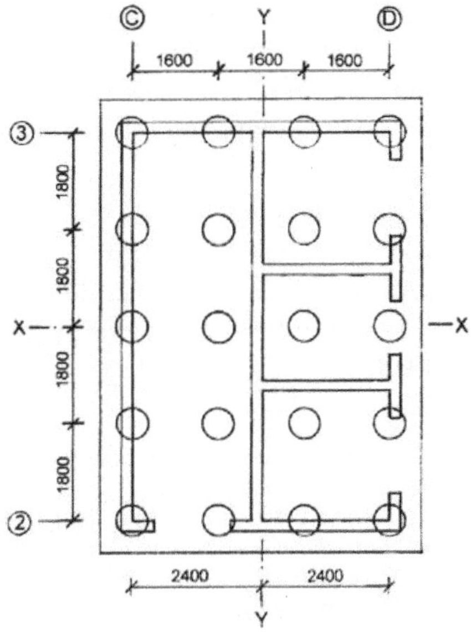 Plan showing layout of piles to internal walls	
	Under service conditions, the maximum bending moments and co-existent vertical load are as follows: $M_x = 5346$ kN m, $M_y = 2700$ kN m, $N = 12{,}374$ kN.	
	Additional characteristic load from area of basement floor slab assumed to extend 3.6 m beyond the gridlines on all sides, and pile cap (plus concrete overlay) is	

Example 2   **Calculation Sheet 26**

Reference	CALCULATIONS	OUTPUT
	$F_{k,add} = 14.4 \times 12.0 \times 11.5 + 8.4 \times 6.0 \times 1.15 \times 25 = 3436$ kN	
	Total co-existent service load $N_s = 12{,}374 + 3436 = 15{,}810$ kN	
	Second moments of area of pile group are as follows:	
	$I_x = 2 \times 4 \times (1.8^2 + 3.6^2) = 129.6$ m^2, $\qquad I_y = 2 \times 5 \times (0.8^2 + 2.4^2) = 64.0$ m^2	
	For wind acting in a direction normal to the long face of the building, maximum load on each pile in outermost row of piles:	
	$N_{p,max} = 15{,}810/20 + 5346 \times 3.6/129.6 = 939$ kN (<944 kN)	
	For wind acting in a direction normal to the short face of the building, maximum load on each pile in outermost row of piles:	
	$N_{p,max} = 15{,}810/20 + 2700 \times 2.4/64.0 = 892$ kN (<944 kN)	
	The maximum design ultimate bending moments and co-existent vertical load are as follows: $M_x = 8020$ kN m, $M_y = 4050$ kN m, $N = 15{,}817$ kN.	
	Corresponding additional load is	
	$F_{k,add} = 14.4 \times 12.0 \times 15.4 + 1.25 \times 1449 = 4473$ kN	
	Total co-existent design ultimate load $N_s = 15{,}817 + 4473 = 20{,}290$ kN	
	For wind acting in a direction normal to the long face of the building, maximum load on each pile in outermost row of piles:	
	$N_{p,max} = 20{,}290/20 + 8020 \times 3.6/129.6 = 1238$ kN	
	Design of pile cap	
	For the piles on the gridlines, the load from the walls will be transferred principally by deep beam action within the walls. For the remaining piles, load transference will occur by bending or truss action in the pile cap.	
	Minimum area of reinforcement required in each direction:	
9.2.1.1 (1)	$A_{s,min} = 0.26 \times (3.0/500) \times 1000 \times 900 = 1404$ mm^2/m (H25-300)	
	For wind acting in a direction normal to the long face of the building, considering the design ultimate loads without the additional load from the basement slab and pile cap, maximum load on each pile in second row of piles:	
	$N_{p,max} = 15{,}817/20 + 8020 \times 1.8/129.6 = 902$ kN	
	Maximum hogging moment in pile cap (at second pile in row);	
	$M_{Ed} = 902 \times 2.4 \times 2/9 = 481$ kN m	
	Area of reinforcement required in transverse tie between piles;	
	$A_s = 481 \times 10^6/(0.87 \times 500 \times 0.95 \times 900) = 1294$ mm^2 (4H25)	Provide H25-300 top and bottom in each direction
	Clearly, the provision of minimum reinforcement throughout will be sufficient.	
	BASEMENT FLOOR AND WALL	
	Analysis	
	The floor slab will be analysed as a series of rectangular strips in two directions at right angles. The width of an internal strip is taken between centrelines of adjacent panels. The width of an edge strip is taken from the centreline of the external wall to the centreline of the first panel. Load on an edge strip is effectively distributed by the basement wall, and there is no need to analyse the strip for structural effects in the direction parallel to the wall. The wall will be considered as a slab propped at the top and continuous at the bottom.	
	The effective span of the wall between ground floor and basement floor is 4.0 m. Two soil conditions will be considered: at-rest pressure with surcharge, and active pressure without surcharge. Ultimate fixed-end moments at bottom of wall are	
	$M_{w,max} = 1.35 \times 0.47 \times (10 \times 4.0^2/8 + 20 \times 4.0^3/15) = 67.0$ kN m/m	
	$M_{w,min} = 1.0 \times 0.30 \times 20 \times 4.0^3/15 = 25.6$ kN m/m	

Example 2   **Calculation Sheet 27**

Reference	CALCULATIONS	OUTPUT
	STRIP ON LINE B	

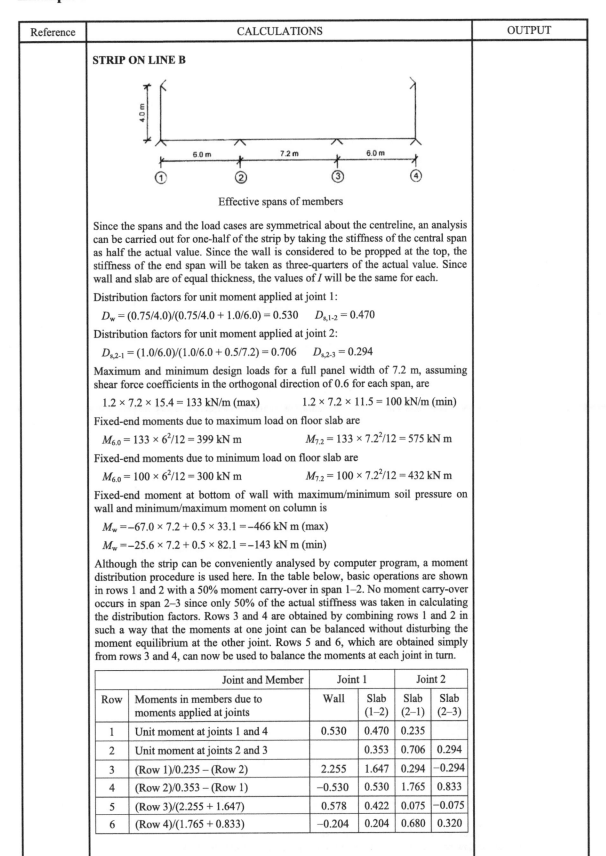

Effective spans of members

Since the spans and the load cases are symmetrical about the centreline, an analysis can be carried out for one-half of the strip by taking the stiffness of the central span as half the actual value. Since the wall is considered to be propped at the top, the stiffness of the end span will be taken as three-quarters of the actual value. Since wall and slab are of equal thickness, the values of I will be the same for each.

Distribution factors for unit moment applied at joint 1:

$D_w = (0.75/4.0)/(0.75/4.0 + 1.0/6.0) = 0.530$ $D_{s,1\text{-}2} = 0.470$

Distribution factors for unit moment applied at joint 2:

$D_{s,2\text{-}1} = (1.0/6.0)/(1.0/6.0 + 0.5/7.2) = 0.706$ $D_{s,2\text{-}3} = 0.294$

Maximum and minimum design loads for a full panel width of 7.2 m, assuming shear force coefficients in the orthogonal direction of 0.6 for each span, are

$1.2 \times 7.2 \times 15.4 = 133$ kN/m (max) $1.2 \times 7.2 \times 11.5 = 100$ kN/m (min)

Fixed-end moments due to maximum load on floor slab are

$M_{6.0} = 133 \times 6^2/12 = 399$ kN m $M_{7.2} = 133 \times 7.2^2/12 = 575$ kN m

Fixed-end moments due to minimum load on floor slab are

$M_{6.0} = 100 \times 6^2/12 = 300$ kN m $M_{7.2} = 100 \times 7.2^2/12 = 432$ kN m

Fixed-end moment at bottom of wall with maximum/minimum soil pressure on wall and minimum/maximum moment on column is

$M_w = -67.0 \times 7.2 + 0.5 \times 33.1 = -466$ kN m (max)

$M_w = -25.6 \times 7.2 + 0.5 \times 82.1 = -143$ kN m (min)

Although the strip can be conveniently analysed by computer program, a moment distribution procedure is used here. In the table below, basic operations are shown in rows 1 and 2 with a 50% moment carry-over in span 1–2. No moment carry-over occurs in span 2–3 since only 50% of the actual stiffness was taken in calculating the distribution factors. Rows 3 and 4 are obtained by combining rows 1 and 2 in such a way that the moments at one joint can be balanced without disturbing the moment equilibrium at the other joint. Rows 5 and 6, which are obtained simply from rows 3 and 4, can now be used to balance the moments at each joint in turn.

	Joint and Member	Joint 1		Joint 2	
Row	Moments in members due to moments applied at joints	Wall	Slab (1–2)	Slab (2–1)	Slab (2–3)
1	Unit moment at joints 1 and 4	0.530	0.470	0.235	
2	Unit moment at joints 2 and 3		0.353	0.706	0.294
3	(Row 1)/0.235 − (Row 2)	2.255	1.647	0.294	−0.294
4	(Row 2)/0.353 − (Row 1)	−0.530	0.530	1.765	0.833
5	(Row 3)/(2.255 + 1.647)	0.578	0.422	0.075	−0.075
6	(Row 4)/(1.765 + 0.833)	−0.204	0.204	0.680	0.320

Example 2   **Calculation Sheet 28**

Reference	CALCULATIONS	OUTPUT
	For each load case, fixed-end moments are determined for each span. The algebraic sum of the moments either side of each internal support can now be balanced and the resulting values obtained as shown in the following table.	

Load Case	Moments (kN m) at Ends of Members for Load Case	Bottom of Wall	Support 1	Support 2 LH	Support 2 RH
1	Maximum pressure on wall and maximum load on all spans of floor				
	Fixed-end moments	−466	−399	399	−575
	(Row 5) × 865	500	365	65	−65
	(Row 6) × 176	−36	36	120	56
	Sum to obtain final moments	−2	2	584	−584
2	As case 1 but minimum load on interior span of floor				
	Fixed-end moments	−466	−399	399	−432
	(Row 5) × 865	500	365	65	−65
	(Row 6) × 33	−7	7	22	11
	Sum to obtain final moments	27	−27	486	−486
3	As case 1 but minimum load on end spans of floor				
	Fixed-end moments	−466	−300	300	−575
	(Row 5) × 766	443	323	58	−58
	(Row 6) × 275	−56	56	187	88
	Sum to obtain final moments	−79	79	545	−545
4	Minimum pressure on wall and maximum load on all spans of floor				
	Fixed-end moments	−143	−399	399	−575
	(Row 5) × 542	313	229	40	−40
	(Row 6) × 176	−36	36	120	56
	Sum to obtain final moments	134	−134	559	−559
5	As case 4 but minimum load on interior span of floor				
	Fixed-end moments	−143	−399	399	−432
	(Row 5) × 542	313	229	40	−40
	(Row 6) × 33	−7	7	22	11
	Sum to obtain final moments	163	−163	461	−461
6	As case 4 but minimum load on end spans of floor				
	Fixed-end moments	−143	−300	300	−575
	(Row 5) × 443	256	187	33	−33
	(Row 6) × 275	−56	56	187	88
	Sum to obtain final moments	57	−57	520	−520

Shear forces at the ends of each span and the maximum sagging moment in the span can now be determined, and the results are given in the following table.

Example 2   **Calculation Sheet 29**

Reference	CALCULATIONS	OUTPUT

Maximum Moments and Shear Forces for Load Cases

Load Case	Moment (kN m) at location				Shear (kN) at location		
	Support 1	Span 1–2	Support 2	Span 2–3	Support 1	Support 2 (LH)	Support 2 (RH)
1	2	343	**− 584**	278	301	**497**	479
2	−27	**363**	−486	162	322	476	360
3	**79**	271	−545	317	196	404	479
4	−134	271	−559	303	328	470	479
5	**−163**	295	−461	187	**349**	449	360
6	−57	192	−520	**342**	223	377	479

5.3.2.2 (4) — The hogging moment at support 2 may be reduced by $\Delta M_{Ed} = F_{Ed,sup}(t/8)$, where $F_{Ed,sup}$ is the design support reaction and t is the breadth of the support.

STRIP ON LINE 2

For the purpose of analysis, the floor slab will be considered fully fixed at line C.

Effective spans of members

Distribution factors for unit moment applied at joint A:

$$D_w = (0.75/4.0)/(0.75/4.0 + 1.0/7.2) = 0.574, \qquad D_{s,A-B} = 0.426$$

Distribution factors for unit moment applied at joint B: $D_{s,B-A} = D_{s,B-C} = 0.500$

Assuming shear force coefficients for the spans in the orthogonal direction of 0.60 for the end span and 0.5 for the interior span, the loaded width for the strip on line 2 is $(0.6 \times 6.0 + 0.5 \times 7.2) = 7.2$ m. Thus, maximum and minimum loads for the full panel width are

$$7.2 \times 15.4 = 111 \text{ kN/m (max)}, \qquad 7.2 \times 11.5 = 83 \text{ kN/m (min)}$$

Fixed-end moments due to maximum and minimum loads on floor slab are

$$M_{max} = 111 \times 7.2^2/12 = 480 \text{ kN m}, \qquad M_{min} = 83 \times 7.2^2/12 = 359 \text{ kN m}$$

Fixed-end moment at bottom of wall with maximum/minimum soil pressure on wall and minimum/maximum moment on column is

$$M_w = -67.0 \times 7.2 + 0.5 \times 53.1 \ = -456 \text{ kN m (max)}$$

$$M_w = -25.6 \times 7.2 + 0.5 \times 113.4 = -128 \text{ kN m (min)}$$

	Joint and Member	Joint A		Joint B		Joint C
Row	Moments in members due to moments applied at joints	Wall	Slab (A–B)	Slab (B–A)	Slab (B–C)	Slab (C–B)
1	Unit moment at joint A	0.574	0.426	0.213		
2	Unit moment at joint B		0.250	0.500	0.500	0.250
3	(Row 1)/0.213 − (Row 2)	2.695	1.750	0.500	−0.500	−0.250
4	(Row 2)/0.250 − (Row 1)	−0.574	0.574	1.787	2.000	1.000
5	(Row 3)/(2.695 + 1.750)	0.606	0.394	0.112	−0.112	−0.056
6	(Row 4)/(1.787 + 2.000)	−0.152	0.152	0.472	0.528	0.264

Example 2   **Calculation Sheet 30**

Reference	CALCULATIONS	OUTPUT

For each load case, fixed-end moments are determined for each span. The algebraic sum of the moments either side of each internal support can now be balanced and the resulting values obtained as shown in the following table.

Load Case	Moments (kN m) at Ends of Members for Load Case	Bottom of Wall	Support A	Support B LH	Support B RH	Support C LH
1	Maximum pressure on wall and maximum load on both spans of floor					
	Fixed-end moments	−456	−480	480	−480	480
	(Row 5) × 936	567	369	105	−105	−53
	Sum to obtain final moments	111	−111	585	−585	427
2	As case 1 but minimum load on span B–C of floor					
	Fixed-end moments	−456	−480	480	−359	359
	(Row 5) × 936	567	369	105	−105	−53
	(Row 6) × −121	19	−19	−57	−64	−32
	Sum to obtain final moments	130	−130	528	−528	274
3	As case 1 but minimum load on span A–B of floor					
	Fixed-end moments	−456	−359	359	−480	480
	(Row 5) × 815	494	321	91	−91	−46
	(Row 6) × 121	−19	19	57	64	32
	Sum to obtain final moments	−19	−19	507	−507	466
4	Minimum pressure on wall and maximum load on both spans of floor					
	Fixed-end moments	−128	−480	480	−480	480
	(Row 5) × 608	368	240	68	−68	−34
	Sum to obtain final moments	240	−240	548	−548	446
5	As case 4 but minimum load on span B–C of floor					
	Fixed-end moments	−128	−480	480	−359	359
	(Row 5) × 608	368	240	68	−68	−34
	(Row 6) × −121	19	−19	−57	−64	−32
	Sum to obtain final moments	259	−259	491	−491	293
6	As case 4 but minimum load on span A–B of floor					
	Fixed-end moments	−128	−359	359	−480	480
	(Row 5) × 487	295	192	54	−54	−27
	(Row 6) × 121	−19	19	57	64	32
	Sum to obtain final moments	148	−148	470	−470	485

Shear forces at the ends of each span and the maximum sagging moment in the span can now be determined, and the results are given in the following table. The hogging moment at support B may be reduced by $\Delta M_{Ed} = F_{Ed,sup}(t/8)$, where $F_{Ed,sup}$ is the design support reaction and t is the breadth of the support.

5.3.2.2 (4)

Example 2

Calculation Sheet 31

Reference	CALCULATIONS	OUTPUT

Maximum Moments and Shear Forces for Load Cases

Load Case	Moment (kN m) at Location					Shear (kN)		At Location	
	Support A	Span A–B	Support B	Span B–C	Support C	Support A (RH)	Support B (LH)	Support B (RH)	Support C (LH)
1	−111	392	**−585**	217	−427	334	**465**	**422**	377
2	−130	**403**	−528	93	−274	344	455	321	277
3	**−19**	303	−507	232	−466	231	367	405	394
4	−240	334	−548	224	−446	357	442	414	385
5	**−259**	348	−491	149	−293	**367**	432	326	272
6	−148	241	−470	**240**	**−485**	254	344	397	**402**

DESIGN OF BASEMENT FLOOR

Flexural design

I.1.2 (3)
Figure I.1

Table I.1

The panels should be notionally divided into column and middle strips, and the bending moments for the full panel width apportioned within specified limits. The width of the column strips on line B and line 2 will be taken as $7200/2 = 3600$ mm. The hogging moments at the internal columns will be allocated in the proportions: 75% on column strips, 25% on middle strips. The sagging moments in the spans, and the hogging moments at the edges of the slab, will be distributed uniformly over the full panel width.

The maximum hogging moments for the panel strips intersecting at support B2, allowing for the reductions due to width of support, are as follows:

Strip on line 2: $M = 585 - (465 + 422) \times 3.6/8 = 186$ kN m

Strip on line B: $M = 584 - (497 + 479) \times 3.6/8 = 144$ kN m

Allowing 50 mm cover (bottom) and 25 mm cover (top) with 16 mm bars in each direction, values of the effective depth are as follows:

Strip on line 2: $d = 215$ mm (top), 190 mm (bottom)

Strip on line B: $d = 200$ mm (top), 175 mm (bottom)

Minimum area of longitudinal tension reinforcement for strip on line 2 (top):

9.3.1.1 (1)

$A_{s,min} = 0.26(f_{ctm}/f_{yk})b_t d = 0.26 \times (3.0/500) \times 1000 \times 215 = 336$ mm²/m

Maximum spacing of principal reinforcement in area of maximum moment:

9.3.1.1 (3)

$2h = 500 \leq 250$ mm.   Elsewhere: $3h = 750 \leq 400$ mm

For minimum reinforcement, provide H12-250 giving 452 mm²/m

According to the values of $M/bd^2 f_{ck}$, appropriate values of $z/d \leq 0.95$ and A_s can be determined (Table A1), and suitable bars can be selected (Table A9).

	Location	Strip	Width	M (kN m)	$M/bd^2 f_{ck}$	z/d	A_s	Bars
Strip on line 2	Support A	panel	7200	259	0.025	0.95	2915	H12-250T
	Span A–B	panel	7200	403	0.049	0.95	5133	H16-250B
	Support B	column	3600	139.5	0.026	0.95	1570	H12-250T
	ditto	middle	3600	46.5	0.009	0.95	524	H12-250T
	Span B–C	panel	7200	240	0.029	0.95	3057	H12-250B
	Support C	panel	7200	485	0.046	0.95	5459	H16-250T
Strip on line B	Support 1	panel	7200	163	0.018	0.95	1973	H12-250T
	Span 1–2	panel	7200	363	0.052	0.95	5020	H16-250B
	Support 2	column	3600	108	0.024	0.95	1307	H12-250T
	ditto	middle	3600	36	0.008	0.95	436	H12-250T
	Span 2–3	panel	7200	342	0.049	0.95	4729	H16-250B

Example 2   **Calculation Sheet 32**

Reference	CALCULATIONS	OUTPUT
	Shear design	
	The critical shear stress occurs at a control perimeter at distance $2d$ from the edge of the pile cap supporting column B2. Mean effective depth for reinforcement at the top of the slab, with 12 mm bars each way, is approximately $d = 210$ mm.	
	The maximum support reaction is obtained for the strip on line B with load case 1 and, allowing for the reduction due to the area of slab supported directly by the pile cap, the maximum shear force $V_{Ed} = (497 + 479) - 133 \times 3.6^2 = 748$ kN.	
	Length of the control perimeter, where b is the side of the pile cap, is	
	$u_1 = 4(b + \pi d) = 4 \times (3600 + \pi \times 210) = 17000$ mm say	
	Average shear stress at the control perimeter is	
	$v_{Ed} = V_{Ed}/(u_1 d) = 748 \times 10^3/(17000 \times 210) = 0.21$ MPa	
	The design shear strength is given by	
6.2.2 (1) Table NA.1	$v_c = 0.12k(100A_{sl}f_{ck}/b_w d)^{1/3} \geq v_{min}$	
	The minimum shear strength, with $k = 1 + (200/d)^{1/2} = 1.97$, is given by	
	$v_{min} = 0.035k^{3/2}f_{ck}^{1/2} = 0.035 \times 1.97^{3/2} \times 32^{1/2} = 0.54$ MPa $(>v_{Ed})$	Shear satisfactory
	Deflection	
7.4.1 (6)	Deflection requirements may be met by limiting the span-effective depth ratio. For the strip on line B, $d = 175$ mm (using 16 mm bars in both directions) and the maximum span/effective depth ratio $= 7200/175 = 41.2$.	
	The design loads are: 100 kN/m (characteristic) and 133 kN/m (ultimate)	
	The maximum service stress in the bottom reinforcement for the 7.2 m span under the characteristic load is given approximately by	
	$\sigma_s = (f_{yk}/\gamma_s)(A_{s,req}/A_{s,prov})[(g_k + q_k)/n]$	
	$= (500/1.15)(4729/5789)(100/133) = 267$ MPa	
	From *Reynolds*, Table 4.21, limiting l/d = basic ratio $\times \alpha_s \times \beta_s$ where:	
	For $100A_s/bd = 100 \times 4729/(7200 \times 175) = 0.38 < 0.1f_{ck}^{0.5} = 0.1 \times 32^{0.5} = 0.56$,	
	$\alpha_s = 0.55 + 0.0075f_{ck}/(100A_s/bd) + 0.005f_{ck}^{0.5}[f_{ck}^{0.5}/(100A_s/bd) - 10]^{1.5}$	
	$= 0.55 + 0.0075 \times 32/0.38 + 0.005 \times 32^{0.5} \times (32^{0.5}/0.38 - 10)^{1.5} = 1.49$	
	$\beta_s = 310/\sigma_s = 310/267 = 1.16$	
7.4.2 Table NA.5	For a flat slab, basic ratio = 24. Since the span does not exceed 8.5 m, there is no need to modify this value and hence	
	Limiting $l/d = 24 \times \alpha_s \times \beta_s = 24 \times 1.49 \times 1.16 = 41.5$ (>actual $l/d = 41.2$)	Check complies
	Cracking	
7.3.2 (2)	Minimum area of reinforcement required in tension zone for crack control:	
	$A_{s,min} = k_c k f_{ct,eff} A_{ct}/\sigma_s$	
	Taking values of $k_c = 0.4$, $k = 1.0$, $f_{ct,eff} = f_{ctm} = 0.3f_{ck}^{(2/3)} = 3.0$ MPa (for general design purposes), $A_{ct} = bh/2$ (for plain concrete section) and $\sigma_s \leq f_{yk} = 500$ MPa	
	$A_{s,min} = 0.4 \times 1.0 \times 3.0 \times 1000 \times (250/2)/500 = 300$ mm^2/m (H12-250)	
	It is reasonable to ignore any crack width requirement based on appearance, since the bottom surface is in contact with the tanking layer, and the top surface will be hidden by the finishes.	
	Curtailment of longitudinal tension reinforcement	
	In the absence of an elastic moment envelope covering all appropriate load cases, the simplified curtailment rules for one-way continuous slabs will be used in each orthogonal direction.	

Example 2   **Calculation Sheet 33**

Reference	CALCULATIONS	OUTPUT
	DESIGN OF BASEMENT WALL	
	Durability	
BS 8500	Since the external surface of the wall is protected by a continuous barrier system, it is reasonable to consider exposure class XC1 for both surfaces of the wall.	
	$c_{min} = 15$ mm $\quad \Delta c_{dev} = 10$ mm, $\qquad c_{nom} = 15 + 10 = 25$ mm	
	Flexural design	
	Allowing for 25 mm cover and 12 mm diameter horizontal bars in the outer layers, for 16 mm diameter vertical bars, $d = 250 - (25 + 12 + 16/2) = 205$ mm.	
	Minimum area of vertical tension reinforcement:	
9.3.1.1 (1)	$A_{s,min} = 0.26 \times (3.0/500) \times 1000 \times 205 = 320$ mm^2/m (H12-250 say)	
	Minimum area of reinforcement required in tension zone for crack control:	
7.3.2 (2)	$A_{s,min} = k_c k f_{ct,eff} A_{ct} / \sigma_s \quad$ where	
	$k_c = 0.4$ for bending, $k = 1.0$ for $h \leq 300$ mm, $f_{ct,eff} = f_{ctm} = 0.3 f_{ck}^{(2/3)} = 3.0$ MPa for general purposes, $A_{ct} = bh/2$ and $\sigma_s \leq f_{ck} = 500$ MPa.	
	$A_{s,min} = 0.4 \times 1.0 \times 3.0 \times 1000 \times 125/500 = 300$ mm^2/m (H12-250 say)	
	Moment of resistance provided by minimum reinforcement is given by	
	$M = A_s (0.87 f_{yk}) z = 452 \times 0.87 \times 500 \times 0.95 \times 205 \times 10^{-6} = 38$ kN m/m	
	For the wall on line 1, from the calculations for the slab strip on line B, maximum sagging/hogging moments at the junction of wall and floor slab are as follows:	
	(Case 5): $M_1 = 163/7.2 = 23$ kN m/m $\quad$ (Case 3): $M_1 = -79/7.2 = -11$ kN m/m	
	The provision of minimum reinforcement (H12-250) is sufficient in both cases.	2.3 m
	The maximum sagging moment at any height in the wall on line 1 occurs for load case 2, where sagging moment at the bottom of the wall, $M_A = 27/7.2 = 4$ kN m/m. In this case, shear force at top of wall is	67 kNm
	$V_{Ed} = 1.35 \times 0.47 \times (10 \times 4.0/2 + 20 \times 4.0^2/6) + 4/4.0 = 47.5$ kN/m	
	If a_o is distance from top of wall to point of zero shear, then	4 kNm
	$V_{Ed} - 1.35 \times 0.47 \times (10a_o + 20a_o^2/2) = 0$ which gives $a_o = 2.3$ m	Wall on line 1
	Hence, maximum sagging moment (at $a_o = 2.3$ m) is	
	$M_{Ed} = V_{Ed} \times a_o - 1.35 \times 0.47 \times (10a_o^2/2 + 20a_o^3/6) = 67$ kN m/m	
	$M/bd^2 f_{ck} = 67 \times 10^6/(1000 \times 205^2 \times 32) = 0.050 \qquad z/d = 0.95$ (max)	
	$A_s = 67 \times 10^6/(0.87 \times 500 \times 0.95 \times 205) = 791$ mm^2/m (H12-125)	
	For the wall on line A, from the calculations for the slab strip on line 2, maximum sagging moment at the junction of wall and floor slab is as follows:	
	(Case 5): $M_A = 259/7.2 = 36.0$ kN m/m.	2.4 m
	The provision of minimum reinforcement (H12-250) is sufficient.	
	Maximum sagging moment at any height in the wall on line A occurs for load case 2, where sagging moment at the bottom of the wall, $M_A = 130/7.2 = 18$ kN m/m. In this case, shear force at top of wall is	75 kNm
	$V_{Ed} = 1.35 \times 0.47 \times (10 \times 4.0/2 + 20 \times 4.0^2/6) + 18/4.0 = 51$ kN/m	
	If a_o is distance from the top of the wall to the point of zero shear, then	18 kNm
	$V_{Ed} - 1.35 \times 0.47 \times (10a_o + 20a_o^2/2) = 0$ which gives $a_o = 2.4$ m	Wall on line A
	Hence, maximum sagging moment (at $a_o = 2.4$ m) is	
	$M_{Ed} = V_{Ed} \times a_o - 1.35 \times 0.47 \times (10a_o^2/2 + 20a_o^3/6) = 75$ kN m/m	Vertical reinforcement H12-250 at outer face H12-125 at inner face
	$M/bd^2 f_{ck} = 75 \times 10^6/(1000 \times 205^2 \times 32) = 0.056 \qquad z/d = 0.948$	
	$A_s = 75 \times 10^6/(0.87 \times 500 \times 0.948 \times 205) = 888$ mm^2/m (H12-125)	

Example 2   **Calculation Sheet 34**

Reference	CALCULATIONS	OUTPUT
	Shear design	
	Maximum shear force occurs at bottom of wall on line 1 where, for the slab strip on line B (case 3):	
	$V_{Ed} = 1.35 \times 0.47 \times (10 \times 4.0/2 + 20 \times 4.0^2/3) + 11/4.0 = 83.2$ kN/m	
	$v_{Ed} = V_{Ed}/(bd) = 83.2 \times 10^3/(1000 \times 200) = 0.42$ MPa	
	Minimum design shear strength, where $k = 2.0$ for $d \leq 200$ mm, is	
6.2.2 (1)	$v_{min} = 0.035k^{3/2}f_{ck}^{1/2} = 0.035 \times 2.0^{3/2} \times 32^{1/2} = 0.56$ MPa $(>v_{Ed})$	
	Cracking due to restrained early thermal contraction	
	Minimum area of horizontal reinforcement, with $f_{ct,eff} = 1.8$ MPa for cracking at age of 3 days, $k_c = 1.0$ for tension and $k = 1.0$ for $h \leq 300$ mm, is given by	
7.3.2 (2)	$A_{s,min} = k_c k f_{ct,eff} A_{ct} / f_{yk} = 1.0 \times 1.0 \times 1.8 \times 1000 \times 125/500 = 450$ mm^2/m (EF)	Horizontal reinforcement H12-200 (EF)
7.3.4	With $c = 35$ mm, $k_1 = 0.8$ for high bond bars, $k_2 = 1.0$ for tension, $h_{c,ef}$ as the lesser of $2.5(h - d)$ and $h/2$, and H12-200 (EF) as minimum reinforcement:	
	$s_{r,max} = 3.4c + 0.425k_1k_2(A_{c,eff}/A_s)\phi$	
	$= 3.4 \times 25 + 0.425 \times 0.8 \times 1.0 \times (2.5 \times 31 \times 1000/565) \times 12 = 645$ mm	
PD 6687 2.16	With $R = 0.8$ for wall on a thick base, $\Delta T = 25°$C for 350 kg/m^3 Portland cement concrete and 250 mm thick wall (*Reynolds*, Table 2.18), and $\alpha = 12 \times 10^{-6}$ per $°$C:	
	$w_k = (0.8R\alpha\Delta T) \times s_{r,max} = 0.8 \times 0.8 \times 12 \times 10^{-6} \times 25 \times 645 = 0.13$ mm	
	Wall as deep beam	
	Provision of H12-200 (EF) horizontally and H12-250 (EF) vertically as minimum reinforcement meets the deep beam requirements, as shown on calculation sheets 21 and 22.	

Bar Marks	Commentary on Bar Arrangement (Drawings 12–13)
01	Bars (shape code 21), supported on tops of piles, with 75 mm cover bottom and ends.
02	Column starter bars (shape code 11) standing on mat formed by bars 01. Height of upstand needed on top of pile cap to allow for 150 mm concrete overlay, 25 mm asphalt tanking, 250 mm floor slab and 75 mm kicker $= 150 + 25 + 250 + 75 = 500$ mm. Projection of bars above top of pile cap to provide lap length above top of kicker $= 1.5 \times 35 \times 32 + 500 = 2200$ mm. Cover to bars in column $= 50$ mm to enable 35 mm cover to links.
03	Closed links (shape code 51) to hold column starter bars in place and reinforce kicker.
04, 05	Bars (shape code 21) with 75 mm cover bottom and ends, and 50 mm cover top.
06, 07	Bars (shape codes 21, 11) lapping with bars 08. Lap length $= 1.5 \times 50 \times 25 \times 1404/1473 = 1800$ mm.
08, 09, 10	Straight bars (maximum length 12 m) with 1800 mm laps.
11	Closed links (shape code 51) with 75 mm cover bottom and sides, and 50 mm cover top.

Example 2: Reinforcement in Pile Caps to Internal Columns and Walls   Drawing 12

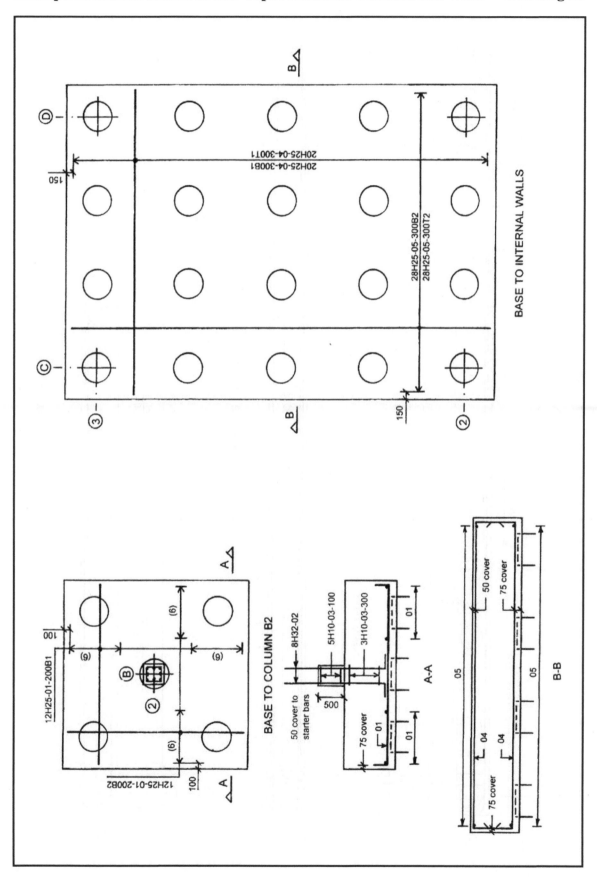

Example 2: Reinforcement in Pile Beams to External Walls   **Drawing 13**

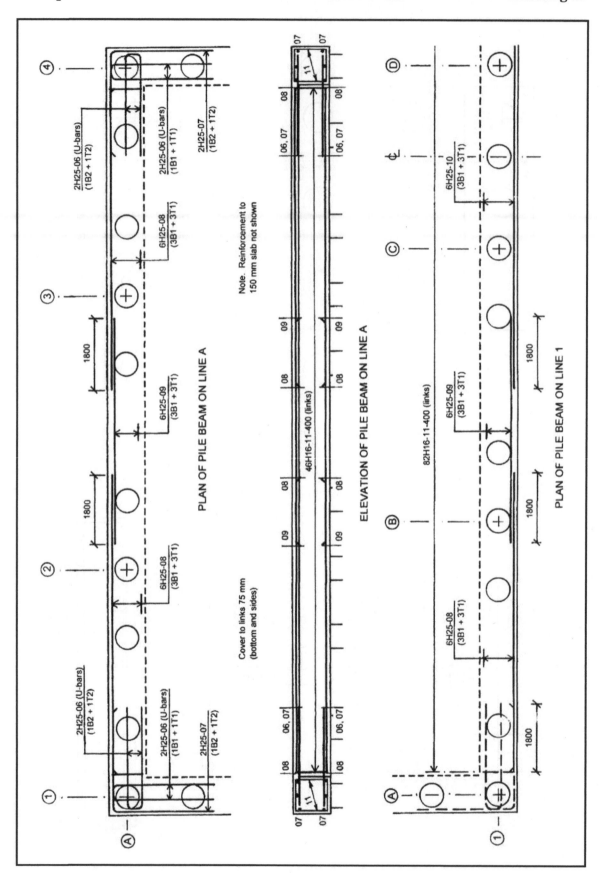

Example 2: Reinforcement in Basement Floor Slab (1)   **Drawing 14**

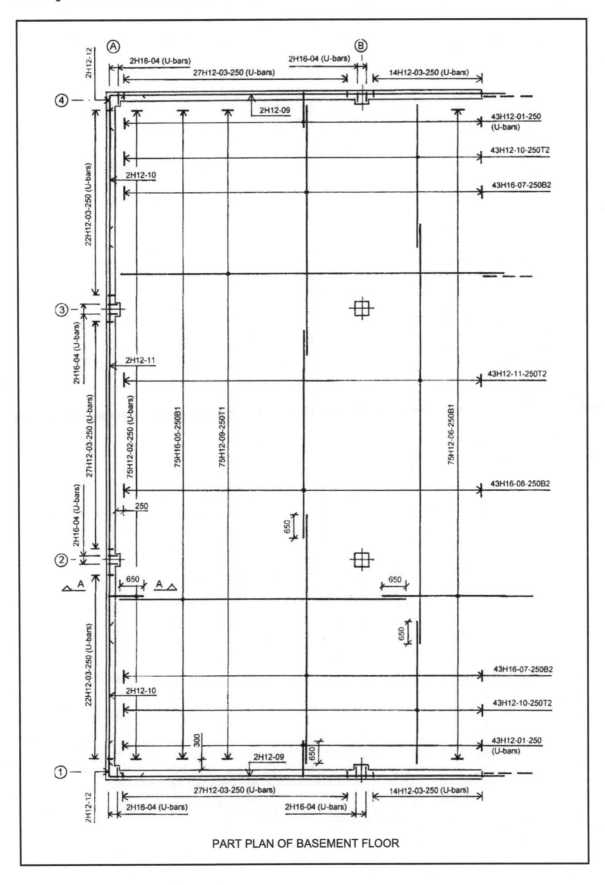

PART PLAN OF BASEMENT FLOOR

Example 2: Reinforcement in Basement Floor Slab (2)   **Drawing 15**

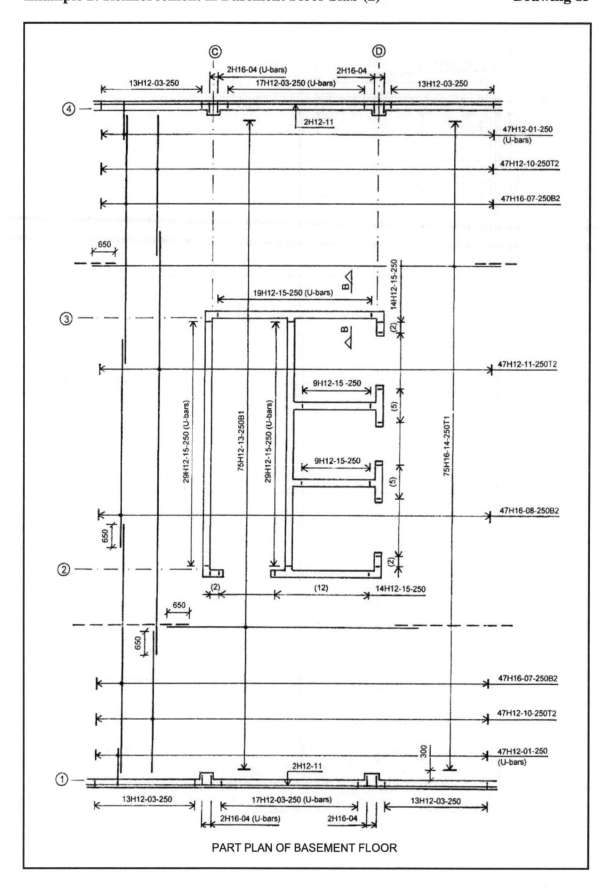

PART PLAN OF BASEMENT FLOOR

Example 2: Reinforcement in Basement Walls   Drawing 16

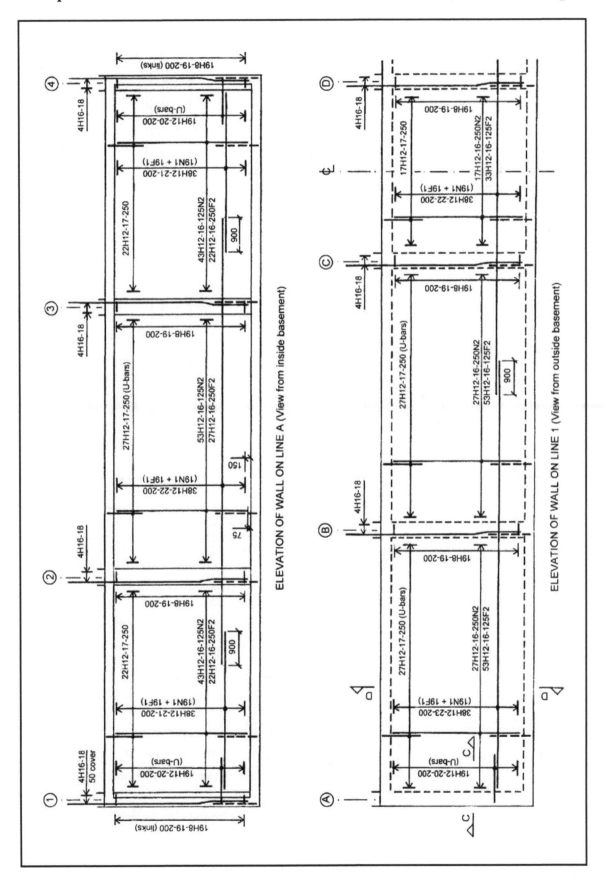

Example 2: Cross-Sections for Basement Floor Slab and Walls   **Drawing 17**

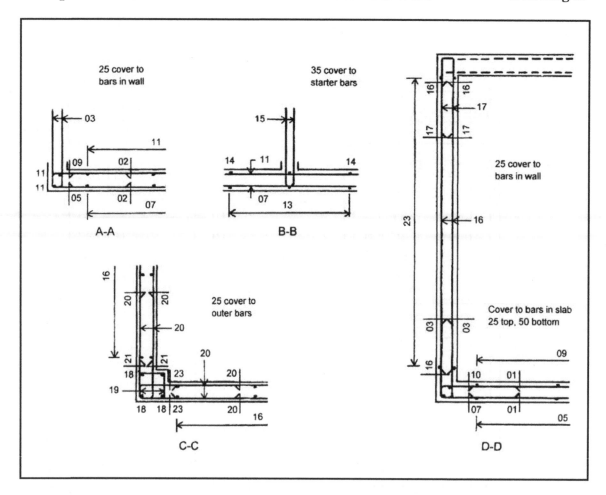

Bar Marks	Commentary on Bar Arrangement (Drawings 14–17)
01, 02	Bars (shape code 21) lapping with bars 07 and 10, and 05 and 09. Lap length = $1.5 \times 35 \times 12 = 650$ mm say. Cover = 50 mm (bottom) and 25 mm (top and ends).
03, 04	Starter bars (shape code 21) to external walls and columns. Projection above basement floor to provide a lap length above kicker = $1.5 \times 35\phi + 75 = 725$ mm for bar 03, and 925 mm for bar 04.
05, 06, 07 08, 09, 10 11, 13, 14	Straight bars (maximum length 12 m) in floor slab, with lap lengths = $1.5 \times 35 \times 12 = 650$ mm.
12	Bars (shape code 11) lapping with bars 09 and 10. Lap length = 650 mm.
15	Starter bars (shape code 21) to internal walls. Projection above basement floor = 725 mm. Cover to bars in wall = 35 mm to enable 25 mm cover to horizontal bars.
16	Straight bars bearing on 75 mm kicker, and curtailed 50 mm below ground floor slab.
17, 20	Bars (shape code 21). Lap length = 650 mm for bar 17, and $1.5 \times 50 \times 12 = 900$ mm for bar 20.
18	Bars (shape code 26) bearing on 75 mm kicker and cranked to fit alongside bars 04. Projection of bars above ground floor level = $1.5 \times 35 \times 16 + 75 = 925$ mm.
19	Closed links (shape code 51), with 25 mm nominal cover, starting above kicker and stopping below ground floor slab.
21, 22, 23	Straight bars (maximum length 12 m) in external walls, with lap lengths = $1.5 \times 50 \times 12 = 900$ mm.

5 Example 3: Free-Standing Cantilever Earth-Retaining Wall

Description

A retaining wall on a spread base is required to support level ground and a footpath adjacent to a road. The existing ground may be excavated as necessary to construct the wall, and the excavated ground behind the wall is to be reinstated by backfilling with a granular material. A graded drainage material will be provided behind the wall, with an adequate drainage system at the bottom. The fill to be retained is 4.0 m high above the top of the base and the surcharge is 5.0 kN/m². For the sub-base, two soil types will be considered: (1) non-cohesive soil, (2) cohesive soil.

Suitable dimensions for the base to a cantilever retaining wall on a spread base can be estimated by means of the following design chart. Stability against overturning is assured over the entire range of the chart, and the maximum bearing pressure under service conditions can be investigated for all types of soil. A uniform surcharge that is small compared to the total forces acting on the wall can be represented by an equivalent height of soil. In this case, l is replaced by $l_e = l + q/\gamma$, where q is the surcharge pressure. In more general cases, $l_e = 3M_h/F_h$ and $\gamma = 2F_h/K_a l_e^2$ can be used, where F_h is the total horizontal force and M_h is the bending moment about the underside of the base due to F_h.

The chart contains two curves denoting conditions where the bearing pressure diagram is uniform, and triangular (reaching zero pressure at the heel), respectively. A uniform bearing condition is important when it is important to avoid tilting, to minimise deflection at the top of the wall. It is generally advisable to maintain ground contact over the full area of the base, especially for clays where the occurrence of ground water beneath the heel could soften the formation.

Values of the dimensional parameters α and β, for the base of an idealised wall of zero thickness as shown below, and for values of $\xi = p_{max}/\gamma l$ and $\psi = \tan \delta_d/\sqrt{K_a}$, where γ is unit weight of soil, δ_d is angle of base friction, and K_a is fully active earth pressure coefficient, can be obtained from the adjoining chart.

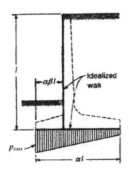

(a) $p_{max} = \dfrac{4F_v}{3(\alpha l - 2e)}$ for $e \geq \dfrac{\alpha l}{6}$

(b) $p_{max} = \dfrac{F_v}{\alpha l}\left(1+\dfrac{6e}{\alpha l}\right)$ for $e \leq \dfrac{\alpha l}{6}$

$e = M_h/F_v + \alpha l/2 - x \leq \alpha l/6$

F_v is resultant of vertical loads including weight of wall, and weight of earth and surcharge on base

M_h is bending moment about underside of base due to horizontal forces acting on full height of wall

x is distance from toe of base to line of action of F_v

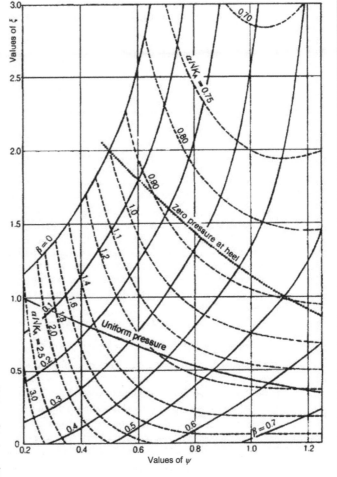

Chart for cantilever wall on spread base

Example 3   **Calculation Sheet 1**

Reference	CALCULATIONS	OUTPUT
	DESIGN PRINCIPLES	

In Eurocode 7, for conventional structures, two combinations of partial factors for actions and soil parameters are considered for the ULS as follows:

Partial Safety Factors for the Ultimate Limit State					
Combination	Safety Factor on Actions[a] γ_F		Safety Factor for Soil Parameters γ_M		
	γ_G	γ_Q	$\gamma_{\varphi'}$	$\gamma_{c'}$	γ_{cu}
1	1.35	1.5	1.0	1.0	1.0
2	1.0	1.3	1.25	1.25	1.4

[a]If the action is favourable, values of $\gamma_G = 1.0$ and $\gamma_Q = 0$ should be used.

Generally, combination 2 determines the size of the structure and combination 1 governs the structural design of the members. Characteristic soil parameters are defined as cautious estimates of the values affecting the occurrence of a limit state.

Thus, for combination 2, design values for soil strength at ULS are given by

$$\tan \phi_d = (\tan \phi)/1.25 \quad \text{and} \quad c'_d = c'/1.25 \qquad \text{where}$$

c' and ϕ are characteristic values of cohesion intercept and angle of shearing resistance (in terms of effective stress), respectively.

Design values for shear resistance at the interface of the base and the sub-soil, for drained (friction) and undrained (adhesion) conditions, respectively, are given by:

$$\tan \delta_d = \tan \phi_d \text{ (cast in-situ concrete)} \quad \text{and} \quad c_{ud} = c_u/1.4 \qquad \text{where}$$

c_u is the undrained shear strength

Walls should be checked for the ULS with regard to overall stability, bearing resistance and sliding. For eccentric loading, bearing pressure is assumed to be uniformly distributed, with the centre of pressure coincident with the line of action of the applied load. The resistance should be checked for both long-term (drained) and short-term (undrained) conditions where appropriate.

The traditional practice of considering characteristic actions and allowable bearing pressures in order to limit ground deformation, and check the bearing resistance, may also be adopted by mutual agreement. With this approach, a linear variation of bearing pressure is assumed for eccentric loading. The ULS still needs to be considered to check sliding and for the structural design.

The partial safety factors for the SLS are given as unity, but it is often prudent to use the ULS values for the active force. In this case, suitable dimensions for the wall can be estimated with the aid of the design chart on the preceding page. Here, the value p_{max} is for a linear variation of bearing pressure and, for a uniform distribution that is coincident with the line of action of the applied load, the contact length is equal to $\delta(\alpha l)$, where δ depends on whether the solution is (a) above, or (b) below, the curve for 'zero pressure at heel' shown on the design chart, as follows:

(a) $\delta = 4(1 - \beta)/3\xi \le 2/3$ with $p = 0.75p_{max}$

(b) $\delta = 4/3 - \xi/3(1 - \beta) > 2/3$ with $p = (1 - \beta)\gamma l/\delta$

For sliding, the chart applies directly to non-cohesive soils. For bases on clay, the long-term condition can be investigated by using ϕ' with $c' = 0$. For the short-term condition, the sliding criterion results in $\alpha = K_a \gamma l/2\delta c_{ud}$, where a minimum value is obtained when $\delta = 1.0$. If the short-term condition is critical, a trial value for δ can be assumed to calculate an initial value for α, for which a corresponding value of β can be obtained from the design chart. The resulting value of δ can then be compared to the assumed value, and the process repeated until parity is obtained.

Ideally, the settlement of spread foundations on clay soils should be checked by calculation but may be taken as satisfactory, in the case of firm-to-stiff clays, if the ratio of design ultimate bearing resistance to service load is at least 3.

Example 3 **Calculation Sheet 2**

Reference	CALCULATIONS	OUTPUT

SOIL PARAMETERS

Properties of the retained soil (well-graded sand and gravel) are as follows:

Unit weight $\gamma = 20$ kN/m^3

Angle of shearing resistance:

$\phi' = 35^\circ \quad \phi'_d = \tan^{-1}[(\tan 35^\circ)/1.25] = 29^\circ$

Coefficient of active earth pressure:

$K_a = (1 - \sin\phi')/(1 + \sin\phi') = 0.27 \quad K_{ad} = (1 - \sin\phi'_d)/(1 + \sin\phi'_d) = 0.35$

Properties of the sub-base soil are as follows:

(1) Medium dense sand:

$\phi' = 30^\circ \quad \phi'_d = \tan^{-1}[(\tan 30^\circ)/1.25] = 25^\circ \qquad \tan\delta_d = \tan\phi'_d = 0.46$

Allowable bearing value: $p_{ba} = 150$ kN/m^2 (kPa)

(2) Firm clay:

$c_u = 35$ kN/m$^2 \quad c_{ud} = c_u/1.4 = 35/1.4 = 25$ kN/m^2

Allowable bearing value: $p_{ba} = 100$ kN/m^2 (kPa)

$\phi' = 25^\circ \quad \phi'_d = \tan^{-1}[(\tan 25^\circ)/1.25] = 20.5^\circ \qquad \tan\delta_d = \tan\phi'_d = 0.37$

WALL DIMENSIONS (Trial values)

Trial dimensions of retaining walls

Taking thickness of wall stem and base as (height of fill)/10 = 4000/10 = 400 mm, height of wall to underside of base is $l = 4.0 + 0.4 = 4.4$ m.

Allowing for surcharge, equivalent height of wall is as follows:

$l_e = l + q/\gamma = 4.4 + 5.0/20 = 4.65$ m.

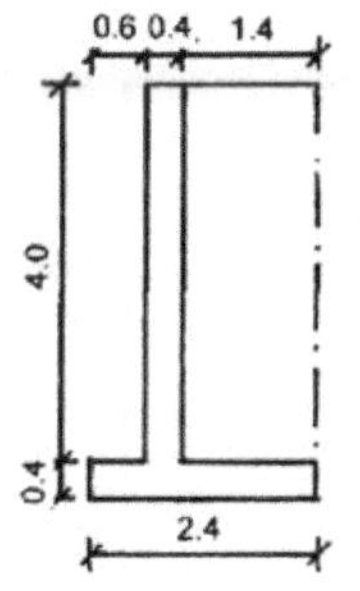

Base on sand

Values of chart parameters and resulting base dimensions, for each sub-base, are as follows:

(1) $\xi = p_{max}/\gamma l_e = 150/(20 \times 4.65) = 1.61 \quad \psi = \tan\delta_d/\sqrt{K_{ad}} = 0.46/\sqrt{0.35} = 0.78$

$\alpha/\sqrt{K_{ad}} = 0.86 \quad \alpha = 0.86\sqrt{0.35} = 0.51 \quad \beta = 0.24$

Base width $= \alpha l_e = 0.51 \times 4.65 = 2.4$ m

Toe length $= \beta(\alpha l_e) = 0.24 \times 2.4 = 0.6$ m

(2) Long-term (drained) condition:

$\xi = p_{max}/\gamma l_e = 100/(20 \times 4.65) = 1.08 \quad \psi = \tan\delta_d/\sqrt{K_{ad}} = 0.37/\sqrt{0.35} = 0.63$

$\alpha/\sqrt{K_{ad}} = 1.07 \quad \alpha = 1.07\sqrt{0.35} = 0.63 \quad \beta = 0.25$

Base width $= 0.63 \times 4.65 = 3.0$ m Toe length $= 0.25 \times 3.0 = 0.75$ m

Short-term (undrained) condition:

Contact length to satisfy the sliding criterion is given by the equation:

$K_{ad}\gamma l_e^2/2c_d = 0.35 \times 20 \times 4.65^2/(2 \times 25) = 3.03$ m

The contact length for a uniform bearing pressure distribution coincident with the line of action of the applied load is given by $\delta(\alpha l_e)$. For solutions that are below the curve for 'zero pressure at heel', $\delta = 4/3 - \xi/3(1 - \beta)$.

Suppose base width = 3.2 m and toe length = 0.9 m

$\alpha = 3.2/4.65 = 0.69, \ \alpha/\sqrt{K_{ad}} = 0.69/\sqrt{0.35} = 1.17, \ \beta = 0.9/3.2 = 0.28$

From design chart, $\xi = 0.83$ and $\delta = 4/3 - 0.83/3(1 - 0.28) = 0.95$

From equation for sliding criterion, $\alpha l_e = 3.03/0.95 = 3.2$ m (check)

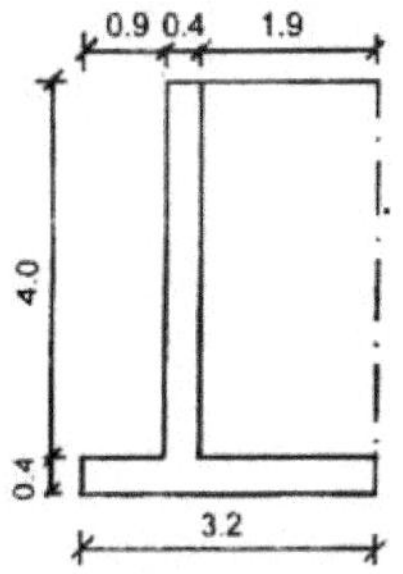

Base on clay

Example 3   **Calculation Sheet 3**

Reference	CALCULATIONS	OUTPUT
	GEOTECHNICAL DESIGN (Combination 2 partial safety factors) **(1) Wall with base on sand** Vertical loads and bending moments about front edge of base	

GEOTECHNICAL DESIGN (Combination 2 partial safety factors)

(1) Wall with base on sand

Vertical loads and bending moments about front edge of base

		Loads (kN)	Lever arm (m)	Moments (kN m)
Surcharge	5×1.4	$= 7.0$	$\times 1.7$	$= 11.9$
Backfill	$20 \times 1.4 \times 4.0$	$= 112.0$	$\times 1.7$	$= 190.4$
Wall stem	$25 \times 0.4 \times 4.0$	$= 40.0$	$\times 0.8$	$= 32.0$
Wall base	$25 \times 0.4 \times 2.4$	$= 24.0$	$\times 1.2$	$= 28.8$
Total		$F_v = 183.0$		$M_v = 263.1$

Horizontal loads and bending moments about bottom of base

		Loads (kN)	Lever arm (m)	Moments (kN m)
Surcharge	$0.35 \times 5 \times 4.4$	$= 7.7$	$\times 4.4/2$	$= 17.0$
Backfill	$0.35 \times 20 \times 4.4^2/2$	$= 67.8$	$\times 4.4/3$	$= 99.4$
Total		$F_h = 75.5$		$M_h = 116.4$

Resultant moment $M_{net} = 263.1 - 116.4 = 146.7$ kN m

Distance from front edge of base to line of action of total vertical force is

$a = M_{net}/F_v = 146.7/183 = 0.80$ m

Eccentricity of vertical force relative to centreline of base is

$e = 2.4/2 - 0.8 = 0.40$ m (edge of middle-third of base width)

Maximum bearing pressure for a triangular distribution over full base width is

$p_{max} = 2 \times 183/2.4 = 152.5$ kN/m^2 ($> p_{ba} = 150$)

Resistance to sliding $= F_v \tan \delta_d = 183 \times 0.46 = 84.2$ kN ($> F_h = 75.5$)

Note: In the design chart, the wall stem and base are taken to be of zero thickness. The actual vertical load is more than assumed, with a consequent increase in the bearing pressure and the resistance to sliding. A small modification can be made.

If the toe length is increased to 0.8 m, the revised values are as follows:

$F_v = 166$ kN, $M_v = 252.4$ kN m, $M_{net} = 136.0$ kN m, $a = 0.82$ m, $e = 0.38$ m

Bearing pressures at front and rear edges, respectively, of base are

$p = (166/2.4) \times (1 \pm 6 \times 0.38/2.4) = 134.9$ kN/m^2 (< 150) and 3.5 kN/m^2

Resistance to sliding $= 166 \times 0.46 = 76.3$ kN ($> F_h = 75.5$)

(2) Wall with base on clay

Vertical loads and bending moments about front edge of base

		Loads (kN)	Lever arm (m)	Moments (kN m)
Surcharge	5×1.9	$= 9.5$	$\times 2.25$	$= 21.4$
Backfill	$20 \times 1.9 \times 4.0$	$= 152.0$	$\times 2.25$	$= 342.0$
Wall stem	$25 \times 0.4 \times 4.0$	$= 40.0$	$\times 1.10$	$= 44.0$
Wall base	$25 \times 0.4 \times 3.2$	$= 32.0$	$\times 1.60$	$= 51.2$
Total		$F_v = 233.5$		$M_v = 458.6$

Horizontal loads and bending moments about bottom of base are as for case (1).

Resultant moment $M_{net} = 458.6 - 116.4 = 342.2$ kN m

Distance from front edge of base to line of action of total vertical force is

$a = M_{net}/F_v = 342.2/233.5 = 1.46$ m ($2a = 2.92$ m < 3.03 m required)

If the base width is increased to 3.3 m, with the toe length increased to 1.0 m, the revised values are as follows:

$F_v = 234.5$ kN, $M_v = 482.0$ kN m, $M_{net} = 365.6$ kN m, $a = 1.56$ m (sufficient)

Eccentricity of vertical force relative to centreline of base is

$e = 3.3/2 - 1.56 = 0.09$ m

OUTPUT

Lateral soil pressure on retaining wall

Final dimensions of retaining walls

Base on sand

Bearing pressure diagram

Base on clay

Example 3   **Calculation Sheet 4**

Reference	CALCULATIONS	OUTPUT
	Long-term (drained) condition	Bearing pressure diagrams
	Bearing pressures at front and rear edges, respectively, of base are	
	$\quad p = (234.5/3.3) \times (1 \pm 6 \times 0.09/3.3) = 82.7$ kN/m^2 (> 100) and 59.5 kN/m^2	
	Resistance to sliding: $\quad F_v \tan \delta_d = 234.5 \times 0.37 = 86.7$ kN	Long-term condition
	Short-term (undrained) condition	
	The contact length is taken as $l_b = 2a = 2 \times 1.56 = 3.12$ m, resulting in a uniform bearing pressure as follows:	
	$\quad p_u = F_v / l_b = 234.5/3.12 = 75.2$ kN/m^2	
	The ultimate bearing resistance is given by the equation:	
	$\quad q_u = (2 + \pi) c_{ud} i_c \quad$ where $i_c = 0.5[1 + \sqrt{1-k}\,]$ and $k = F_h/c_{ud} l_b$	
	$\quad i_c = 0.5[1 + \sqrt{1 - 75.5/(25 \times 3.12)}\,] = 0.59$	
	$\quad q_u = (2 + \pi) \times 25 \times 0.59 = 75.8$ kN/m^2 (> 75.2)	
	Resistance to sliding: $\quad c_{ud} l_b = 25 \times 3.12 = 78.0$ kN ($> F_h = 75.5$)	Short-term condition
	STRUCTURAL DESIGN (Combination 1 partial safety factors)	
	For the ULS, $\gamma_G = 1.35$ and $\gamma_M = 1.0$. In this case, all the forces calculated for the geotechnical design are multiplied by 1.35, but the coefficient of active earth pressure is taken as 0.27 instead of 0.35.	
	(1) Wall with base on sand	
	The revised vertical load and bending moments are as follows:	
	$\quad F_v = 1.35 \times 166 = 224.1$ kN/m $\quad M_v = 1.35 \times 252.4 = 340.8$ kN m/m	
	$\quad M_h = 1.35 \times (0.27/0.35) \times 116.4 = 121.2$ kNm/m $\quad M_{net} = 219.6$ kN m/m	
	$\quad a = 219.6/224.2 = 0.98$ m $\quad e = 1.2 - 0.98 = 0.22$ m	
	Bearing pressures at front and rear edges, respectively, of base are	
	$\quad p = (224.1/2.4) \times (1 \pm 6 \times 0.22/2.4) = 144.8$ kN/m^2 and 42.0 kN/m^2	
	Pressure at 0.8 m from edge of toe $= 144.8 - 102.8 \times 0.8/2.4 = 110.5$ kN/m^2	
	Bending moment in base at 0.8 m from edge of toe (i.e., face of wall) is	
	$\quad M = (110.5 - 1.35 \times 25 \times 0.4) \times 0.8^2/2 + 34.3 \times 0.8^2/3 = 38.4$ kNm/m	Bearing pressure diagram (base on sand)
	Pressure at 0.8 m from edge of heel $= 42.0 + 102.8 \times 0.8/2.4 = 76.3$ kN/m^2	
	Bending moment in base at 0.8 m from edge of heel (i.e., bottom of splay) is	
	$\quad M = [42.0 - 1.35 \times (5.0 + 20 \times 4.0 + 25 \times 0.4)] \times 0.8^2/2 + 34.3 \times 0.8^2/6$	
	$\quad\quad = -24.0$ kN m/m	
	(2) Wall with base on clay	
	The revised vertical load and bending moments are as follows:	
	$\quad F_v = 1.35 \times 234.5 = 316.6$ kN/m $\quad M_v = 1.35 \times 482.0 = 650.7$ kN m/m	
	$\quad M_h = 1.35 \times (0.27/0.35) \times 116.4 = 121.2$ kN m/m $\quad M_{net} = 529.5$ kN m/m	
	$\quad a = 529.5/316.6 = 1.67$ m $\quad e = 1.65 - 1.67 = -0.02$ m	
	Bearing pressures at rear and front edges, respectively, of base are:	
	$\quad p = (316.6/3.3) \times (1 \pm 6 \times 0.02/3.3) = 99.4$ kN/m^2 and 92.5 kN/m^2	
	Pressure at 1.0 m from edge of toe $= 92.5 + 6.9 \times 1.0/3.3 = 94.6$ kN/m^2	
	Bending moment in base at 1.0 m from edge of toe (i.e., face of wall) is	
	$\quad M = (92.5 - 1.35 \times 25 \times 0.4) \times 1.0^2/2 + 2.1 \times 1.0^2/6 = 39.9$ kN m/m	Bearing pressure diagram (base on clay)
	Pressure at 1.5 m from edge of heel $= 99.4 - 6.9 \times 1.5/3.3 = 96.3$ kN/m^2	

Example 3 **Calculation Sheet 5**

Reference	CALCULATIONS	OUTPUT
	Bending moment in base at 1.5 m from edge of heel (i.e., bottom of splay) is $M = [96.3 - 1.35 \times (5.0 + 20 \times 4.0 + 25 \times 0.4)] \times 1.5^2/2 + 3.1 \times 1.5^2/3$ $\quad = -33.6$ kN m/m	
	Durability	
BS8500 4.4.1.3 Table NA.1	For the buried parts of the wall, assuming non-aggressive soil conditions, exposure class XC2 applies. For the visible surface, exposure classes XD1 and XF1 apply. Concrete of minimum strength class C28/35 is recommended for both conditions, with cover $c_{min} = 35$ mm. If concrete strength class C32/40 is used, $c_{min} = 30$ mm and $c_{nom} = 40$ mm. For concrete cast against blinding, $c_{nom} = 50$ mm.	Concrete strength class C32/40 with covers: 50 mm (bottom of base), and 40 mm (other surfaces)
	Flexural design	
	Since the junction between wall stem and base is an 'opening corner', a 400×400 strengthening splay will be provided and the stem thickness reduced to 350 mm.	Reduce thickness of stem to 350 mm and introduce 400×400 splay between stem and base.
	For the vertical bars in the stem, allowing for 40 mm cover and 16 mm bars, with the horizontal bars in the outer layers, $d = 350 - (40 + 16 + 16/2) = 280$ mm say.	
	Minimum area of vertical tension reinforcement in wall stem:	
9.2.1.1	$A_{s,min} = 0.26 \times (3.0/500) \times 1000 \times 280 = 437$ mm^2/m (H12-200 say)	
	Design bending moment in wall at top of splay is	
	$M = 1.35 \times 0.27 \times (5.0 \times 3.6^2/2 + 20 \times 3.6^3/6) = 68.5$ kN m/m	
	Hence, from Table A1:	
	$M/bd^2f_{ck} = 68.5 \times 10^6/(1000 \times 280^2 \times 32) = 0.032 \qquad z/d = 0.95$ (maximum)	
	$A_s = 68.5 \times 10^6/(0.87 \times 500 \times 0.95 \times 280) = 592$ mm^2/m (H16-300)	Effective depth for reinforcement in splay
	Design bending moment in wall at bottom of splay is	
	$M = 1.35 \times 0.27 \times (5.0 \times 4.0^2/2 + 20 \times 4.0^3/6) = 92.4$ kN m/m	
	For the inclined reinforcement, effective depth at bottom of splay is	Vertical reinforcement in stem and diagonal reinforcement in splay H12-200, except for bars at bottom of stem
	$d = 400\sqrt{2} - (40 + 16 + 16/2) = 500$ mm $\qquad\qquad$ Hence,	
	$M/bd^2f_{ck} = 92.4 \times 10^6/(1000 \times 500^2 \times 32) = 0.012 \qquad z/d = 0.95$ (maximum)	
	$A_s = 92.4 \times 10^6/(0.87 \times 500 \times 0.95 \times 500) = 448$ mm^2/m (H12-200 say)	
	For the transverse reinforcement in the base to the wall;	
	$d = 400 - (40 + 16/2) = 350$ mm say.	
	Minimum area of reinforcement in base to wall:	Reinforcement in base H12-200 throughout
	$A_{s,min} = 0.26 \times (3.0/500) \times 1000 \times 350 = 546$ mm^2/m (H12-200)	
	The transverse bending moments in the base are small in magnitude and minimum reinforcement, top and bottom, will suffice.	
	Shear design	
	Design shear force in wall at top of splay is as follows:	
	$V_{Ed} = 1.35 \times 0.27 \times (5.0 \times 3.6 + 20 \times 3.6^2/2) = 53.8$ kN/m $\qquad$ Hence,	
	$v_{Ed} = V_{Ed}/(bd) = 53.8 \times 10^3/(1000 \times 280) = 0.20$ MPa	
	Minimum design shear strength, where $k = 1 + (200/d)^{1/2} = 1.84$, is	
6.2.2 (1)	$v_{min} = 0.035k^{3/2}f_{ck}^{1/2} = 0.035 \times 1.84^{3/2} \times 32^{1/2} = 0.49$ MPa ($> v_{Ed}$)	
	Cracking due to flexure	
	Minimum area of reinforcement required in tension zone for crack control:	
7.3.2 (2)	$A_{s,min} = k_c k f_{ct,eff} A_{ct}/\sigma_s \quad$ where	
	$k_c = 0.4$ for bending, $k = 0.97$ for $h = 350$ mm, $f_{ct,eff} = f_{ctm} = 0.3f_{ck}^{(2/3)} = 3.0$ MPa for general design purposes, $A_{ct} = bh/2$ and $\sigma_s \leq f_{yk} = 500$ MPa.	
	$A_{s,min} = 0.4 \times 0.97 \times 3.0 \times 1000 \times 175/500 = 408$ mm^2/m ($< A_s$ provided)	

Example 3   **Calculation Sheet 6**

Reference	CALCULATIONS	OUTPUT
7.3.3 (2) Table 7.2 Table 7.3	For the bars at the bottom of the wall stem, since all the loads are permanent, the reinforcement stress under service loading is given approximately by $\sigma_s = (0.87f_{yk}/\gamma_s) \times (A_{s,req}/A_{s,prov}) = (0.87 \times 500/1.35) \times 592/670 = 285$ MPa The crack width criterion can be satisfied by limiting either the bar size or the bar spacing. From *Reynolds*, Table 4.24, with $w_k = 0.3$ mm and $\sigma_s = 285$ MPa, the recommended maximum values are $\phi^*_s = 12$ mm or bar spacing = 140 mm say. Clearly, the criterion is not satisfied and the reinforcement stress must be reduced. If the reinforcement is increased to H16-200, the following values are obtained; $\sigma_s = 285/1.5 = 190$ MPa and maximum bar spacing = 260 mm (< 200). Where the reinforcement is reduced to H12-200, the value of ϕ^*_s is given by $\phi^*_s = \phi_s (2.9/f_{ct,eff})[2(h-d)/(k_c h_{cr})] = 12 \times 2.9/3.0 \times 2 \times 70/(0.4 \times 175) = 24$ mm From *Reynolds*, Table 4.24, with a bar spacing of 200 mm, maximum value of service stress $\sigma_s = 240$ MPa Maximum design ultimate moment is then given by $M = A_s (1.35\sigma_s)z = 565 \times 1.35 \times 240 \times 0.95 \times 280 \times 10^{-6} = 48.7$ kN m/m Bending moment at distance a from top of wall is given by $M = 1.35 \times 0.27 \times (5.0 \times a^2/2 + 20 \times a^3/6) = 48.7$ kN m/m where $a = 3.2$ m Here, H12-200 is sufficient but bars to be curtailed should continue for a distance $a_1 = d = 280$ mm. Projection of bars above top of splay = 0.4 + 0.28 = 0.68 m.	Vertical reinforcement at bottom of wall stem H16-200 (earth face)
7.3.2 (2) 7.3.4 PD 6687 2.16	**Cracking due to restrained early thermal contraction** Minimum area of horizontal reinforcement, with $f_{ct,eff} = 1.8$ MPa for cracking at age of 3 days, $k_c = 1.0$ for tension and $k = 0.97$ for $h = 350$ mm, is given by $A_{s,min} = k_c k f_{ct,eff} A_{ct}/f_{yk} = 1.0 \times 0.97 \times 1.8 \times 1000 \times 350/500 = 1222$ mm²/m With $c = 40$ mm, $k_1 = 0.8$ for high bond bars, $k_2 = 1.0$ for tension, $h_{c,ef}$ as the lesser of 2.5($h - d$) and $h/2$, and H16-300 (EF) as minimum reinforcement: $s_{r,max} = 3.4c + 0.425k_1k_2(A_{c,eff}/A_s)\phi$ $= 3.4 \times 40 + 0.425 \times 0.8 \times 1.0 \times (2.5 \times 48 \times 1000/670) \times 16 = 1110$ mm With $R = 0.8$ for wall on a thick base, $\Delta T = 28°$C for 350 kg/m³ Portland cement concrete and 350 mm thick wall (*Reynolds*, Table 2.18), and $\alpha = 12 \times 10^{-6}$ per °C: $w_k = (0.8R\alpha\Delta T) \times s_{r,max} = 0.8 \times 0.8 \times 12 \times 10^{-6} \times 28 \times 1110 = 0.24$ mm In this case, since any early thermal cracking should be properly controlled by the reinforcement, it would be reasonable to provide movement joints at 12 m centres to accommodate long-term movements due to temperature and moisture change.	Horizontal bars in wall stem H16-300 (EF)
3.1.3 (2) Table 3.1 3.1.4 (2) Figure 3.1 7.4.3 (5)	**Deflection** For the purpose of the calculation, the stiffening effect of the splay at the bottom of the wall will be ignored, but the effect of base tilting as a result of the variation of bearing pressure will be considered. For the SLS, bending moment at bottom of splay is $M = 92.4/1.35 = 68.5$ kN m/m. Secant modulus of elasticity of concrete at 28 days: $E_{cm} = 22[(f_{ck} + 8)/10]^{0.3} = 22 \times 4^{0.3} = 33.3$ GPa Final creep coefficient, for a C32/40 concrete with normally hardening cement in outside conditions (RH = 85%), for a member of notional thickness 350 mm and loaded at 28 days, is $\varphi(\infty,t_0) = 1.5$ say. Effective modulus of elasticity for long-term deformation is $E_{c,eff} = E_{cm}/[1 + \varphi(\infty,t_0)] = 33.3/2.5 = 13.3$ GPa Second moment of area values (uncracked and cracked sections) for a 350 mm thick section reinforced with H12-200 (EF), where $\alpha_e = E_s/E_{c,eff} = 200/13.3 = 15$ and $A_s = A'_s$, can be obtained as follows:	

Example 3

Calculation Sheet 7

Reference	CALCULATIONS	OUTPUT
	$I_o = bh^3/12 + 2(\alpha_e - 1)A_s(d - 0.5h)^2$ $= 1000 \times 350^3/12 + 2 \times 14 \times 565 \times (280 - 175)^2 = 3747 \times 10^6$ mm^4/m $I_c = bx^3/3 + \{\alpha_e(d-x)^2 + (\alpha_e - 1)(x - d')^2\}A_s$ where $x/d = \{[(2\alpha_e - 1)\rho]^2 + 2[\alpha_e + (\alpha_e - 1)(d'/d)]\rho\}^{0.5} - (2\alpha_e - 1)\rho$ and $\rho = \rho' = A_s/bd = 565/(1000 \times 280) = 0.002$ Hence, $x/d = \{(29 \times 0.002)^2 + 2 \times (15 + 14 \times 70/280) \times 0.002\}^{0.5} - 29 \times 0.002$ $= 0.220$ $x = 0.22 \times 280 = 62$ mm $I_c = 1000 \times 62^3/3 + \{15 \times (280 - 62)^2 + 14 \times (62 - 70)^2\} \times 565 = 483 \times 10^6$ mm^4	
	Moment to cause cracking of section, where $f_{ctm} = 3.0$ MPa, is given by $M_{cr} = f_{ctm} I_o/(h/2) = 3.0 \times 3747/175 = 64.2$ kN m/m	
	For sections that are expected to be cracked, the curvature may be determined by the relationship:	
7.4.3 (3)	$\dfrac{1}{r_b} = \dfrac{M}{EI_c}\left[\zeta + (1-\zeta)\dfrac{I_c}{I_o}\right]$ where $\zeta = 1 - \beta\left(\dfrac{M_{cr}}{M}\right)^2$	
	With $\beta = 0.5$ for sustained loading, $\zeta = 1 - 0.5 \times (64.2/68.5)^2 = 0.56$, which gives $1/r_b = \{(68.5 \times 10^6)/(13.3 \times 10^3 \times 483 \times 10^6)\} \times (0.56 + 0.44 \times 483/3747)$ $= 6.6 \times 10^{-6}$ (5.56×10^{-6} from backfill and 1.04×10^{-6} from surcharge)	
	The deflection at the top of wall can be estimated form the relationship: $a = \sum Kl^2(1/r_b) + Kl^2(1/r_{cs})$	
	For curvatures due to backfill (triangular load), surcharge (uniform load) and concrete shrinkage (uniform moment), K = 0.2, 0.25 and 0.5, respectively. Since the section is symmetrically reinforced, and the design moment only just exceeds the cracking moment, the effect of concrete shrinkage will be ignored. Then, $a = (0.2 \times 5.56 + 0.25 \times 1.04) \times 4000^2 \times 10^{-6} = 22$ mm	
	Note: Since the effect of the splay has been ignored, and the design moment only just exceeds the cracking moment, the actual deflection is likely to be somewhat less than the calculated value.	
	From the distributions of bearing pressure shown on calculation sheet 4, it can be seen that the deflection of the wall will be increased by base tilting, for the base on sand. The additional deflection can be estimated from the relationship: $a_{add} = [(p_1 - p_2)/k_s] \times l/b$ where p_1 and p_2 are bearing pressures at the front and rear edges of the base for the SLS, k_s is a modulus of subgrade reaction, l is the wall height and b is the base width. For k_s in the range 10 to 80 MN/m^3 (see Table B1): $a_{add} = (144.8 - 42.0)/(1.35 \times k_s \times 10^3) \times 4400/2.4 = 14$ to 2 mm	

Bar Marks	Commentary on Bar Arrangement (Drawing 1)
01, 03	Bars (shape code 11) with 50 mm cover bottom and ends (for concrete cast against blinding). Vertical legs to project above kicker (100 mm high above top of splay) to provide lap length = 1.5 × 35 × 12 = 650 mm say.
02	Bars (shape code 25) with nominal end projections = 200 mm say, providing length beyond top and bottom of splay not less than an anchorage length = 35 × 12 = 420 mm.
04, 05	Straight bars with 50 mm end cover, and 40 mm top cover to bars 04.
06, 08	Straight bars with bars 06 bearing on top of kicker, and bars 08 with 40 mm side cover and 50 mm at ends.
07	Bars (shape code 21) with nominal lap = 300 say with bars 06.

Example 3: Reinforcement in Typical Wall Panel and Base

Drawing 1

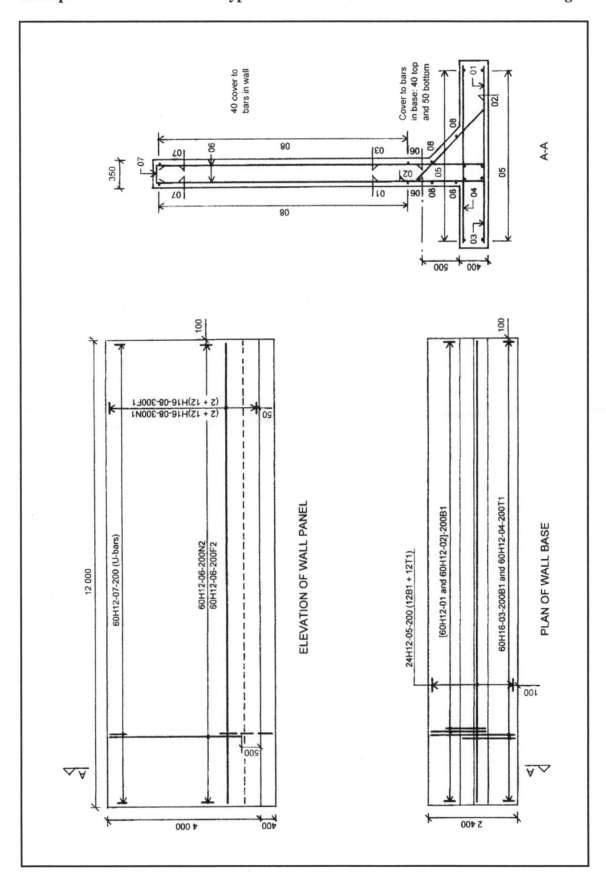

6 Example 4: Underground Service Reservoir

Description

A reservoir is required to contain 6500 m³ of potable water with a freeboard, when the reservoir is full, of 250 mm. At a depth of 5 m below the existing ground level, a firm clay stratum exists with presumed values 150 kN/m² for allowable bearing pressure, and 12 MN/m³ for modulus of subgrade reaction. The reservoir will be constructed in open excavation, and the ground behind the wall reinstated with excavated material. A graded drainage material will be provided behind the wall, with perimeter drains at the bottom of the wall, and a connecting system of drains below the floor. The roof will be covered with topsoil over a drainage material and waterproof membrane.

The reservoir wall will be provided with movement joints to accommodate differential settlements and minimise restraint to the effects of temperature change. The base to the perimeter wall will be separated from the rest of the floor, and the wall will be constructed in discontinuous lengths. The reservoir roof will be in the form of a continuous flat slab supported on internal columns, and provide a propped connection to the perimeter wall.

Since the perimeter wall is in the form of separate elements, individual units will need to be checked with regard to overall stability, bearing resistance and sliding, and there will be no separating layer between the base and the blinding. To minimise tilting, the wall base will be extended inwards to support the first row of columns.

Consider a wall of thickness 400 mm, and height 6 m above the base. The base thickness will be taken as 400 mm, with the top of the base level with the top of the floor. Allowing for the freeboard, the maximum depth of water is 5.75 m. Required internal area of the reservoir is $6500/5.75 = 1130$ m². For a square plan form, the internal dimension = $\sqrt{1130} = 33.6$ m.

For the roof slab, the total distance between centres of bearing at the top of the wall is $33.6 + 0.4 = 34$ m. Taking a group of 36 columns spaced at 5 m centres in each direction, the distance from the centre of the perimeter wall to the first column = $(34 - 5 \times 5)/2 = 4.5$ m. The slab thickness will be taken as 200 mm, and the column size as 300 mm diameter. The column head will be enlarged to avoid the need for shear reinforcement in the slab.

The floor slab will be constructed as a series of 5 m wide continuous strips with separation joints between the strips. Each strip will support a centrally placed line of columns. The slab thickness will be taken as 200 mm, and the bottom of each column enlarged to avoid the need for shear reinforcement in the slab. A separating layer of 1000 gauge polyethylene will be provided between the slab and the blinding concrete. The proposed arrangement is shown in Drawing 1.

Note: For structures of this type, care needs to be taken to minimise the effect of any thermal expansion of the roof on the perimeter walls. This will normally be achieved by ensuring that the roof covering is applied before the soil is placed behind the wall. Alternatively, restraint may be minimised by inserting a durable compressible filler material between the wall and the surrounding soil. This will prevent the build-up of large passive pressures in the upper portion of the soil and allow the wall to deflect as a long flexible cantilever.

Example 4: General Arrangement (Plan and Cross-Section)   **Drawing 1**

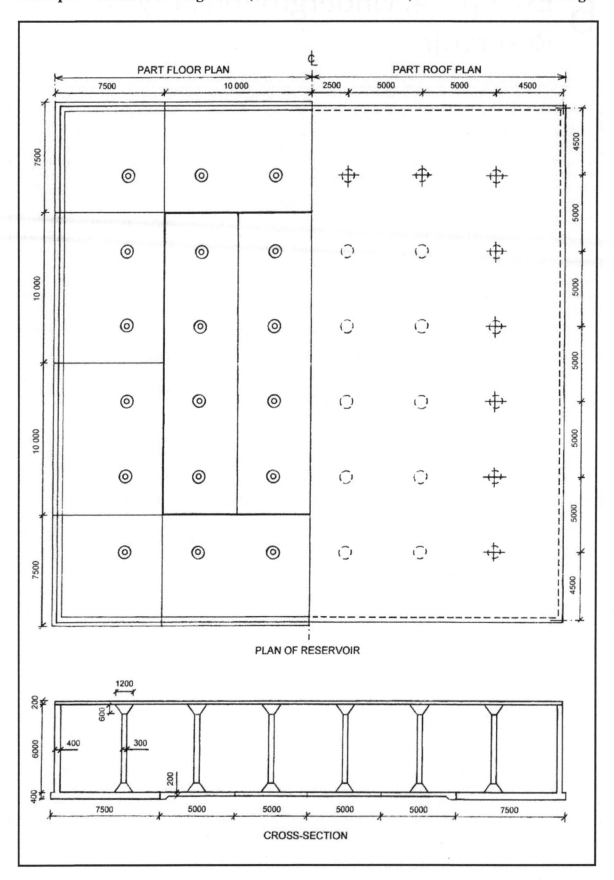

PART FLOOR PLAN   ₡   PART ROOF PLAN

PLAN OF RESERVOIR

CROSS-SECTION

Example 4

Calculation Sheet 1

Reference	CALCULATIONS	OUTPUT
	DESIGN PRINCIPLES	

In **Eurocode 7**, for conventional structures, two combinations of partial factors for actions and soil parameters are considered for the ULS as follows:

Partial safety factors for the ULS					
Combination	Safety factor on actions[a] γ_F		Safety factor for soil parameters γ_M		
	γ_G	γ_Q	$\gamma_{\varphi'}$	$\gamma_{c'}$	γ_{cu}
1	1.35	1.5	1.0	1.0	1.0
2	1.0	1.3	1.25	1.25	1.4

[a]If the action is favourable, values of $\gamma_G = 1.0$ and $\gamma_Q = 0$ should be used.

For combination 2, design values for soil strength at ULS are given by:

$$\tan \varphi'_d = (\tan \varphi')/1.25 \text{ and } c'_d = c'/1.25 \qquad \text{where}$$

c' and φ' are characteristic values of cohesion intercept and angle of shearing resistance (in terms of effective stress), respectively.

Design values for shear resistance at the interface of the base and the sub-soil, for drained (friction) and undrained (adhesion) conditions, respectively, are given by:

$$\tan \delta_d = \tan \varphi'_d \text{ (cast } in\text{-}situ \text{ concrete) } \text{ and } c_{ud} = c_u/1.4 \qquad \text{where}$$

c_u is the undrained shear strength.

Walls should be checked for the ULS with regard to overall stability, bearing resistance and sliding. For eccentric loading, bearing pressure is assumed to be uniformly distributed, with the centre of pressure coincident with the line of action of the applied load. The resistance should be checked for both long-term (drained) and short-term (undrained) conditions where appropriate.

The traditional practice of considering characteristic actions and allowable bearing pressures to limit ground deformation, and check the bearing resistance, may also be adopted by mutual agreement. With this approach, a linear variation of bearing pressure is assumed for eccentric loading. The ULS still needs to be considered to check sliding and for the structural design.

Ideally, the settlement of spread foundations on clay soils should be checked by calculation but may be taken as satisfactory, in the case of firm-to-stiff clays, if the ratio of design ultimate bearing resistance to service load is at least 3.

BS EN 1991-4 NA.A.2.1

In **Eurocode 1: Part 4**, the recommendations in Annexes A and B are replaced by those in the UK National Annex. For liquid induced loads, $\gamma_Q = 1.2$ may be taken for the ULS. The liquid level will be taken up to the top of the walls, assuming the liquid outlets are blocked. For the SLS, $\gamma_Q = 1.0$, and it is reasonable to take the liquid level to the maximum operational level.

BS EN 1992-3

In **Eurocode 2: Part 3**, for serviceability, structures are classified in relation to a required degree of protection against leakage. Class 1 refers to structures where leakage should be limited to a small amount but some surface staining or damp patches are acceptable. In this case, the width of any cracks that can be expected to pass through the full thickness of the section should be limited to w_{kl} given by:

$$0.05 \text{ mm} \leq w_{kl} = 0.225(1 - z_w/45\,h) \leq 0.2 \text{ mm}$$

where h is the wall thickness, and z_w is the liquid depth, at the section considered.

In situations where cracks are not expected to pass through the full thickness of the section, and the depth of the compression zone is at least equal to the lesser of $0.2\,h$ or 50 mm for all design conditions, the requirements of Eurocode 2: Part 1 may be applied. In Eurocode 2: Part 3, it is implied although not clearly stated, that it is sufficient to check for cracking under quasi-permanent loading. In this example, characteristic loading will be taken, as explained in Chapter 1.

Example 4   **Calculation Sheet 2**

Reference	CALCULATIONS	OUTPUT
	SOIL PARAMETERS	

Properties of the retained soil (well-graded sand and gravel) are as follows:

$$\gamma = 20 \text{ kN/m}^3 \qquad \varphi' = 35° \qquad \varphi'_d = \tan^{-1}[(\tan 35°)/1.25] = 29°$$

Coefficient of at-rest earth pressure:

$$K_o = 1 - \sin \varphi' = 0.43 \qquad K_{od} = 1 - \sin \varphi'_d = 0.52$$

Properties of the sub-base soil (firm clay) are as follows:

$$\gamma = 18 \text{ kN/m}^3 \qquad c_u = 50 \text{ kN/m}^2 \qquad c_{ud} = c_u/1.4 = 50/1.4 = 35 \text{ kN/m}^2$$

$$\varphi' = 27° \qquad \qquad \tan \delta = \tan \varphi' = 0.50$$

$$\varphi'_d = \tan^{-1}[(\tan 27°)/1.25] = 22° \qquad \tan \delta_d = \tan \varphi'_d = 0.40$$

Coefficient of passive earth pressure: $\quad K_p = (1 + \sin \varphi'_d)/(1 - \sin \varphi'_d) = 2.5$

PERIMETER WALL AND BASE

The reservoir must be designed for both full and empty conditions. The maximum water level will be taken at 6.0 m for ULS and 5.75 m for SLS, as explained in the design principles. The structure will be designed for the effects of earth pressures based on at-rest conditions when empty, but no relief will be given for beneficial earth pressures when full. Liquid-retaining structures are generally filled to test for water-tightness before any soil is placed against the walls.

The 200 mm thick roof slab will be covered with a waterproof membrane, 100 mm drainage material, and 200 mm topsoil. Allowance will be made for an additional 3.0 kN/m^2 superimposed load on the roof, and on the backfill to the walls.

Characteristic vertical loads

Roof load:

Concrete	$0.200 \times 25 =$	5.0
Overlay	$0.300 \times 18 =$	5.4
		10.4 kN/m^2

$q_k = 3.0 \text{ kN/m}^2$

Line load at bottom of wall:

Concrete	$5.0 \times 2.0 + 0.4 \times 6.0 \times 25 =$	70.0
Overlay	5.4×2.0	$= 10.8$
		80.8 kN/m

$q_k = 3.0 \times 2.0 = 6.0 \text{ kN/m}$

Line load at bottom of columns (spaced at 5.0 m centres):

Concrete	$5.0 \times 5.0 + 0.070 \times 6.0 \times 25/5 =$	27.1
Overlay	5.4×5.0	$= 27.0$
		54.1 kN/m

$q_k = 3.0 \times 5.0 = 15.0 \text{ kN/m}$

Analysis

The base, subjected to vertical loading from the wall and the first row of columns, will be considered initially as a beam with free ends bearing on an elastic soil. The moment required to restrain the resulting rotation at the junction with the wall will then be determined. The wall will be considered initially as a beam propped at the top and fixed at the bottom, and analysed to determine the moment at the base due to lateral pressure, for the full and empty conditions. The out of balance moment at the joint will then be distributed according to the relative stiffness of the members, and the resulting effects determined.

The base will be extended beyond the centre of the wall by 0.5 m and beyond the first line of columns by 2.5 m. Then, the base length $L = 7.5$ m, and the distances measured from the outer edge are approximately $0.07L$ to the centre of the wall, and $0.67L$ to the line of columns. For simplicity, the rotation at the junction of the wall and the base will be taken as the value at the end of the base. The resulting small error is of little consequence in the context of these calculations.

Example 4   **Calculation Sheet 3**

Reference	CALCULATIONS	OUTPUT
	The end slopes for beams on elastic foundations can be determined from the data in Table B2, where $\lambda L = (3k_s L^4/E_c h^3)^{1/4}$. With $E_c = 32$ GN/m^2 for C28/35 concrete, $L = 7.5$ m and $h = 0.4$ m: $\lambda L = [3 \times 12 \times 10^3 \times 7.5^4/(32 \times 10^6 \times 0.4^3)]^{1/4} = 2.75$ say In Table B2, the coefficients do not vary linearly between successive values of λL, and the values used in the following calculations have been derived from the basic equations. However, values obtained by linear interpolation could still be used. For a concentrated moment M_0 at end A, and concentrated loads, F_1 at $a/L = 0.07$ and F_2 at $a/L = 0.67$, the slopes at end A are $\theta_M = 83.59 M_0/(k_s BL^3)$ and $\theta_F = (-10.14F_1 + 3.19F_2)/(k_s BL^2)$ Thus, the fixed-end moment at A required to offset the slope θ_F is $\theta M_0 = -(k_s BL^3/83.59) \times \theta_F = 0.121F_1 L - 0.038F_2 L$ Flexural stiffness values of the base and wall ($l = 6.3$ m) are $K_b = k_s BL^3/83.59 = 12 \times 10^3 \times 1.0 \times 7.5^3/83.59 = 0.0606 \times 10^6$ kN m/m $K_w = E_c Bh_w^3/4l = 32 \times 10^6 \times 1.0 \times 0.4^3/(4 \times 6.3) = 0.0838 \times 10^6$ kN m/m Distribution factors: $D_b = 0.0606/(0.0606 + 0.0838) = 0.420$, $D_w = 0.580$	Dimensions of wall and combined base
	GEOTECHNICAL DESIGN (Combination 2 partial safety factors) **(1) Reservoir full (no earth loading on wall or roof)** Loading due to the water in the reservoir can be represented by a uniform load over the entire area of the base, modified by upward loads due to the absence of water on the outer nib of the base, and the displacement of water by concrete in the line loads at each end of the base. For simplicity, the upward load due to the absence of water on the nib will be added to the upward load acting on the line of the wall. The uniform loads due to the weight of the water and base are transferred directly to the ground, and have no structural affect on the base or the wall. Design ultimate values of $\gamma_G = 1.0$ for concrete and $\gamma_Q = 1.2$ for water apply, with the maximum depth of water taken as 6.0 m. The design ultimate loads acting on the wall and column lines, respectively, are $F_1 = 70.0 - 1.2 \times 0.7 \times 6.0 \times 9.81 = 20.6$ kN/m $F_2 = 27.1 - 1.2 \times 2.1 \times 9.81/25 = 26.1$ kN/m Fixed-end moments at junction of base and wall are approximately: Base: $M_b = (0.121 \times 20.6 - 0.038 \times 26.1) \times 7.5 = 11$ kN m/m Wall: $M_w = 1.2 \times 9.81 \times 6.0^2 \times 6.3/15 = 178$ kN m/m Resulting moments at junction, after releasing fixed-end moments, are $M_w = 178 - 0.580 \times (178 + 11) = 68$ kN m/m $\quad M_b = -68$ kN m/m Horizontal force at junction of base and wall is $F_h = 1.2 \times 9.81 \times 6.0^2/3 + 68/6.3 = 152$ kN/m Vertical force at underside of base, including weight of water and base, is $F_v = (1.0 \times 0.4 \times 25 + 1.2 \times 6.0 \times 9.81) \times 7.5 + 20.6 + 26.1 = 651$ kN/m Resistance to sliding for the short-term and long-term conditions are Short-term (undrained) condition (contact length $l_b = 7.5$ m): $c_{ud} l_b = 35 \times 7.5 = 262$ kN/m ($>F_h = 152$) Long-term (drained) condition: $F_v \tan \delta_d = 651 \times 0.4 = 260$ kN/m ($>F_h = 152$) The characteristic loads acting on the wall and column lines, respectively, with the maximum depth of water taken as 5.75 m, are	Hydrostatic loading for ultimate condition

Example 4   **Calculation Sheet 4**

Reference	CALCULATIONS	OUTPUT
	$F_1 = 70.0 - 0.7 \times 5.75 \times 9.81 = 30.5$ kN/m	

$F_2 = 27.1 - 0.07 \times 5.75 \times 9.81/5 = 26.3$ kN/m

Fixed-end moments at junction of base and wall are approximately:

  Base: $M_b = (0.121 \times 30.5 - 0.038 \times 26.3) \times 7.5 = 20$ kN m/m

  Wall: $M_w = 9.81 \times 5.75^2 \times 6.3/15 = 136$ kN m/m

Resulting moments at junction, after releasing fixed-end moments, are

  $M_w = 136 - 0.580 \times (136 + 20) = 46$ kN m/m $M_b = -46$ kN m/m

From Tables B3 and B7, bearing pressure (at $x/L = 0$) due to M_b, F_1 and F_2 is

  $q = -15.20(M_b/BL^2) + 4.589(F_1/BL) - 0.334(F_2/BL)$

    $= 15.20 \times 46/7.5^2 + (4.589 \times 30.5 - 0.334 \times 26.3)/7.5 = 30$ kN/m^2

Maximum bearing pressure, including weight of water and base, is

  $q = 0.4 \times 25 + 5.75 \times 9.81 + 30 = 97$ kN/m^2

Hydrostatic loading for service condition

(2) Reservoir empty (earth loading on wall and roof)

Depth of surcharge on backfill at mid-depth of roof = $0.4 + 3.0/20 = 0.55$ m. Load due to earth on the outer nib of the base will be added to the line load on the wall.

Design ultimate values of $\gamma_G = 1.0$ for concrete and soil, and $\gamma_Q = 0$ for live load on the roof, will be taken. Design ultimate loads acting on the wall and column lines, respectively, are:

  $F_1 = 0.3 \times 6.65 \times 20 + 80.8 = 39.9 + 80.8 = 120.7$ kN/m $F_2 = 54.1$ kN/m

Fixed-end moments at junction of base and wall, with $K_{od} = 0.52$, are

  Base: $M_b = (0.121 \times 120.7 - 0.038 \times 54.1) \times 7.5 = 94$ kN m/m

  Wall: $M_w = -0.52 \times 20 \times (6.3^3/15 + 0.55 \times 6.3^2/8) = -202$ kN m/m

Resulting moments at junction, after releasing fixed-end moments, are

  $M_w = -202 + 0.580 \times (202 - 94) = -140$ kN m/m $M_b = 140$ kN m/m

Horizontal force at underside of base is

  $F_h = 0.52 \times 20 \times (6.3^2/3 + 0.55 \times 6.3/2 + 6.95 \times 0.2) + 140/6.3 = 192$ kN/m

Total vertical force at underside of base, including weight of base, is

  $F_v = 0.4 \times 25 \times 7.5 + 120.7 + 54.1 = 250$ kN/m

Resultant moment of forces about outer edge of base is

  $M = 75 \times 3.75 + 39.9 \times 0.15 + 80.8 \times 0.5 + 54.1 \times 5.0 + 140 = 738$ kN m/m

Distance to centre of vertical force = $M/F_v = 738/250 = 2.95$ m, which is within the middle third of the base. Average bearing pressure = $250/7.5 = 33$ kN/m

Earth loading for geotechnical design

Design resistances to sliding for undrained and drained conditions are as follows:

  (a) Short-term (undrained) condition (contact length $l_b = 2 \times 2.95 = 5.9$ m):

    $c_{ud}l_b = 35 \times 5.9 = 206$ kN/m ($> F_h = 192$)

  (b) Long-term (drained) condition:

    $F_v \tan \delta_d = 250 \times 0.40 = 100$ kN/m ($< F_h = 192$)

Since the resistance to sliding could become inadequate in the long-term, it would be prudent to introduce a shear key below the base, in line with the stem of the wall, to mobilise a passive earth resistance $\geq (192 - 100) = 92$ kN/m.

Average bearing pressure behind shear key = $33 - 0.35 \times 42/7.5 = 31$ kN/m^2

Shear key below base

The design passive resistance, where q is the bearing pressure below the base and z is the depth of the shear key, is given by $K_p(\gamma z^2/2 + qz)$. For a 1.0 m deep shear key with $q = 31$ kN/m^2, the passive resistance is:

  $K_p(\gamma z^2/2 + qz) = 2.5 \times (18 \times 1.0^2/2 + 31 \times 1.0) = 100$ kN/m (> 92)

Bearing pressure

Example 4   **Calculation Sheet 5**

Reference	CALCULATIONS	OUTPUT
	The earth pressure acting on the shear key causes an additional bending moment that affects the moment equilibrium at the junction of the wall and the base. The passive earth resistance acts at a depth of $(18/3 + 31/2)/(18/2 + 31) = 0.54$ m, and the moment about the mid-depth of the base $M_p = -92 \times 0.74 = -68$ kN m/m.	
	Thus, the equilibrium moments at the junction of the wall and the base are:	
	$\quad M_w = -202 + 0.580 \times (202 + 68 - 94) = -100$ kN m/m $\qquad M_b = 168$ kN m/m	
	The modified horizontal force at the underside of the base is:	
	$\quad F_h = 0.52 \times 20 \times (6.3^2/3 + 0.55 \times 6.3/2 + 6.95 \times 0.2) + 100/6.3 = 186$ kN/m	
	Thus, the required passive resistance $= 186 - 100 = 86$ kN/m (< 100)	
	STRUCTURAL DESIGN (Combination 1 partial safety factors)	
	(1) Reservoir full (no earth loading)	
	In this case, the values used and the results obtained for the geotechnical design apply for the structural design also.	
	(2) Reservoir empty (full earth loading on wall and roof)	
	(a) Short-term (undrained) condition:	
	Design ultimate values of $\gamma_G = 1.35$ for concrete, retained soil and soil on roof, and $\gamma_Q = 1.5$ for live load on roof, will be taken to obtain maximum moment in wall.	
	Design ultimate loads acting on lines of wall and columns respectively are	$l = 6.3$
	$\quad F_1 = 1.35 \times 120.7 + 1.5 \times 6.0 = 172.0$ kN/m	
	$\quad F_2 = 1.35 \times 54.1 + 1.5 \times 15.0 = 95.5$ kN/m	
	Fixed-end moments at junction of base and wall are	$1.35 K_0\, \gamma\, (0.55 + l)$
	$\quad$ Base: $\ M_b = (0.121 \times 172.0 - 0.038 \times 95.5) \times 7.5 = 129$ kN m/m	Earth loading for structural design
	$\quad$ Wall: $\ M_w = -1.35 \times 0.43 \times 20 \times (6.3^3/15 + 0.55 \times 6.3^2/8) = -225$ kN m/m	
	Resulting moments at junction, after releasing fixed-end moments, are	
	$\quad M_w = -225 + 0.580 \times (225 - 129) = -169$ kN m $\qquad M_b = 169$ kN m	
	Horizontal force at junction of base and wall is	
	$\quad F_h = 1.35 \times 0.43 \times 20 \times (6.3^2/3 + 0.55 \times 6.3/2) + 169/6.3 = 201$ kN/m	
	The service loads acting on the wall and column lines, respectively, are	
	$\quad F_1 = 120.7 + 0.6 \times 6.0 = 124.3$ kN/m $\qquad F_2 = 54.1 + 0.6 \times 15.0 = 63.1$ kN/m	
	Fixed-end moments at junction of base and wall are	
	$\quad$ Base: $\ M_b = (0.121 \times 124.3 - 0.038 \times 63.1) \times 7.5 = 95$ kN m/m	
	$\quad$ Wall: $\ M_w = -0.43 \times 20 \times (6.3^3/15 + 0.55 \times 6.3^2/8) = -167$ kN m/m	
	Final service moments at junction, after releasing fixed-end moments, are	
	$\quad M_w = -167 + 0.580 \times (167 - 95) = -125$ kN m $\qquad M_b = 125$ kN m	
	(b) Long-term (drained) condition:	
	Design ultimate values of $\gamma_G = 1.35$ for retained soil, $\gamma_G = 1.0$ for concrete and soil on roof, and $\gamma_Q = 0$ for live load on roof, will be taken to obtain maximum moment in base. Fixed-end moments at junction of base and wall are $M_b = 94$ kN m/m and $M_w = -225$ kN m/m. Ignoring the effect of the shear key gives the following:	
	Resulting moments at junction, after releasing fixed-end moments, are	
	$\quad M_w = -225 + 0.580 \times (225 - 94) = -149$ kN m $\qquad M_b = 149$ kN m	
	Horizontal force at junction of base and wall is	
	$\quad F_h = 1.35 \times 0.43 \times 20 \times (6.3^2/3 + 0.55 \times 6.3/2) + 149/6.3 = 198$ kN/m	

Example 4   Calculation Sheet 6

Reference	CALCULATIONS	OUTPUT
	Resistance to sliding provided by base friction (with $\gamma_{\varphi'} = 1.0$) is $\quad F_v \tan\delta = 250 \times 0.50 = 125$ kN/m ($< F_h = 198$) Resulting passive earth force acting on shear key $= 198 - 125 = 73$ kN/m Moment about the mid-depth of the base $M_p = -73 \times 0.74 = -54$ kN m/m. Modified equilibrium moments at junction of wall and base are $\quad M_w = -225 + 0.580 \times (225 + 54 - 94) = -118$ kN m/m $\qquad M_b = 172$ kN m/m *Note*: By comparison with the values obtained in the geotechnical design, the force on the shear key is less but the moment in the base is slightly more. **Durability**	
BS 8500	For the outer face of the wall, assuming non-aggressive soil conditions, exposure class XC2 applies. For the underside of the roof and the upper portions of the inner face of the wall, exposure class XC3/XC4 applies. Concrete of minimum strength class C28/35 will be specified, with covers $c_{min} = 30$ mm and $c_{nom} = 40$ mm.	Concrete strength class C28/35 with 40 mm cover to both faces
9.2.1.1	**WALL STEM** **Flexural design** Allowing for 40 mm cover, 12 mm horizontal bars in the outer layers, and 16 mm vertical bars, $d = 400 - (40 + 12 + 16/2) = 340$ mm. Minimum area of vertical tension reinforcement in wall stem: $\quad A_{s,min} = 0.26 \times (3.0/500) \times 1000 \times 340 = 531$ mm^2/m (H12-200) (1) Reservoir full Maximum design moment at junction of wall and base is $M_{Ed} = 68$ kN m/m Hence, from Table A1: $\quad M_{Ed}/bd^2f_{ck} = 68 \times 10^6/(1000 \times 340^2 \times 28) = 0.021 \qquad z/d = 0.95$ (max) $\quad A_s = 68 \times 10^6/(0.87 \times 500 \times 0.95 \times 340) = 484$ mm^2/m (H12-200) Shear force at top of wall is $\quad V_t = 1.2 \times 9.81 \times 6.0^2/6 - 68/6.3 = 60$ kN/m If a_o is distance from top of wall to point of zero shear, then $\quad V_t - 1.2 \times 9.81 \times a_o^2/2 = 0$ which gives $a_o = 3.19$ m Hence, maximum sagging moment (at $a_o = 3.19$ m) is $\quad M_{Ed} = V_t \times a_o - 1.2 \times 9.81 \times a_o^3/6 = 128$ kNm/m $\quad A_s = 128 \times 10^6/(0.87 \times 500 \times 0.95 \times 340) = 911$ mm^2/m (H16-200) (2) Reservoir empty Maximum design moment at junction of wall and base is $M_{Ed} = 169$ kN m/m $\quad M_{Ed}/bd^2f_{ck} = 169 \times 10^6/(1000 \times 340^2 \times 28) = 0.052 \qquad z/d = 0.95$ (max) $\quad A_s = 169 \times 10^6/(0.87 \times 500 \times 0.95 \times 340) = 1203$ mm^2/m (H16-150) In the long-term (drained) condition, the passive earth pressure acting on the shear key reduces the moment at the bottom of the wall to $M_w = 118$ kN m/m. Shear force at top of wall is $\quad V_t = 1.35 \times 0.43 \times 20 \times (6.3^2/6 + 0.55 \times 6.3/2) - 118/6.3 = 78$ kN/m If a_o is distance from top of wall to point of zero shear, then $\quad V_t - 1.35 \times 0.43 \times 20 \times (a_o^2/2 + 0.55a_o) = 0$ which gives $a_o = 3.16$ m Hence, maximum sagging moment (at $a_o = 3.16$ m) is	60 kN 3.19 128 kNm 152 kN 68 kNm (1) Reservoir full 78 kN 3.16 154 kNm 118 kNm 193 kN (2) Reservoir empty

Example 4   Calculation Sheet 7

Reference	CALCULATIONS	OUTPUT
	$M_{Ed} = V_t \times a_o - 1.35 \times 0.43 \times 20 \times (a_o^3/6 + 0.55a_o^2/2) = 154$ kN m/m $A_s = 154 \times 10^6/(0.87 \times 500 \times 0.95 \times 340) = 1096$ mm^2/m (H16-175) **Shear design**	
6.2.2 (1) Table NA.1	The design shear strength of a flexural member without shear reinforcement is given by $$v_c = \left(\frac{0.18k}{\gamma_c}\right)\left(\frac{100A_{sl}f_{ck}}{b_w d}\right)^{1/3} \geq v_{min} = 0.035k^{3/2}f_{ck}^{1/2}$$ where $k = 1 + \sqrt{\dfrac{200}{d}} \leq 2.0$, $\left(\dfrac{100A_{sl}}{b_w d}\right) \leq 2.0$ and $\gamma_c = 1.5$ The values of v_c and v_{min} may be increased by $0.15N_{Ed}/A_c$ where N_{Ed} is the axial force on the section due to loading (positive for compression). With $k = 1 + (200/330)^{1/2} = 1.78$, $v_{min} = 0.035 \times 1.78^{3/2} \times 28^{1/2} = 0.44$ MPa (1) Reservoir full Design shear force at the junction of the wall and the base is $V_b = 152$ kN/m. Since the junction is an 'opening corner', no reduction of this shear force will be taken. $v_{Ed} = V_{Ed}/b_w d = 152 \times 10^3/(1000 \times 340) = 0.45$ MPa $N_{Ed} = 70$ kN/m $v_{min} + 0.15N_{Ed}/A_c = 0.44 + 0.15 \times 70 \times 10^3/(1000 \times 400) = 0.46$ MPa $(> v_{Ed})$ (2) Reservoir empty Design shear force at the junction of the wall and the base is $V_b = 201$ kN/m. Since the junction is a 'closing corner', the critical section for shear may be taken at a distance d above the top of the base (say 0.5 m above the centre of the base). $V_{Ed} = 201 - 1.35 \times 0.43 \times 20 \times 6.6 \times 0.5 = 163$ kN/m $v_{Ed} = V_{Ed}/b_w d = 163 \times 10^3/(1000 \times 340) = 0.48$ MPa $N_{Ed} = 1.35 \times 80.8 + 1.5 \times 6.0 = 118$ kN/m With $100A_{sl}/b_w d = 100 \times 1571/(1000 \times 340) = 0.46$ $v_c = (0.18 \times 1.78/1.5) \times (0.46 \times 28)^{1/3} = 0.50$ MPa $v_c + 0.15N_{Ed}/A_c = 0.50 + 0.15 \times 118 \times 10^3/(1000 \times 400) = 0.54$ MPa $(> v_{Ed})$ When the reservoir is empty, in order to mobilise the required resistance to sliding in the long-term (drained) condition, the resulting shear force on the key below the base is $V_{Ed} = 92$ kN/m (see calculation sheet 4). Hence, $v_{Ed} = V_{Ed}/b_w d = 92 \times 10^3/(1000 \times 340) = 0.27$ MPa $(\leq v_{min})$ **Shear at junction of wall and roof** Suppose that the top of the wall is chamfered at the inner edge and provided with a waterstop at the outer edge, so that the contact width between the roof and the wall is $b_i = 360$ mm. Maximum design shear force at top of wall is $V_{Ed} = 78$ kN/m. Design shear resistance at the interface between concretes cast at different times is given by:	
6.2.5	$V_{Rdi} = v_{Rdi}\, b_i = [cf_{ctd} + \mu(\sigma_n + \rho f_{yd})]b_i$ where c and μ are factors that depend on the roughness of the interface, with $c = 0.35$ and $\mu = 0.6$ for a free surface left without further treatment after vibration	
Table 3.1	$f_{ctd} = f_{ctk,0.05}/\gamma_c = 1.92/1.5 = 1.28$ MPa $f_{yd} = f_{yk}/\gamma_s = 500/1.15 = 435$ MPa With H10-300 U-bars at the interface, $\rho = A_s/A_i = 524/(1000 \times 360) = 0.0014$ σ_n is the normal stress across the interface due to the minimum vertical load. For the characteristic load due to the roof slab and soil: $\sigma_n = 20.8/360 = 0.058$ MPa	

Example 4   Calculation Sheet 8

Reference	CALCULATIONS	OUTPUT
	In this instance, movement of the roof slab during construction could invalidate any contribution from the tensile strength of the concrete and $c = 0$ will be taken. $V_{\text{Rdi}} = 0.6 \times (0.058 + 0.0014 \times 435) \times 360 = 144\ \text{kN/m}\ (> V_{\text{Ed}})$ **Cracking due to loading**	
BS EN 1992-3 7.3.1	The requirements of Eurocode 2: Part 1 may be applied, provided the depth of the compression zone is at least equal to the lesser of $0.2h$ or 50 mm in all conditions. With minimum reinforcement, $100A_s/bd = 100 \times 565/(1000 \times 340) = 0.166$. From Table A6, $x/d = 0.200$ (for $\alpha_e = 15$) and $x = 0.200 \times 340 = 68$ mm (≥ 50 mm). Minimum area of reinforcement required in tension zone for crack control:	
7.3.2 (2)	$A_{s,\min} = k_c k f_{ct,eff} A_{ct}/\sigma_s$ where $k_c = 0.4$ for bending, $k = 0.93$ for $h = 400$ mm, $f_{ct,eff} = f_{ctm} = 0.3 f_{ck}^{(2/3)} = 2.8$ MPa for general design purposes, $A_{ct} = bh/2$ and $\sigma_s \leq f_{yk} = 500$ MPa. $A_{s,\min} = 0.4 \times 0.93 \times 2.8 \times 1000 \times 200/500 = 417\ \text{mm}^2/\text{m}\ (< A_s\ \text{provided})$ The reinforcement stress under characteristic loading is given approximately by $\sigma_s = (0.87 f_{yk}) \times (M_k/M_u) \times (A_{s,req}/A_{s,prov})$ where M_k and M_u are moments due to characteristic and ultimate loads, respectively.	
7.3.3 (2) Table 7.3	Crack width criterion can be met by limiting the bar spacing. For section at bottom of wall, values obtained from *Reynolds,* Table 4.24 for $w_k = 0.3$ mm, are (1) Reservoir full (H12-200) $\sigma_s = 0.87 \times 500 \times 46/68 \times 484/565 = 250$ MPa Bar spacing ≤ 185 mm (2) Reservoir empty (H16-150) $\sigma_s = 0.87 \times 500 \times 125/169 \times 1203/1340 = 280$ MPa Bar spacing ≤ 150 mm It can be seen that the bar spacing requirement is satisfied for condition (2) but not for condition (1). The requirement is also unlikely to be satisfied in the case of the sagging moments for conditions (1) and (2). The vertical reinforcement will be increased to H16-150 for all cases.	For vertical bars: Provide H16-150 (EF)
	Cracking due to restrained early thermal contraction	
BS EN 1992-3 7.3.1	For cracks that can be expected to pass through the full thickness of the section, the crack width limit is given by: $w_{k,\lim} = 0.225(1 - z_w/45h) \leq 0.2$ mm where z_w is depth of water at section Hence, for $h = 0.4$ m: $w_{k,\lim} = 0.225(1 - z_w/18) \leq 0.2$ mm from which $w_{k,\lim} = 0.2$ mm for $z_w \leq 1.67$ m decreasing linearly to 0.15 mm at $z_w = 6.0$ m Minimum area of horizontal reinforcement, with $f_{ct,eff} = 1.8$ MPa for cracking at age of 3 days, $k_c = 1.0$ for tension and $k = 0.93$ for $h = 400$ mm, is given by	
7.3.2 (2) 7.3.4	$A_{s,\min} = k_c k f_{ct,eff} A_{ct}/f_{yk} = 1.0 \times 0.93 \times 1.8 \times 1000 \times 400/500 = 1340\ \text{mm}^2/\text{m}$ With $c = 40$ mm, $k_1 = 0.8$ for high bond bars, $k_2 = 1.0$ for tension, $h_{c,ef}$ as the lesser of $2.5(h-d)$ and $h/2$, and H12-150 (EF) as minimum reinforcement: $s_{r,\max} = 3.4c + 0.425 k_1 k_2 (A_{c,eff}/A_s)\varphi$ $= 3.4 \times 40 + 0.425 \times 0.8 \times 1.0 \times (2.5 \times 46 \times 1000/754) \times 12 = 758$ mm	
PD 6687 2.16	With $R = 0.8$ for wall on a thick base, $\Delta T = 30°C$ for 350 kg/m^3 Portland cement concrete and 400 mm thick wall (*Reynolds,* Table 2.18), and $\alpha = 12 \times 10^{-6}$ per °C: $w_k = (0.8 R \alpha \Delta T) \times s_{r,\max} = 0.8 \times 0.8 \times 12 \times 10^{-6} \times 30 \times 758 = 0.18$ mm Similarly, with H12-125 (EF): $s_{r,\max} = 655$ mm and $w_k = 0.15$ mm	For horizontal bars: Provide H12-150 (EF) for depth of 3.6 m and H12-125 (EF) below

Example 4

Reference	CALCULATIONS	OUTPUT
	Thus, H12-150 (EF) is sufficient for values of $z_w \leq 18(1 - 0.18/0.225) = 3.6$ m, and H12-125 (EF) for values of $z_w > 3.6$ m.	

Corner panels

At the corner of the reservoir, the vertical edges of the wall panels are fixed at one end and free at the other. The negative moment at the fixed edge will be taken as the value obtained for a long rectangular panel fixed at both ends. From the tables in *Appendix C*, for panel 1 and $l_x/l_z = 4.0$, the following values are obtained:

Assuming 12 mm horizontal bars, $d = 400 - (40 + 12/2) = 350$ mm say.

(1) Reservoir full (Table C2)

$M = 0.037 \times 1.2 \times 9.81 \times 6.0^2 \times 6.3 = 98.8$ kN m/m

$A_s = 98.8 \times 10^6/(0.87 \times 500 \times 0.95 \times 350) = 683$ mm²/m (H12-150)

(2) Reservoir empty (Tables C2 and C3)

$M = 1.35 \times 0.43 \times 20 \times (0.037 \times 6.3^3 + 0.081 \times 0.55 \times 6.3^2) = 128$ kN m/m

$A_s = 128 \times 10^6/(0.87 \times 500 \times 0.95 \times 350) = 885$ mm²/m (H12-125)

It can be seen that an arrangement of H12-125 (EF) will be sufficient to meet the requirements for bending and cracking due to restrained early thermal contraction.

OUTPUT: For corner panels H12-125 (EF) at all levels

WALL BASE

Flexural design

Moments due to loads F_1 (at $a/L = 0.07$) and F_2 (at $a/L = 0.67$) can be determined from Tables B7 and B9. Hence, design ultimate bending moments at junction of wall and base ($x/L = 0.07$) due to moment M_b, and loads F_1 and F_2, are as follows:

(1) Reservoir full ($M_b = -68$ kN m/m, $F_1 = 20.6$ kN/m, $F_2 = 26.1$ kN/m)

$M_{Ed} = -68 + (0.010 \times 20.6 - 0.001 \times 26.1) \times 7.5 = -67$ kN m/m

Provide H12-150 to align with bars in wall stem.

(2) Reservoir empty ($M_b = 169$ kN m/m, $F_1 = 172.0$ kN/m, $F_2 = 95.5$ kN/m)

$M_{Ed} = 169 + (0.010 \times 172.0 - 0.001 \times 95.5) \times 7.5 = 181$ kN m/m

With 50 mm cover, $d = 400 - (50 + 16/2) = 340$ mm say

$M_{Ed}/bd^2f_{ck} = 181 \times 10^6/(1000 \times 340^2 \times 28) = 0.056$ $z/d = 0.948$

$A_s = 181 \times 10^6/(0.87 \times 500 \times 0.948 \times 340) = 1291$ mm²/m (H16-150)

OUTPUT: Provide H12-150 (T) and H16-150 (B) in transverse direction, H12-200 (T and B) in longitudinal direction

Suppose the base is constructed in 10 m long panels, with each panel supporting two columns arranged symmetrically. All the columns are to be enlarged to 1.2 m diameter at the bottom (calculation sheet 13). Consider each column supported on a square area of side 5.0 m, with the load applied over an equivalent square area of side $1.2 \times (\pi/4)^{1/2} = 1.0$ m say. Column load $N_{Ed} = 95.5 \times 5.0 = 478$ kN.

Total bending moment for a 5 m wide strip at face of loaded area is

$M = (478/25) \times 5.0 \times 2.0^2/2 = 192$ kN m

Assuming 12 mm bars in second layer, $d = 400 - (50 + 16 + 12/2) = 320$ mm

$A_s = 192 \times 10^6/(0.87 \times 500 \times 0.95 \times 320) = 1452$ mm² (H12-200 minimum)

OUTPUT: Part plan showing column layout

Shear at junction of wall and base

(2) Reservoir empty

From the equations used to derive the coefficients in Tables B3 and B7, maximum design ultimate shear force at $x/L = 0.07$, is

$V_{Ed} = -0.837(M_0/L) - 0.717F_1 + 0.095F_2$

$= -0.837 \times 169/7.5 - 0.717 \times 172.0 + 0.095 \times 95.5 = -133$ kN/m

$v_{Ed} = V_{Ed}/b_w d = 133 \times 10^3/(1000 \times 340) = 0.39$ MPa ($< v_{min}$)

Example 4   Calculation Sheet 10

Reference	CALCULATIONS	OUTPUT
	Punching shear at columns	
	Mean effective depth: $d_{av} = 400 - (50 + 15) = 335$ mm.	See plan on sheet 9 for basic control perimeter
	Length of the basic control perimeter, at distance $2d_{av}$ from the face of the 1.2 m diameter enlargement at the bottom of the column, is:	
	$u_1 = 2\pi \times (600 + 2 \times 335) = 7980$ mm	
	Conservatively, taking $V_{Ed} = N_{Ed}$, the shear stress at the control perimeter is:	
	$v_{Ed} = V_{Ed}/(u_1 d) = 478 \times 10^3/(7980 \times 335) = 0.18$ MPa $(< v_{min})$	

ROOF SLAB

Analysis

The roof is considered to be divided in two orthogonal directions into strips of slab on knife-edge supports. The widths of the strips are taken as 5 m for the interior strips and 2 m for the edge strips. The effective spans between centres of supports are 5 m for the interior spans and 4.5 m for the end spans.

For a 7-span continuous slab on knife-edge supports, where the length of each end span is 0.9 × the length of an interior span, bending moments and shear forces can be obtained from the following tables:

Moment Coefficients for 7-Span Slab Where End Span is 0.9 × Interior Span							
Spans loaded	Span A–B	Support B	Span B–C	Support C	Span C–D	Support D	Span D–E
All	0.060	−0.093	0.038	−0.081	0.043	−0.084	0.041
Odds	0.084	−0.037	−0.040	−0.043	0.083	−0.042	−0.042
Evens	−0.024	−0.056	0.078	−0.038	−0.040	−0.042	0.083

M = coefficient × FL, where F is total load on interior span L

Shear Coefficients for 7-Span Slab Where End Span is 0.9 × Interior Span							
Spans Loaded	Support A	Support B		Support C		Support D	
		LH	RH	LH	RH	LH	RH
All	0.347	0.553	0.512	0.488	0.497	0.503	0.500
Odds	0.409	0.491	−0.006	0.006	0.501	0.499	0
Evens	−0.062	0.062	0.518	0.482	−0.004	0.004	0.500

V = coefficient × F, where F is total load on interior span

The design ultimate loads (see calculation sheet 2 for characteristic loads) are:

Permanent: $1.35 \times 10.4 = 14.0$ kN/m^2 Variable: $1.5 \times 3.0 = 4.5$ kN/m^2

The maximum design bending moments and shear forces for a 5 m wide panel are given in the following tables:

Location	Bending Moment for 5 m Wide Panel	kN m
Span A–B	$(0.060 \times 14.0 + 0.084 \times 4.5) \times 5^2 \times 5$ =	152
Support B	$-0.093 \times (14.0 + 4.5) \times 5^2 \times 5$ =	−215
Span C–D	$(0.043 \times 14.0 + 0.083 \times 4.5) \times 5^2 \times 5$ =	122
Support D	$-0.084 \times (14.0 + 4.5) \times 5^2 \times 5$ =	−195

Location	Shear Force for 5 m Wide Panel	kN
Support A	$0.409 \times (14.0 + 4.5) \times 5^2$ =	189
Support B	$(0.553 + 0.512) \times (14.0 + 4.5) \times 5^2$ =	493

Example 4 **Calculation Sheet 11**

Reference	CALCULATIONS	OUTPUT
	When the reservoir is full, the roof slab is also subjected to direct tensile forces in each direction, equal in magnitude to the shear force at the top of the perimeter wall. The slab will be designed for the combined effects of the bending moments resulting from the roof load, and the direct tension resulting from the hydrostatic pressure acting on the wall. From calculation sheet 5, when the reservoir is full, the design shear force at the top of the wall is $V_t = 60$ kN/m.	
	Design for combined flexure and tension	
	A rectangular section that is subjected to a bending moment M and a tensile force N acting at the mid-depth of the section, where $M/N \geq (d - 0.5h)$, can be designed for a reduced moment $M_1 = M - N(d - 0.5h)$ and a tensile force N acting at the level of the tension reinforcement. Design tensile force = 60 kN/m.	Tensile force acting at mid-depth of section
	Allowing for 40 mm cover and 12 mm bars in each direction, for the second layer of bars, $d = 200 - (40 + 12 + 12/2) = 140$ mm say.	
I.1.2 (3) Figure I.1 Table I.1	The panels will be notionally divided into 2.5 m wide column and middle strips, and the moments for the full panel width apportioned between specified limits. The hogging moments at the columns will be allocated in the proportions: 70% on column strips and 30% on middle strips. The sagging moments in the spans will be allocated in the proportions: 50% on column strips and 50% on middle strips.	Tensile force acting at level of reinforcement
	Support B (Column strip)	
	$M = 0.7 \times 215/2.5 = 60.2$ kNm/m	
	$M_1 = 60.2 - 60 \times (0.140 - 0.100) = 60.2 - 2.4 = 57.8$ kNm/m	
	$M_1/bd^2f_{ck} = 57.8 \times 10^6/(1000 \times 140^2 \times 28) = 0.106 \quad z/d = 0.896$	
	$A_s = 57.8 \times 10^6/(0.87 \times 500 \times 0.896 \times 140) + 60 \times 10^3/(0.87 \times 500)$	
	$\quad = 1060 + 138 = 1198$ mm^2/m (H12-200 + H16-200)	
	Support B (Middle strip)	
	$M = 0.3 \times 215/2.5 = 25.8$ kNm/m $\quad M_1 = 25.8 - 2.4 = 23.4$ kNm/m	
	$M_1/bd^2f_{ck} = 23.4 \times 10^6/(1000 \times 140^2 \times 28) = 0.043 \quad z/d = 0.95$ (max)	
	$A_s = 23.4 \times 10^6/(0.87 \times 500 \times 0.95 \times 140) + 138 = 543$ mm^2/m (H12-200)	
	Support D (Column strip)	Provide H12-200 (T) throughout and add H16-200 at the outer columns and H12-200 at all other columns
	$M = 0.7 \times 195/2.5 = 54.6$ kNm/m $\quad M_1 = 54.6 - 2.4 = 52.2$ kNm/m	
	$M_1/bd^2f_{ck} = 52.2 \times 10^6/(1000 \times 140^2 \times 28) = 0.095 \quad z/d = 0.908$	
	$A_s = 52.2 \times 10^6/(0.87 \times 500 \times 0.908 \times 140) + 138 = 1082$ mm^2/m (H12-100)	
	Span A–B (Column strip and middle strip)	
	$M = 152/5.0 = 30.4$ kNm/m $\quad M_1 = 30.4 - 2.4 = 28.0$ kNm/m	
	$M_1/bd^2f_{ck} = 28.0 \times 10^6/(1000 \times 140^2 \times 28) = 0.051 \quad z/d = 0.95$ (max)	
	$A_s = 28.0 \times 10^6/(0.87 \times 500 \times 0.95 \times 140) + 138 = 622$ mm^2/m (H16-200)	
	Span C–D (Column strip and middle strip)	Provide H16-200 (B) in the end spans and H12-200 (B) elsewhere
	$M = 122/5.0 = 24.4$ kNm/m $\quad M_1 = 24.4 - 2.4 = 22.0$ kNm/m	
	$A_s = 22.0 \times 10^6/(0.87 \times 500 \times 0.95 \times 140) + 138 = 519$ mm^2/m (H12-200)	
	Shear at perimeter wall	
	Conservatively, taking the reaction at the centre of support A: $\quad V_{Ed} = 189$ kN	
	Assuming the reaction to be uniformly distributed across the panel width:	
	$v_{Ed} = V_{Ed}/b_w d = 189 \times 10^3/(5000 \times 140) = 0.27$ MPa	
6.2.2 (1)	With $v_{min} = 0.035k^{3/2}f_{ck}^{1/2}$, where $k = 2.0$ for $d \leq 200$ mm, and $\sigma_{cp} = N_{Ed}/A_c$:	
	$v_{min} + k_1 \sigma_{cp} = 0.035 \times 2^{3/2} \times 28^{1/2} - 0.15 \times 60/200 = 0.48$ MPa	

Example 4　　　　　　　　　　　　　　　　　　**Calculation Sheet 12**

Reference	CALCULATIONS	OUTPUT
	Punching shear at columns	
	For the top reinforcement in the column strips, with 16 mm bars in two orthogonal directions, the mean effective depth $d_{av} = 200 - (40 + 16) = 144$ mm.	
6.4.2 (1) Figure 6.13	If the column is provided with a 1.2 m diameter head, length of the basic control perimeter at distance $2d_{av}$ from the face: $u_1 = 2\pi \times (600 + 2 \times 144) = 5580$ mm	Provide 1.2 m diameter head to all columns
6.4.3 Figure 6.21N Equation 6.42	Taking $\beta = 1.15$ for an internal column, modified to take account of the diameters of the column and the column head, and $V_{Ed} = 493$ kN at support B: 　$\beta = 1.0 + 0.15 \times (300 + 4 \times 144)/(1200 + 4 \times 144) = 1.075$ 　$v_{Ed} = \beta V_{Ed}/(u_1 d) = 1.075 \times 493 \times 10^3/(5580 \times 144) = 0.66$ MPa The mean reinforcement percentage with 1570 mm^2/m in each direction is 　$100A_{sl}/b_w d = 100 \times 1570/(1000 \times 144) = 1.09$ Hence, with $k = 2.0$ for $d \leq 200$ mm, 　$v_{Rd,c} = (0.18 \times 2.0/1.5) \times (1.09 \times 28)^{1/3} - 0.15 \times 60/200 = 0.70$ MPa $(\geq v_{Ed})$	
6.4.4 (1)	**Deflection**	
	Deflection requirements may be met by limiting the span-effective depth ratio. For the interior spans, the actual span/effective depth ratio = 5000/140 = 36.	
7.4.1 (6)	The quasi-permanent load, taking $\psi_2 = 0.3$, for a full panel width is given by 　$g_k + \psi_2 q_k = 10.4 + 0.3 \times 3.0 = 11.3$ kN/m^2	
BS EN 1990 Table NA.A1.1	For spans C-D, $A_{s,req} = 516$ mm^2/m, $A_{s,prov}/A_{s,req} = 754/516 = 1.46$ (≤ 1.5), and the corresponding stress under quasi-permanent load is given approximately by 　$\sigma_s = (f_{yk}/\gamma_s)(A_{s,req}/A_{s,prov})[(g_k + \psi_2 q_k)/n]$ 　　$= (500/1.15)(516/754)(11.3/18.5) = 182$ MPa	
	From *Reynolds*, Table 4.21, limiting l/d = basic ratio $\times \alpha_s \times \beta_s$ where: For $100A_s/bd = 100 \times 516/(1000 \times 140) = 0.37 < 0.1f_{ck}^{0.5} = 0.1 \times 28^{0.5} = 0.53$, 　$\alpha_s = 0.55 + 0.0075f_{ck}/(100A_s/bd) + 0.005f_{ck}^{0.5}[f_{ck}^{0.5}/(100A_s/bd) - 10]^{1.5}$ 　　$= 0.55 + 0.0075 \times 28/0.37 + 0.005 \times 28^{0.5} \times (28^{0.5}/0.37 - 10)^{1.5} = 1.35$ 　$\beta_s = 310/\sigma_s = 310/182 = 1.70$	
	For a flat slab, the basic ratio = 24. Hence, 　Limiting $l/d = 24 \times \alpha_s \times \beta_s = 24 \times 1.35 \times 1.70 = 55$ ($>$ actual $l/d = 36$)	Check complies
7.4.2 Table NA.5	**Cracking due to loading**	
	Minimum area of reinforcement required in tension area for crack control: 　$A_{s,min} = k_c k f_{ct,eff} A_{ct}/\sigma_s$　　where, at points of contra-flexure,	
7.3.2 (2)	$k_c = 1.0$ for tension, $k = 1.0$ for $h \leq 300$ mm, $f_{ct,eff} = f_{ctm} = 0.3f_{ck}^{(2/3)} = 2.8$ MPa for general design purposes, $A_{ct} = bh$ and $\sigma_s \leq f_{yk} = 500$ MPa. 　$A_{s,min} = 1.0 \times 1.0 \times 2.8 \times 1000 \times 200/500 = 1120$ mm^2/m　　H12-200 (EF)	
	No other specific measures are necessary provided overall depth does not exceed 200 mm, and detailing requirements are observed.	Check complies
7.3.3 (1)	**Cracking due to restrained early thermal contraction**	
	Minimum area of horizontal reinforcement, with $f_{ct,eff} = 1.8$ MPa for cracking at age of 3 days, $k_c = 1.0$ for tension and $k = 1.0$ for $h \leq 300$ mm, is given by 　$A_{s,min} = k_c k f_{ct,eff} A_{ct}/f_{yk} = 1.0 \times 1.0 \times 1.8 \times 1000 \times 200/500 = 720$ mm^2/m	
7.3.2 (2)	With $c = 52$ mm, $k_1 = 0.8$ for high bond bars, $k_2 = 1.0$ for tension, $h_{c,ef}$ as the lesser of $2.5(h - d)$ and $h/2$, and H12-200 (EF) as minimum reinforcement:	

Example 4   **Calculation Sheet 13**

Reference	CALCULATIONS	OUTPUT
	$s_{r,max} = 3.4c + 0.425k_1k_2(A_{c,eff}/A_s)\varphi$	
	$\quad = 3.4 \times 52 + 0.425 \times 0.8 \times 1.0 \times (100 \times 1000/565) \times 12 = 899$ mm	
PD 6687 2.16	Taking $R = 0.8$ for infill bays, $\Delta T = 15°$C for 325 kg/m³ Portland cement concrete and slab thickness ≤ 300 mm (*Reynolds,* Table 2.18), and $\alpha = 12 \times 10^{-6}$ per °C:	
	$w_k = (0.8R\alpha\Delta T) \times s_{r,max} = 0.8 \times 0.8 \times 12 \times 10^{-6} \times 15 \times 899 = 0.10$ mm	
	Detailing requirements	
	Minimum area of longitudinal tension reinforcement:	
9.3.1.1 (1)	$A_{s,min} = 0.26(f_{ctm}/f_{yk})b_td = 0.26 \times (3.0/500)\ b_td = 0.00156bd \geq 0.0013\ b_td$	
	$\quad = 0.00156 \times 1000 \times 140 = 219$ mm²/m	
9.3.1.1 (3)	Maximum spacing of principal reinforcement in area of maximum moment:	
	$2h = 400 \leq 250$ mm Elsewhere: $3h = 600$ mm ≤ 400 mm	
9.4.1 (2)	At internal columns, the top reinforcement in the column strip should normally be placed with two-thirds of the required area in the central half of the strip. It is reasonable to waive this requirement when enlarged column heads are provided.	
9.4.1 (3)	Bottom reinforcement (≥ 2 bars) in each orthogonal direction should be provided to pass through each column.	
	Curtailment of longitudinal tension reinforcement	
	The following simplified curtailment rules for one-way continuous slabs will be used in each orthogonal direction.	
	For bottom reinforcement, continue 50% onto support for a distance $\geq 10\phi$ from the face, and 100% to within a distance from centre of support:	
	$\leq 0.1 \times$ span $= 450$ mm at end supports	
	$\leq 0.2 \times$ span $= 1000$ mm at interior supports	
	For top reinforcement, continue for distance beyond face of interior support:	
	100% for $\geq 0.2 \times$ span $= 1000$ mm ($\geq l_{b,rqd} + d = 35 \times 16 + 140 = 700$ mm)	
	50% for $\geq 0.3 \times$ span $= 1500$ mm	
	FLOOR SLAB	
	Analysis	
	The floor slab is to be constructed as a series of 5 m wide continuous strips with each strip supporting a centrally placed line of columns. The slab is 200 mm thick, and the bottom of each column will be enlarged to 1.2 m diameter. The maximum design load at the bottom of support D is	
	$N_{Ed} = 1.03 \times 18.5 \times 5^2 + 1.35 \times 0.07 \times 6.0 \times 25 = 490$ kN	
	Each column will be considered as supported on a 5 m × 5 m square area, with the column load applied over a square area of side $1.2 \times (\pi/4)^{1/2} = 1.0$ m say.	
	Total bending moment for a 5 m wide strip at face of loaded area is	
	$M = (490/25) \times 5.0 \times 2.0^2/2 = 196$ kN m	
	Design for flexure	
	Allowing for 50 mm cover and 12 mm bars in each direction, for the second layer of bars, $d = 200 - (50 + 12 + 12/2) = 130$ mm say.	
	Taking 70% of the total moment on the middle half of the width:	In both directions provide H12-100 (B) at each column and H12-200 (B) elsewhere
	$M_{Ed}/bd^2f_{ck} = 0.7 \times 196 \times 10^6/(2500 \times 130^2 \times 28) = 0.116 \quad z/d = 0.884$	
	$A_s = 0.7 \times 196 \times 10^6/(0.87 \times 500 \times 0.884 \times 130) = 2745$ mm² (H12-100)	
	Clearly, for the outer quarters of the width, H12-200 will be sufficient.	

Example 4   **Calculation Sheet 14**

Reference	CALCULATIONS	OUTPUT
	Punching shear at columns	
	Mean effective depth for bottom reinforcement, $d_{av} = 130 + 12 = 142$ mm	
	Radius of the basic control perimeter at distance $2d_{av}$ from the face of the 1.2 m diameter foot to the column = $600 + 2 \times 142 = 884$ mm.	
	If the floor slab is considered as a flexible plate bearing on an elastic soil, the radius of relative stiffness r_k, where ν is Poisson's ratio, is given by	
	$r_k = [E_c h^3 / 12 (1 - \nu^2) k_s]^{1/4} = [E_c h^3 / 11.52 k_s]^{1/4}$ for $\nu = 0.2$	
	$\quad = [32 \times 10^6 \times 0.2^3 / (11.52 \times 12 \times 10^3)]^{1/4} = 1.2$ m say	
	From Table B11, with $r_x / r_k = 884/1200 = 0.74$, by interpolation:	
	$V_{Ed} = 0.18 (F/r_k) = 0.18 \times 490/1.2 = 73.5$ kN/m	
	Assuming a uniform distribution of shear stress (i.e. no moment transfer):	
	$v_{Ed} = V_{Ed}/(u_1 d) = 73.5 \times 10^3/(1000 \times 142) = 0.52$ MPa	
	With H12-100 in two orthogonal directions, mean reinforcement percentage:	
	$100 A_{sl}/b_w d = 100 \times 1131/(1000 \times 142) = 0.80$	
6.2.2 (1) Table NA.1	Hence, with $k = 2.0$ for $d \leq 200$ mm, $v_{Rd,c} = (0.18 \times 2.0/1.5) \times (0.80 \times 28)^{1/3} = 0.67$ MPa $(> v_{Ed})$	Shear satisfactory
	Cracking due to restrained early thermal contraction	
	Since a separation layer is to be provided between the slab and the blinding, and movement joints are provided between the strips, cracking is unlikely. However, as a precaution, the recommended minimum reinforcement area will be provided.	
7.3.2 (2)	$A_{s,min} = k_c k f_{ct,eff} A_{ct}/f_{yk} = 1.0 \times 1.0 \times 1.8 \times 1000 \times 200/500 = 720$ mm^2/m	
	The provision of H12-200 (bottom) and A393 fabric (top) meets this requirement, and also caters for the small moments that occur in the areas between the columns.	Provide A393 fabric at top of slab
	COLUMNS	
	Effective length and slenderness	
	Using the simplified method for braced columns given in Concise Eurocode 2, for condition 2 (a monolithic connection to members on each side that are shallower than the overall depth of the column but generally not less than half the column depth) at both top and bottom of the column, the effective length:	
	$l_0 = 0.85l = 0.85 \times 6.0 = 5.1$ m	
	Radius of gyration of a circular concrete section: $i = h/4 = 0.3/4 = 0.075$ m	
5.8.3.2 (1)	Slenderness ratio: $\lambda = l_0/i = 5.1/0.075 = 68$	
5.8.3.1 (1)	Slenderness criterion: $\lambda_{lim} = 20(A \times B \times C)/\sqrt{n}$ where $n = N_{Ed}/A_c f_{cd}$	
	At support B (mid-height), $N_{Ed} = 493 + 1.35 \times 0.07 \times 3.0 \times 25 = 500$ kN	
	$n = 500 \times 10^3/(\pi \times 150^2 \times 0.85 \times 28/1.5) = 0.45$	
	Taking $A = 0.7$, $B = 1.1$, $C = 0.7$ (moments predominately due to imperfections)	
	$\lambda_{lim} = 20 \times 0.7 \times 1.1 \times 0.7/\sqrt{0.45} = 16$	
	Since $\lambda > \lambda_{lim}$, second order effects will need to be considered.	
	Design moments	
6.1 (4)	Minimum design moment, with $e_0 = h/30 = 300/30 \geq 20$ mm:	
	$M_{min} = N_{Ed} e_0 = 500 \times 0.02 = 10$ kN m	
5.8.2 (9)	First-order moment due to imperfections (simplified procedure):	
	$M_i = N l_0/400 = 500 \times 5.1/400 = 6.4$ kN m	

Example 4   **Calculation Sheet 15**

Reference	CALCULATIONS	OUTPUT
5.8.8.2 (3) 5.8.8.3 (1) 5.8.8.3 (3)	Using the method based on nominal curvature, the nominal second-order moment $M_2 = N_{Ed} e_2$ where $e_2 = (1/r) l_0^2/10$ and $1/r = K_r K_\varphi (f_{yd}/E_s)/0.45d$ $K_r = (n_u - n)/(n_u - n_{bal}) \leq 1$ where $n_{bal} = 0.4$ may be used	
	Since n is only slightly more than n_{bal}, $K_r = 1.0$ will be taken.	
5.8.8.3 (4)	$K_\varphi = 1 + \beta \varphi_{ef} \geq 1.0$ where $\beta = 0.35 + f_{ck}/200 - \lambda/150 = 0.35 + 28/200 - 68/150 = 0.04$	
5.8.4 (2)	$\varphi_{ef} = \varphi(\infty, t_0) \times (M_{0Eqp}/M_{0Ed}) = 0.6 \times \varphi(\infty, t_0)$ say	
3.1.4 (2) Figure 3.1	For outside conditions, with $t_0 \geq 30$ days, $h_0 = h/2 = 150$ mm and C28/35 concrete: $\varphi(\infty, t_0) = 2.0$ say $\varphi_{ef} = 0.6 \times 2.0 = 1.2$ $K_\varphi = 1 + 0.04 \times 1.2 = 1.05$ $1/r = 1.0 \times 1.05 \times 500/(1.15 \times 200 \times 10^3 \times 0.45 \times 0.24) = 0.0212$ $M_2 = 500 \times 0.0212 \times 5.1^2/10 = 27.6$ kNm	
5.8.9	Considering second-order moments about two perpendicular axes, and first order moment due solely to imperfections about one axis, the equivalent design moment about one axis for a circular cross-section is $M_{Ed} = M_i + \sqrt{2}M_2 = 6.4 + \sqrt{2} \times 27.6 = 45.5$ kNm ($\geq M_{min} = 10$ kN m)	
	Design of cross-section	
	Allowing for 40 mm cover, 8 mm links, and 16 mm main bars, diameter of the circle through centres of bars: $h_s = 300 - 2 \times (40 + 8 + 16/2) = 188$ mm $h_s/h = 188/300 = 0.6$ say $M/h^3 f_{ck} = 45.5 \times 10^6/(300^3 \times 28) = 0.060$ $N/h^2 f_{ck} = 500 \times 10^3/(300^2 \times 28) = 0.20$ From Table A7: $A_s f_{yk}/A_c f_{ck} = 0.20$ and $K_r = 0.80$	
	Since $K_r < 1.0$, the calculated second order effects can be reduced. Hence, With $K_r = 0.80$: $1/r = 0.017$ and $M_2 = 22.1$ kN m $M_{Ed} = 6.4 + \sqrt{2} \times 22.1 = 37.7$ kN m and $M/h^3 f_{ck} = 0.050$ From Table A7: $A_s f_{yk}/A_c f_{ck} = 0.10$ and $K_r = 0.80$ $A_s = 0.10 \times \pi \times 150^2 \times 28/500 = 396$ mm^2 (6H12)	6H12

Bar Marks	Commentary on Bar Arrangement (Drawing 2)
01, 03	Bars (shape code 21) with 50 mm cover against blinding (bottom and ends), and 40 mm cover (top).
02	Straight bars with 50 mm end cover. Bars in wall stem are placed in outer layers with 40 mm cover.
04	Bars (shape code 21) projecting from shear key to provide starter bars for wall stem. Projection above top of base to provide a lap length above 150 mm kicker = $1.5 \times 38 \times 16 + 150 = 1075$ mm say.
05	Straight bars bearing on wall kicker and stopping below roof slab.
06	Bars (shape code 21) with lap length = $1.5 \times 38 \times 10 = 575$ mm.
07	Column starter bars (shape code 11) standing on mat formed by bars 01 and 02. Projection of bars above top of base to provide a lap length above kicker on top of column foot = $1.5 \times 38 \times 12 + 600 + 100 = 1400$ mm say. Cover to bars in column = 50 mm to enable 40 mm cover to links.
08	Circular links (shape code 75) to hold column starter bars in place during construction.

Example 4: Reinforcement in Reservoir Wall and Base   **Drawing 2**

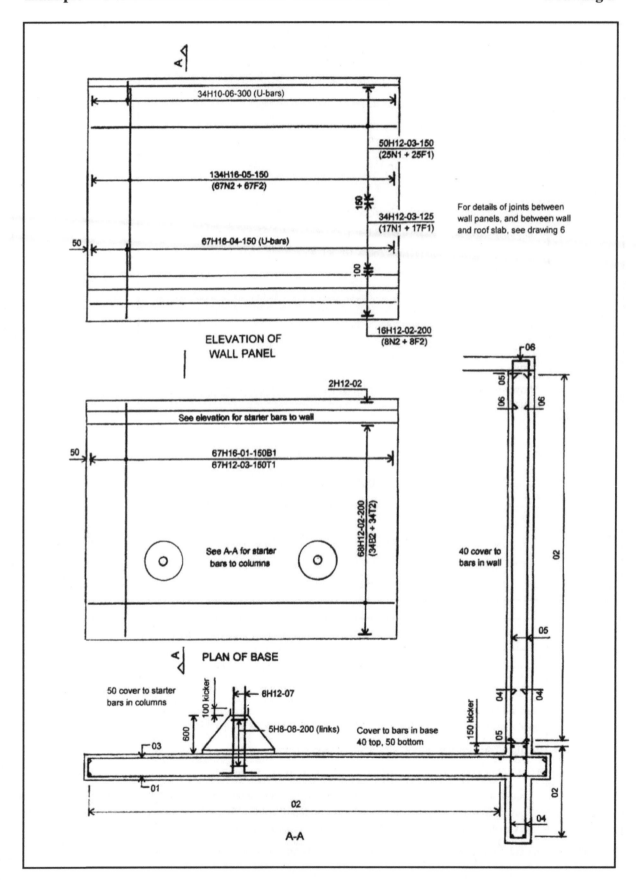

ELEVATION OF
WALL PANEL

For details of joints between
wall panels, and between wall
and roof slab, see drawing 6

PLAN OF BASE

A-A

Example 4: **Reinforcement in Reservoir Floor and Column** Drawing 3

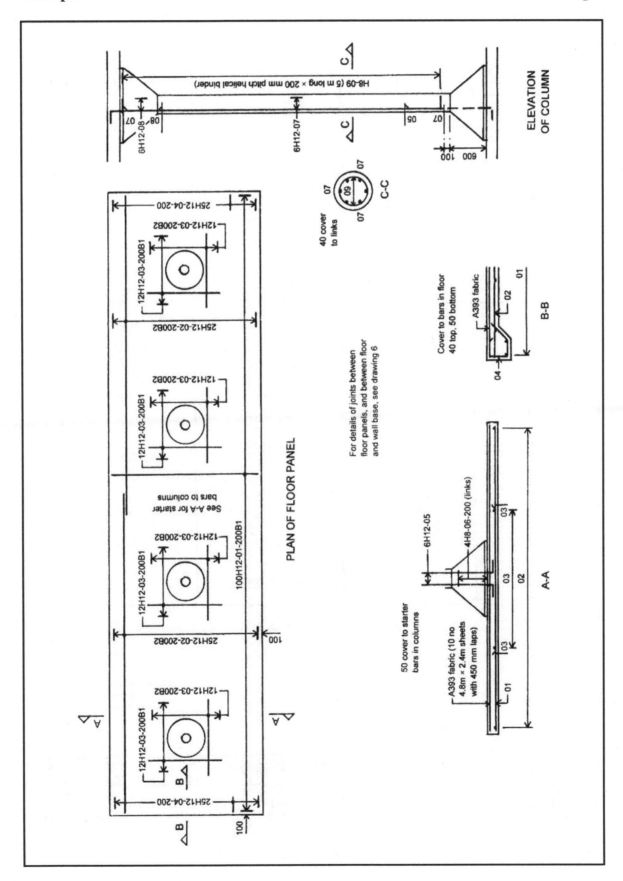

Example 4: Bottom Reinforcement in Reservoir Roof   **Drawing 4**

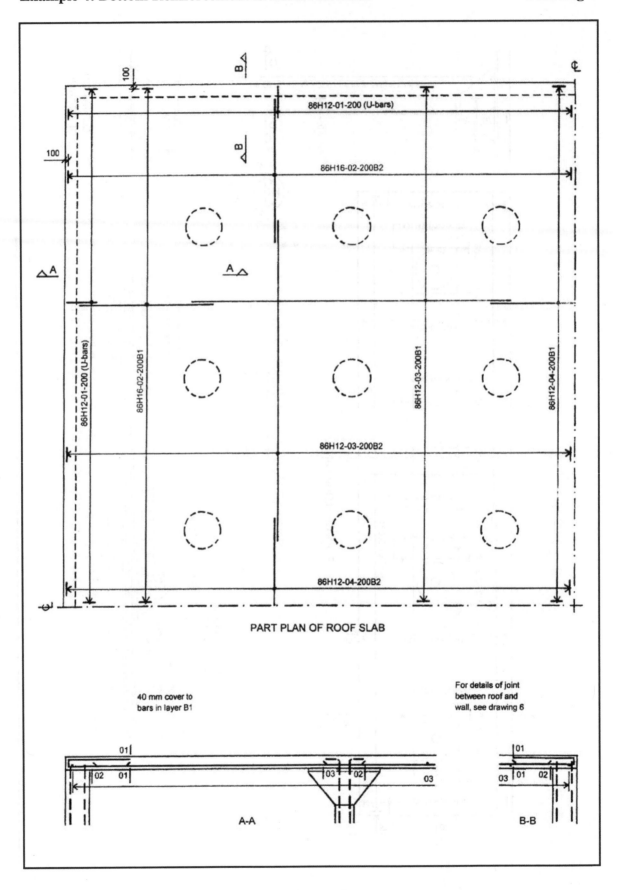

PART PLAN OF ROOF SLAB

40 mm cover to
bars in layer B1

For details of joint
between roof and
wall, see drawing 6

A-A

B-B

Example 4: Top Reinforcement in Reservoir Roof   Drawing 5

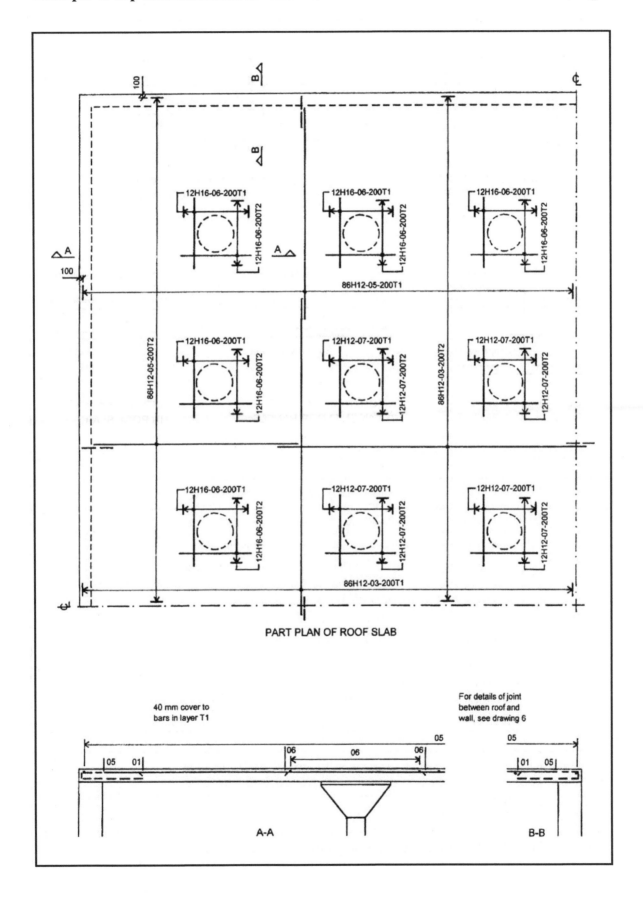

PART PLAN OF ROOF SLAB

40 mm cover to
bars in layer T1

For details of joint
between roof and
wall, see drawing 6

A-A   B-B

Example 4: Joint Details   **Drawing 6**

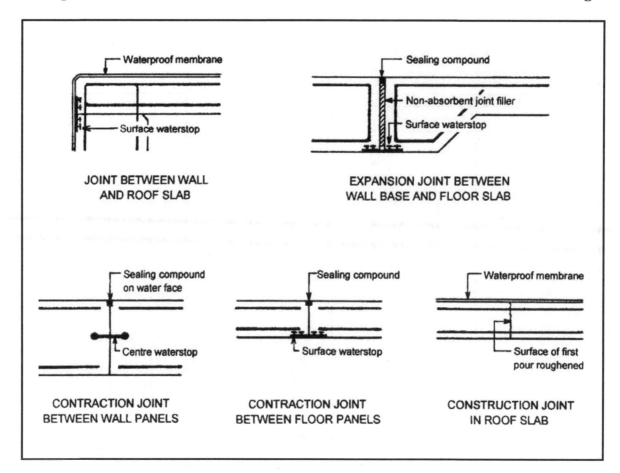

Bar Marks	Commentary on Bar Arrangement (Drawing 3)
01, 02, 03	Straight bars with 50 mm cover generally. For bars 02, lap length = 1.5 × 38 × 12 = 700 mm say.
04	Bars (shape code 36) in edge thickening with 50 mm cover generally.
05	Column starter bars (shape code 11) standing on mat formed by bars 03. Projection of bars above top of base to provide a lap length above kicker on top of column foot = 1.5 × 38 × 12 + 600 + 100 = 1400 mm say. Cover to bars in column = 50 mm to enable 40 mm cover to links.
06	Circular links (shape code 75) to hold column starter bars in place during construction.
07	Straight bars bearing on column kicker and stopping below roof slab.
08	Bars (shape code 11) with lap length = 1.5 × 38 × 12 = 700 mm say.
09	Helical binder (shape code 77), with 40 mm cover, starting above kicker and stopping below the roof slab. Pitch of binder should not exceed 20 × 12 = 240 mm generally. Immediately above base and below slab, main bars are further restrained by column thickening.

Bar Marks	Commentary on Bar Arrangement (Drawings 4 and 5)
01	Bars (shape code 21), with covers: 40 mm bottom and 55 mm top in one direction, and 55 mm bottom and 40 mm top in other direction.
02, 03, 04 05, 06, 07	Straight bars, with lap lengths = 1.5 × 38 × 12 = 700 mm say. Bars 06 and 07 are arranged to alternate with bars 03 and 05, to provide a top mat of bars at 100 mm centres at each column.

7 Example 5: Open-Top Rectangular Tank

Description

An open-top rectangular tank is required to contain 250 m^3 of non-potable water with a minimum freeboard of 100 mm. Allowance is to be made for the water table rising to the ground level, with the underside of the base located at a depth of 1.5 m below ground level. The bearing stratum is medium dense sand, with the following presumed values: 100 kN/m^2 allowable bearing pressure and 20 kN/m^3 modulus of subgrade reaction.

Consider a tank whose internal dimensions are $11.7 \times 5.7 \times 3.85$ m^3 deep, where the thickness of the wall and the base is 300 mm. The capacity of the tank, allowing for freeboard, is $11.7 \times 5.7 \times 3.75 = 250$ m^3.

The tank will be constructed without movement joints, so that continuity is obtained in both horizontal and vertical planes. Maximum values of bending moments and shear forces on vertical and horizontal strips of unit width can be determined from the tables in Appendix C, for tank walls that are either hinged or fixed at the bottom. In considering the horizontal spans, shear forces at the vertical edges of one wall result in axial forces in the adjacent walls. Thus, for internal hydrostatic loading, the shear force at the end of one wall is equal to the tensile force in the adjacent walls, and vice versa.

Note: At the bottom of the wall, a hinged condition can be created by using a narrow footing tied into the floor slab, or by adopting a reinforced hinge detail. A fixed condition can be created by widening the footing until a uniform distribution of bearing pressure is obtained. This will result when the width of the base is such that the moment about the centreline of the wall, due to the weight of water on the base is equal, to the fixed-edge moment at the bottom of the wall. When the tank wall is continuous with the floor slab, the deformation of the floor is complex and dependent on the assumed ground conditions. The effect of the resulting edge rotation on the moment at the bottom of the wall will be examined in this example.

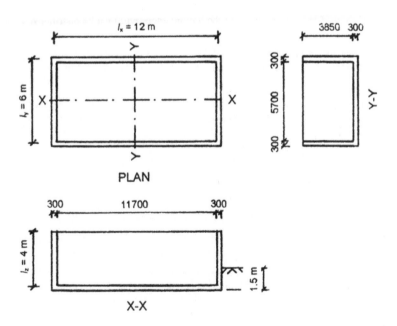

PLAN

X-X

Note: In the UK National Annex to Eurocode 1: Part 4, $\gamma_Q = 1.2$ is recommended for liquid-induced loads at the ultimate limit state. In Eurocode 2: Part 3, it is implied although not clearly stated, that it is sufficient to check for cracking under quasi-permanent loading. In this example, characteristic loading will be taken, as explained in Chapter 1.

Example 5   Calculation Sheet 1

Reference	CALCULATIONS	OUTPUT
BS EN 1997 2.4.7.4	**GEOTECHNICAL DESIGN** **Stability check against uplift due to groundwater** Design stabilising force due to weight of empty tank, with $\gamma_{G,stb} = 0.9$, $G_{dst,d} = = [(12.3 \times 6.3 + 36.0 \times 3.85) \times 0.3 \times 25] \times 0.9 = 1458$ kN Design destabilising vertical force due to 1.5 m head of water, with $\gamma_{G,dst} = 1.0$, $V_{dst,d} = [12.3 \times 6.3 \times 1.5 \times 9.81] \times 1.0 = 1140$ kN $(< G_{dst,d} = 1458$ kN$)$	
	STRUCTURAL DESIGN **(1) Tank full (no external loading)** The internal liquid level will be taken to the tops of the walls assuming that the liquid outlets are blocked. Vertical loading due to water in the tank can be taken as a uniform load over the entire area of the floor, modified by an upward line load due to the displacement of water by concrete in the perimeter wall loads. Uniform loading due to the weight of water and floor slab will be transferred directly to the ground, with no structural effect on the floor or the walls. **Analysis** The following is an approximate analysis in which the continuity of middle strips of unit width is examined in each direction. The floor slab is considered initially as a series of beams with free ends bearing on an elastic soil. The moments needed to restrain the resulting rotations at the junctions with the walls are then determined. The walls are considered as rectangular panels fixed at the bottom, and analysed to determine the vertical moment at the middle of the bottom edge due to hydrostatic loading. The out-of-balance moments at the bottom edges will then be distributed according to the stiffness of the members and the resulting effects determined. The end slopes for beams on elastic foundations can be determined from the data in Table B2, where $\lambda L = (3k_s L^4/E_c h^3)^{1/4}$. With $E_c = 32$ GN/m^2 and $h = 0.3$ m, $\lambda L = [3 \times 20 \times 10^3 /(32 \times 10^6 \times 0.3^3)]^{1/4} \times L = 0.5L$ say The end slopes for beams of unit width with loads $0.5F$ at each end $(a/L = 0)$, and the moments needed to offset the end slopes, can be determined as follows: Section X-X, $L = 12$ m, $\lambda L = 6.0$, $\theta_A = -\theta_B = -36.1F/k_s L^2$, $M_A = -M_B = -(k_s L^3/861.1) \times \theta_A = 0.0419FL$ Section Y-Y, $L = 6$ m, $\lambda L = 3.0$, $\theta_A = -\theta_B = -8.75F/k_s L^2$, $M_A = -M_B = -(k_s L^3/117.6) \times \theta_A = 0.0744FL$ Stiffness values and moment distribution factors can be determined as follows: Floor: $K_f = k_s L^3/c_\theta$, where c_θ is the end slope coefficient obtained from Table B2 for concentrated moments at both ends, according to the value of λL. Wall: $K_z = \alpha_{kz} D/l_z$, where α_{kz} is the panel stiffness coefficient at middle of panel length obtained from Table C1 for a rectangular panel with top edge free and the other edges fixed, according to the value of l_x/l_z (or l_y/l_z). $$D = E_c h^3/12(1-\upsilon^2) = 32 \times 10^6 \times 0.3^3/(12 - 0.2^2) = 0.075 \times 10^6$$	

Section	Panel	Dimensions	Stiffness (kN m/m)
X-X	Floor wall	$L = 12$ m, $\lambda L = 6$ $l_y/l_z = 6/4 = 1.5$	$K_f = 20 \times 10^3 \times 12^3/861.1 = 0.0401 \times 10^6$ $K_z = 5.2 \times 0.075 \times 10^6/4 = 0.0975 \times 10^6$
		$D_f = 0.0401/(0.0401 + 0.0975) = 0.291$, $D_w = 0.709$	
Y-Y	Floor wall	$L = 6$ m, $\lambda L = 3$ $l_x/l_z = 12/4 = 3$	$K_f = 20 \times 10^3 \times 6^3/117.6 = 0.0367 \times 10^6$ $K_z = 2.0 \times 0.075 \times 10^6/4 = 0.0375 \times 10^6$
		$D_f = 0.0367/(0.0367 + 0.0375) = 0.495$, $D_w = 0.505$	

Example 5   **Calculation Sheet 2**

Reference	CALCULATIONS	OUTPUT
	Design ultimate values of $\gamma_G = 1.35$ for concrete, and $\gamma_Q = 1.2$ for water, apply with the maximum water depth taken as 3.85 m.	

Total design ultimate vertical load on floor beam due to load at both ends is

$F = 2 \times (1.35 \times 25 - 1.2 \times 9.81) \times 3.85 \times 0.3 = 50.8$ kN/m.

Fixed-edge moments at junctions of floor and walls can be determined as follows:

Floor: $M_f = 0.0419FL$ (section X-X) and $M_f = 0.0744FL$ (section Y-Y)

Wall: $M_w = \alpha_{mz}(1.2\gamma)l_z^3$, where α_{mz} is the relevant vertical moment coefficient obtained from Table C6 for case (2), that is, $\alpha_{mz,y}$ for X-X and $\alpha_{mz,x}$ for Y-Y.

OUTPUT: Fixed-edge moments

$M_w = 38.4$

$M_f = 25.6$

Section X-X

$M_w = 98.7$

$M_f = 22.7$

Section Y-Y

Section	Panel	Dimensions	Fixed-Edge Moments (kN m/m)
X-X	Floor	$L = 12$ m, $\lambda L = 6$	$M_f = 0.0419 \times 50.8 \times 12 = 25.6$
	wall	$l_x/l_z = 3$, $l_y/l_z = 1.5$	$M_w = 0.051 \times 1.2 \times 9.81 \times 4.0^3 = 38.4$
Y-Y	Floor	$L = 6$ m, $\lambda L = 3$	$M_f = 0.0744 \times 50.8 \times 6 = 22.7$
	wall	$l_x/l_z = 3$, $l_y/l_z = 1.5$	$M_w = 0.131 \times 1.2 \times 9.81 \times 4.0^3 = 98.7$

Resulting moments at the junctions, after releasing the fixed-end moments, are:

Section X-X: $M_f = 25.6 - 0.291 \times (25.6 + 38.4) = +7.0$, $M_w = -7.0$ kN m/m

Section Y-Y: $M_f = 22.7 - 0.495 \times (22.7 + 98.7) = -37.4$, $M_w = +37.4$ kN m/m

It can be seen that these values are considerably less than the fixed-edge moments at the bottom of the walls. For section X-X, the moment has even changed sign. A reasonable approach will be to design the rest of the tank on the basis of a hinged condition at the bottoms of the walls.

Maximum moments in the walls at other positions can be determined as follows:

$M = \alpha_m(1.2\gamma)l_z^3 = (1.2 \times 9.81 \times 4.0^3)\alpha_m = 754\alpha_m$, where α_m is the relevant moment coefficient obtained from Table C7 for case (4). The value of z/l_z at which the maximum value occurs can be obtained from Table C3.

Moment Considered	z/l_z	α_m	$M = 754\alpha_m$ (kN m/m)
Negative moment at corners	0.9	0.132	99.5
Positive moment for span l_x	1.0	0.080	60.3
Positive moment for span l_y	0.6	0.012	9.1
Positive moment for span $l_{z,x}$	0.4	0.053	40.0
Positive moment for span $l_{z,y}$	0.3	0.021	15.8

Maximum shear forces in the walls can be determined as follows:

$V = \alpha_v(1.2\gamma)l_z^2 = (1.2 \times 9.81 \times 4.0^2)\alpha_v = 188.5\alpha_v$, where α_v is an appropriate shear force coefficient estimated from the values in Table C2.

For the short wall, α_v values given for panel type 4 will be used. For the long wall, where there is approximately 40% partial fixity at the bottom, α_v values obtained by interpolation between those given for panel types 2 and 4 will be used. Also, an allowance will be made for the effect of continuity of the horizontal spans on the shear forces at the side edges (reduced for long wall, increased for short wall).

Thus, for the short wall, $\alpha_v = 0.26$ at bottom edge and 0.37 at side edge (before adjustment for continuity). For the long wall, by interpolation, $\alpha_v = 0.42$ at bottom edge and 0.60 at side edge (before adjustment for continuity).

Shear considered	Short Wall		Long Wall	
	α_v	$V = 188.5\alpha_v$	α_v	$V = 188.5\alpha_v$
Shear force at side edge	0.40	75.4	0.54	101.8
Shear force at bottom edge	0.26	49.0	0.42	79.2

Example 5   **Calculation Sheet 3**

Reference	CALCULATIONS	OUTPUT
	Bending moments at the middle of the floor can be determined as follows:	

$M = c_o M_f + c_1 FL$, where c_o and c_1 are coefficients obtained from Tables B3 and B5, for values of $x/L = 0.5$ and $a/L = 0$, respectively.

Section	Dimensions	Moment at Middle of Floor (kNm/m)
X-X	$L = 12$ m, $\lambda L = 6$	$M = -0.084 \times 7.0 - 0.001 \times 50.8 \times 12 = -1.2$
Y-Y	$L = 6$ m, $\lambda L = 3$	$M = 0.492 \times (-37.4) - 0.070 \times 50.8 \times 6 = -39.8$

Maximum shear force at the edges of the floor is as follows:

$V = 0.5F = 0.5 \times 50.8 = 25.4$ kN/m

(2) Tank empty (1.5 m head of groundwater)

An analysis similar to that for the tank full condition will be used, but the pressure distribution under the floor will be uniform.

Stiffness values can be determined as follows:

Floor: $K_z = \alpha_k D/l_z$, where values of coefficient α_k are obtained from Table C1 for a rectangular panel with all edges fixed, and $l_x/l_z = 12.0/6.0 = 2.0$.

Wall: $K_z = \alpha_{kz} D/l_z$, where values of coefficient α_{kz} are as obtained for case (1)

Moment distribution factors are as follows:

Section X-X, $D_f = (3.0/6.0)/(3.0/6.0 + 5.2/4.0) = 0.278$, $D_w = 0.722$

Section Y-Y, $D_f = (2.5/6.0)/(2.5/6.0 + 2.0/4.0) = 0.455$, $D_w = 0.545$

Uniform upward pressure due to maximum design weight of walls is

$p = 1.35 \times 36 \times 3.85 \times 0.3 \times 25/(12.3 \times 6.3) = 18.1$ kN/m^2

Fixed-edge moments for the floor can be determined as follows:

$M_f = \alpha_m pl_z^2$, where values of coefficient α_m are obtained from Table C5 for a rectangular panel with all edges fixed, and $l_x/l_z = 12.0/6.0 = 2.0$.

Section X-X, $M_f = 0.057 \times 18.1 \times 6.0^2 = 37.2$ kN m/m

Section Y-Y, $M_f = 0.083 \times 18.1 \times 6.0^2 = 54.1$ kN m/m

Fixed-edge moments for the walls due to earth pressure and groundwater will be taken as for a vertical cantilever, where effective height = 1.35 m. Considering design minimum load due to active earth pressure and groundwater:

$M_w = -(0.3 \times 20 + 9.81) \times 1.35^3/6 = -6.5$ kNm/m

Resulting moments at edge of floor, after releasing fixed-end moments, are

Section X-X, $M_f = 37.2 - 0.278 \times (37.2 - 6.5) = 28.7$ kN m/m

Section Y-Y, $M_f = 54.1 - 0.455 \times (54.1 - 6.5) = 32.5$ kN m/m

Hogging moments at middle of floor can be determined from Table C5 as follows:

Section X-X, $M_f = 0.016 \times 18.1 \times 6.0^2 + (37.2 - 28.7) = 19.0$ kN m/m

Section Y-Y, $M_f = 0.042 \times 18.1 \times 6.0^2 + (54.1 - 32.5) = 49.0$ kN m/m

Shear forces at edges of floor can be determined from Table C5 as follows:

$V_f = \alpha_v pl_z$, where values of α_v are taken for all edges hinged and $l_x/l_z = 2.0$

Section X-X, $V_f = 0.37 \times 18.1 \times 6.0 = 40.2$ kN/m

Section Y-Y, $V_f = 0.50 \times 18.1 \times 6.0 = 54.3$ kN/m

Durability

External and internal surfaces of the tank are likely to be exposed to cyclic wet and dry conditions, class XC4, and moderate water saturation without de-icing agents, class XF1 (*Reynolds*, Table 4.5). Concrete of minimum strength class C28/35 will be specified, with covers $c_{min} = 30$ mm and $c_{nom} = 40$ mm.

OUTPUT column:

Fixed-edge moments

$M_w = 6.5$

$M_f = 37.2$

Section X-X

$M_w = 6.5$

$M_f = 54.1$

Section Y-Y

Reference: BS 8500

Concrete strength class C28/53 with 40 mm cover to both faces

Example 5

Calculation Sheet 4

Reference	CALCULATIONS	OUTPUT
	WALLS	
	In the horizontal direction, the walls are subjected to a combination of bending and direct tension (when the tank is full), where the direct tension in one wall is equal in magnitude to the shear force at the end of the adjacent wall.	Tensile force acting at mid-depth of section
	Design for combined flexure and tension	
	A rectangular section that is subjected to a bending moment M and a tensile force N acting at the mid-depth of the section, where $M/N \geq (d - 0.5h)$, can be designed for a reduced moment $M_1 = M - N(d - 0.5h)$ and a tensile force N acting at the level of the tension reinforcement. The required tensile reinforcement is given by:	Tensile force acting at level of reinforcement
	$A_s = M_1/(0.87f_{yk}z) + N/0.87f_{yk}$ where z/d can be determined from Table A1	
	Allowing for 40 mm cover and 16 mm diameter horizontal bars in the outer layers, $d = 300 - (40 + 16/2) = 250$ mm say.	
	Maximum design horizontal moment at corner of tank, and coexistent tensile force in short wall, are $M_{Ed} = 99.5$ kN m/m, and $N_{Ed} = 101.8$ kN/m, respectively.	
	$M_1 = 99.5 - 101.8 \times (0.250 - 0.150) = 99.5 - 10.2 = 89.3$ kN m/m	
	$M_1/bd^2f_{ck} = 89.3 \times 10^6/(1000 \times 250^2 \times 28) = 0.051$ $z/d = 0.95$ (max)	
	$A_s = 89.3 \times 10^6/(0.87 \times 500 \times 0.95 \times 250) + 101.8 \times 10^3/(0.87 \times 500)$	
	$= 865 + 234 = 1099$ mm^2/m (H16-150)	
	Maximum design moment, and coexistent tensile force, at mid-point of long wall are $M_{Ed} = 60.3$ kNm/m, and $N_{Ed} = 75.4$ kN/m, respectively.	
	$M_1 = 60.3 - 75.4 \times (0.250 - 0.150) = 60.3 - 7.6 = 52.7$ kN m/m	
	$A_s = 52.7 \times 10^6/(0.87 \times 500 \times 0.95 \times 250) + 75.4 \times 10^3/(0.87 \times 500)$	
	$= 510 + 174 = 684$ mm^2/m (H12-150)	
	For the vertical bars, $d = 300 - (40 + 16 + 12/2) = 235$ mm say.	
	Maximum design vertical moment at the bottom of the wall is $M_{Ed} = 37.4$ kN m/m. Higher up the wall, the sagging moment of 40 kNm/m, which was obtained on the basis of a hinged condition, will be reduced due to the partial fixity at the bottom.	
	$A_s = 37.4 \times 10^6/(0.87 \times 500 \times 0.95 \times 235) = 386$ mm^2/m (H12-200)	
	Design for shear	
	Maximum design shear force and coexistent tensile force at ends of long wall are $V_{Ed} = 101.8$ kN/m, and $N_{Ed} = 75.4$ kN/m, respectively.	
6.2.2 (1)	$v_{Ed} = V_{Ed}/b_wd = 101.8 \times 10^3/(1000 \times 250) = 0.41$ MPa	
	With $v_{min} = 0.035k^{3/2}f_{cu}^{1/2}$, where $k = 1 + (200/250)^{1/2} = 1.89$ and $\sigma_{cp} = N_{Ed}/A_c$:	
	$v_{min} + k_1\sigma_{cp} = 0.035 \times 1.89^{3/2} \times 28^{1/2} - 0.15 \times 75.4/300 = 0.44$ MPa $(> v_{Ed})$	
	Maximum design shear force at bottom edge is $V_{Ed} = 79.2$ kN/m. Hence,	
	$v_{Ed} = 79.2 \times 10^3/(1000 \times 235) = 0.41$ MPa $(< v_{min} = 0.48$ MPa)	
	Cracking due to loading	
BS EN 1992-3 7.3.1	The requirements of Eurocode 2: Part 1 may be applied, provided the depth of the compression zone is at least equal to the lesser of $0.2h$ or 50 mm in all conditions.	
	For elastic analysis of the section, values of x/d can be determined, according to the value of $100M_1/bd^2\sigma_s$, from Table A6. Assuming the ratio M_1/σ_s is the same for ULS and SLS, and allowing for the reduction in σ_s due to $A_{s,prov} > A_{s,req}$, the following values are obtained:	
	For the horizontal spans and the section at the corner of the tank	
	$100M_1/bd^2\sigma_s = 100 \times 89.3 \times 10^6/(1000 \times 250^2 \times 435 \times 1099/1340) = 0.400$	
	$x/d = 0.305$ (for $\alpha_e = 15$) and $x = 0.305 \times 250 = 76$ mm $(\geq 50$ mm$)$	

Example 5   **Calculation Sheet 5**

Reference	CALCULATIONS	OUTPUT
	Similarly, for the section at the mid-span of the long wall	
	$100M_1/bd^2\sigma_s = 100 \times 52.7 \times 10^6/(1000 \times 250^2 \times 435 \times 684/754) = 0.214$	
	$x/d = 0.232$ (for $\alpha_e = 15$) and $x = 0.232 \times 250 = 58$ mm (≥ 50 mm)	
	For the vertical section at the bottom of the long wall (flexure only)	
	$100A_s/bd = 100 \times 565/(1000 \times 235) = 0.240$	
	$x/d = 0.235$ (for $\alpha_e = 15$) and $x = 0.235 \times 235 = 55$ mm (≥ 50 mm)	
	Minimum area of reinforcement required in tension zone for crack control:	
7.3.2 (2)	$A_{s,min} = k_c k f_{ct,eff} A_{ct}/\sigma_s$ where	
	$k = 1.0$ for $h = 300$ mm, $f_{ct,eff} = f_{ctm} = 0.3f_{ck}^{(2/3)} = 2.8$ MPa for general design purposes, and $\sigma_s \leq f_{yk} = 500$ MPa.	
	For horizontal spans, at points of zero moment (pure tension), $k_c = 1.0$, $A_{ct} = bh$	
	$A_{s,min} = 1.0 \times 1.0 \times 2.8 \times 1000 \times 300/500 = 1680$ mm²/m H12-125 (EF)	
	For vertical spans (pure bending), $k_c = 0.4$, $A_{ct} = bh/2$	
	$A_{s,min} = 0.4 \times 1.0 \times 2.8 \times 1000 \times 150/500 = 336$ mm²/m (H12-300)	
	The reinforcement stress under characteristic loading is given approximately by:	
	$\sigma_s = (0.87f_{yk}/1.2) \times (A_{s,req}/A_{s,prov}) = 363 \times A_{s,req}/A_{s,prov}$	
7.3.3 (2) Table 7.3	Crack width criterion can be met by limiting the bar spacing. For section subjected to bending, with $w_k = 0.3$ mm, values obtained from *Reynolds*, Table 4.24 are	
	For horizontal spans, and section at corner of tank (H16-150)	
	$\sigma_s = 363 \times 1099/1340 = 298$ MPa Maximum bar spacing = 125 mm	
	For section at mid-span of long wall (H12-150)	
	$\sigma_s = 363 \times 684/754 = 330$ MPa Maximum bar spacing = 90 mm	
	For vertical section at bottom of long wall (H12-200)	Horizontal bars as follows: at corners H16-125 (inside face), elsewhere H12-125 Vertical bars: H12-150 throughout
	$\sigma_s = 363 \times 386/565 = 248$ MPa Maximum bar spacing = 190 mm	
	It can be seen that the bar spacing requirements are not met. In order to comply, the horizontal bars will be changed to H16-125 at the corners and H12-125 in the span. In the latter case, $\sigma_s = 363 \times 684/905 = 275$ MPa, which is acceptable. The vertical bars will be changed to H12-150(EF) throughout.	
BS EN 1992-3 7.3.3	For section subjected to pure tension in short wall, with H12-125 (EF),	
	$\sigma_s = N_{Ed}/(1.2A_s) = 101.8 \times 10^3/(1.2 \times 1810) = 47$ MPa	
	From the chart given in *Reynolds*, Table 4.25, it can be implied that $w_k < 0.05$ mm.	
BS EN 1992-3 7.3.1	**Cracking due to restrained early thermal contraction**	
	For cracks that can be expected to pass through the full thickness of the section, the crack width limit is given by	
	$w_{k,lim} = 0.225(1 - z_w/45h) \leq 0.2$ mm where z_w is depth of water at section	
	Hence, for $h = 0.3$ m: $w_{k,lim} = 0.225(1 - z_w/13.5) \leq 0.2$ mm from which	
	$w_{k,lim} = 0.2$ mm for $z_w \leq 1.5$ m decreasing linearly to 0.16 mm at $z_w = 3.85$ m	
7.3.2 (2) 7.3.4	Minimum area of horizontal reinforcement, with $f_{ct,eff} = 1.8$ MPa for cracking at age of 3 days, $k_c = 1.0$ for tension and $k = 1.0$ for $h = 300$ mm, is given by	
	$A_{s,min} = k_c k f_{ct,eff} A_{ct}/f_{yk} = 1.0 \times 1.0 \times 1.8 \times 1000 \times 300/500 = 1080$ mm²/m	
	With $c = 40$ mm, $k_1 = 0.8$ for high bond bars, $k_2 = 1.0$ for tension, $h_{c,ef}$ as the lesser of $2.5(h - d)$ and $h/2$, and H12-125 (EF) as minimum reinforcement:	
	$s_{r,max} = 3.4c + 0.425k_1k_2(A_{c,eff}/A_s)\varphi$	
	$= 3.4 \times 40 + 0.425 \times 0.8 \times 1.0 \times (2.5 \times 46 \times 1000/905) \times 12 = 655$ mm	

Example 5   **Calculation Sheet 6**

Reference	CALCULATIONS	OUTPUT
PD 6687 2.16	With $R = 0.8$ for wall on a thick base, $\Delta T = 25°C$ for 350 kg/m³ Portland cement concrete and 300 mm thick wall (*Reynolds*, Table 2.18), and $\alpha = 12 \times 10^{-6}$ per °C: $w_k = (0.8 R \alpha \Delta T) \times s_{r,max} = 0.8 \times 0.8 \times 12 \times 10^{-6} \times 25 \times 655 = 0.13$ mm **Curtailment of horizontal reinforcement** The publication referred to in Table C1 provides moment coefficients at intervals of one-tenth of the height and length of each wall. The moments in the horizontal spans reduce rapidly from the maximum value that occurs at a corner of the tank. At distances from the corner of $0.05 l_x$ (by interpolation), and $0.1 l_y$, the moment coefficient is 0.080. Thus, for the short wall: $M_{Ed} = 754 \times 0.08 = 60.3$ kNm/m $N_{Ed} = 101.8$ kN/m (constant) $M_1 = 60.3 - 101.8 \times (0.250 - 0.150) = 60.3 - 10.2 = 50.1$ kN m/m $A_s = 50.1 \times 10^6 /(0.87 \times 500 \times 0.95 \times 250) + 101.8 \times 10^3 /(0.87 \times 500)$ $\qquad = 485 + 234 = 719$ mm²/m (H12-125)	
9.2.1.3 (2)	Hence, the bars required at the corner of the tank can be reduced to H12-125 at a distance of 600 mm from the corner. However, the curtailed bars should extend for a further minimum distance $a_l = d = 250$ mm. It is also necessary to ensure that the bars extend for a distance not less than $(a_l + l_{bd})$ beyond the adjacent wall.	Corner bars to extend for a distance beyond face of adjacent wall not less than 1125 mm
8.4.3 (2)	From *Reynolds*, Table 4.30, for poor bond conditions, $l_{b,rqd} = 54\phi$. Taking $l_{bd} = l_{b,rqd}$ for simplicity, the curtailed bars should extend beyond the face of the adjacent wall for a distance not less than $(a_l + l_{bd}) = 250 + 54 \times 16 = 1125$ mm. For the lapping bars, from *Reynolds*, Table 4.31, the design lap length:	
8.7.3 (1)	$l_0 = \alpha_6 \times (54\phi) = 1.5 \times 54 \times 12 = 1000$ mm say.	Lap length = 1000 mm
	Deflection The maximum deflection at the top of the walls occurs at the middle of the long wall. A rough estimate of the deflection under service loading can be made from the relationship given in Table C3: $a = \alpha_d \gamma l_z^5 /D$ where, for a rectangular panel with top edge free and $l_x /l_z = 3.0$, $\qquad \alpha_d = 0.0487$ (bottom edge hinged) and 0.0184 (bottom edge fixed). *Note*: More accurate values can be obtained from the publication referred to in Table C1 where, for a rectangular tank with $l_x /l_z = 3.0$ and $l_y /l_z = 1.5$, $\alpha_d = 0.0635$ (bottom edge hinged) and 0.0197 (bottom edge fixed). It can be seen that these values exceed those obtained for a rectangular panel with fixed vertical edges. On the other hand, an interpolated value allowing for 40% partial fixity at the bottom of the long wall would be close to the value given for the rectangular panel with bottom edge hinged. This value will be used in the following calculations. The flexural rigidity $D = E_c h^3 /12(1-\upsilon^2)$ for an uncracked plain concrete section, where the moment required to cause cracking, with $f_{ctm} = 2.8$ MPa, is $M_{cr} = f_{ctm}(bh^2/6) = 2.8 \times 10^3 \times (1.0 \times 0.3^2/6) = 42$ kN m/m Since the service moments in the vertical direction are all less than this value, it is reasonable to estimate the deflection on the basis of an uncracked section. An allowance for the effect of creep can be made by using an effective long-term modulus of elasticity as follows:	
3.1.3 (2) Table 3.1	Secant modulus of elasticity of concrete at 28 days: $E_{cm} = 22[(f_{ck} + 8)/10]^{0.3} = 22 \times 3.6^{0.3} = 32.3$ GPa	
3.1.4 (2) Figure 3.1	Final creep coefficient, for a C28/35 concrete with normally hardening cement in outside conditions (RH = 85%), for a member of notional thickness 300 mm and loaded at 28 days, is $\varphi(\infty,t_0) = 1.5$ say.	
7.4.3 (5)	Effective modulus of elasticity for long-term deformation is $E_{c,eff} = E_{cm}/[1 + \varphi(\infty,t_0)] = 32.3/2.5 = 13.0$ GPa	

Example 5   Calculation Sheet 7

Reference	CALCULATIONS	OUTPUT
	Hence, with $\upsilon = 0.2$, $D = 13.0 \times 10^6 \times 0.3^3/11.52 = 30.5 \times 10^3$ kN m. Then,	
	$a = \alpha_d \gamma l_z^5/D = 0.0487 \times 9.81 \times 4.0^5/30.5 = 16$ mm	
	FLOOR	
	When the tank is full, the slab is subjected to bending moments combined with direct tensions equal in magnitude to the shear forces at the bottom of the walls. When the tank is empty, there are no direct tensions, but the slab is subjected to larger bending moments due to the effect of the groundwater.	
	Design for combined flexure and tension	
	In the short span direction, with tank full, maximum design moment at middle of floor is $M_{Ed} = 39.8$ kNm/m. Co-existent tensile force is $N_{Ed} = 79.2$ kN/m.	
	$M_1 = 39.8 - 79.2 \times (0.250 - 0.150) = 31.9$ kN m/m	
	$M_1/bd^2 f_{cu} = 31.9 \times 10^6/(1000 \times 250^2 \times 28) = 0.019$ $z/d = 0.95$ (max)	
	$A_s = 31.9 \times 10^6/(0.87 \times 500 \times 0.95 \times 250) + 79.2 \times 10^3/(0.87 \times 500)$	
	$\quad = 309 + 182 = 491$ mm²/m (H12-150 say)	
	In the short span direction, with tank empty, maximum design moment at middle of floor is $M_{Ed} = 49.0$ kNm/m.	
	$M/bd^2 f_{cu} = 49.0 \times 10^6/(1000 \times 250^2 \times 28) = 0.028$ $z/d = 0.95$ (max)	
	$A_s = 49.0 \times 10^6/(0.87 \times 500 \times 0.95 \times 250) = 475$ mm²/m (H12-150 say)	
	Design for shear	
	In the short span direction, with tank empty, maximum design shear force at edge of floor is $V_{Ed} = 54.3$ kN/m.	
6.2.2 (1)	$v_{Ed} = V_{Ed}/b_w d = 54.3 \times 10^3/(1000 \times 250) = 0.22$ MPa ($< v_{min} = 0.48$ MPa)	
	Cracking due to loading	
	In the short span direction, with tank full, assuming the ratio M_1/σ_s is the same for ULS and SLS, and allowing for the reduction in σ_s due to $A_{s,prov} > A_{s,req}$:	
	$100 M_1/bd^2 \sigma_s = 100 \times 31.9 \times 10^6/(1000 \times 250^2 \times 435 \times 491/754) = 0.180$	
	$x/d = 0.214$ (for $\alpha_e = 15$) and $x = 0.214 \times 250 = 53$ mm (≥ 50 mm)	
7.3.3 (2) Table 7.3	Stress in reinforcement, and maximum bar spacing for $w_k = 0.3$ mm, are	Provide H12-150 (EW) throughout
	$\sigma_s = 363 \times 491/754 = 237$ MPa Maximum bar spacing = 200 mm	

Bar Marks	Commentary on Bar Arrangement (Drawings 1 and 2)
01, 02	Bars (shape code 21) lapping with bars 04 and 05, respectively. Lap length = $1.5 \times 38 \times 12 = 700$ mm say. Cover to outer bars = 50 mm (bottom), and 40 mm (top and ends).
03	Bars (shape code 21) to lap with vertical bars in walls. Projection above floor to provide a lap length above kicker = $1.5 \times 38 \times 12 + 100 = 800$ mm say. Cover = 55 mm, to allow for 40 mm cover to horizontal bars.
04	Straight bars curtailed 200 mm from face of wall, and lapping 700 mm with bar 01.
05	Straight bars in 6 m lengths, with 700 mm lap in floor and 1000 mm lap in walls (see comment for bar 06).
06	Bars (shape code 21) lapping with bars 05 and 09. Cover = 40 mm. Assuming poor bond conditions, lap length = $1.5 \times 54 \times 12 = 1000$ mm say.
07	Straight bars bearing on kicker and curtailed 150 mm below top of wall.
08	Bars (shape code 21) lapping 700 mm with bars 07.
09	Straight bars curtailed 350 mm from face of adjacent wall, and lapping 1000 mm with bar 06.

Example 5: Reinforcement in Tank Floor   Drawing 1

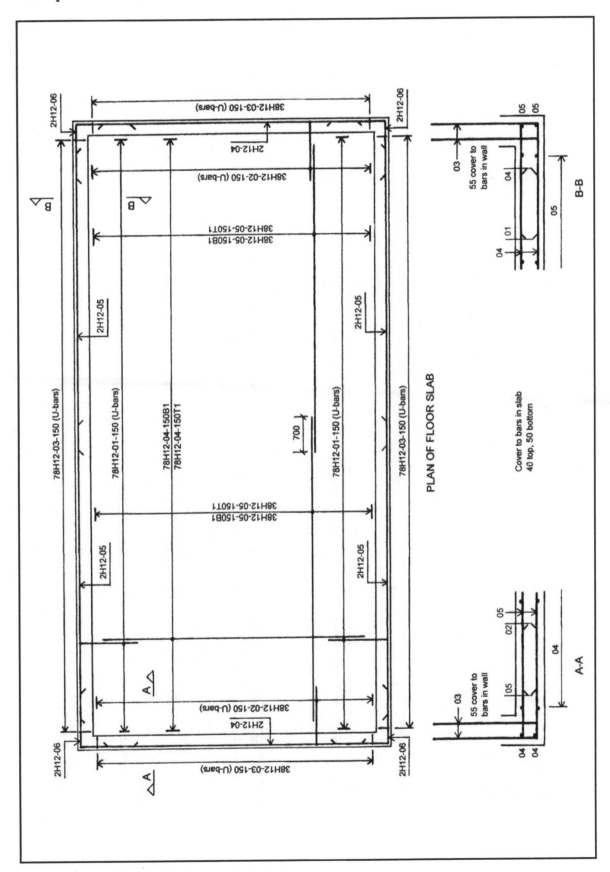

PLAN OF FLOOR SLAB

Cover to bars in slab
40 top, 50 bottom

Example 5: Reinforcement in Tank Walls   **Drawing 2**

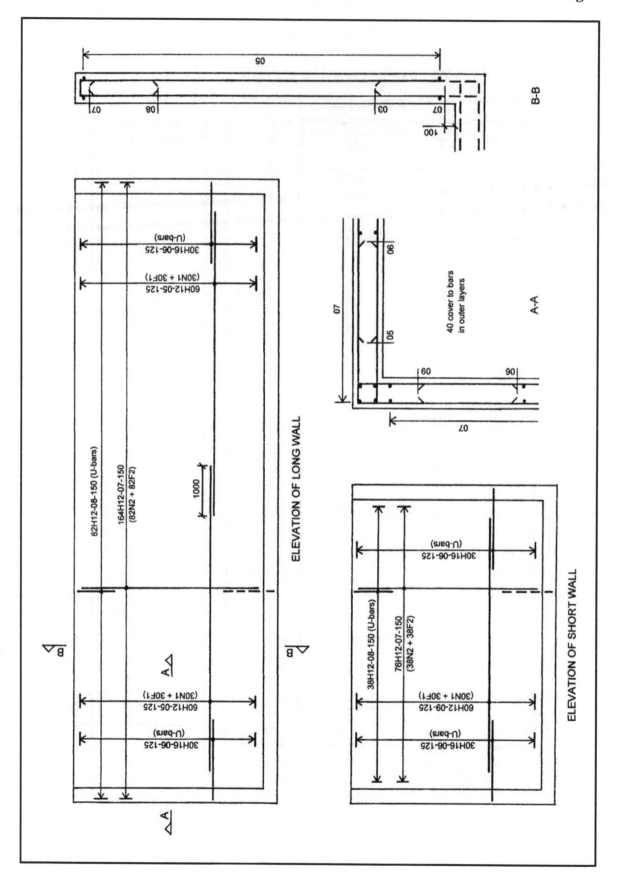

8 Example 6: Open-Top Cylindrical Tank

Description

An open-top cylindrical tank is required to contain 600 m³ of non-potable water with a minimum freeboard of 125 mm. The bearing stratum is firm clay, with the following presumed values: 150 kN/m² allowable bearing pressure and 18 kN/m³ modulus of subgrade reaction.

Consider a tank whose internal dimensions are 11.75 m diameter × 5.875 m deep, where the thickness of the wall and the base is 250 mm. Capacity of tank, allowing for freeboard, is $(\pi/4) \times 11.75^2 \times 5.75 = 620$ m³.

The tank will be constructed without movement joints, so that continuity is obtained between the wall and the floor. Values of circumferential forces, vertical moments and radial shears can be determined from the tables in Appendix C, for tank walls that are either hinged or fixed at the bottom. In this example, where the wall and the base are continuous, the effect of the resulting edge rotation on the forces and moments in the wall will be examined.

The following conditions will be considered: (1) tank with water at ambient temperature; (2) tank subjected to hydrostatic and thermal actions, with ambient temperature in the range 0–20°C, and water temperature in the range 0–40°C.

Note: In the UK National Annex to Eurocode 1: Part 4, $\gamma_Q = 1.2$ is recommended for liquid-induced loads at the ULS. In Eurocode 2: Part 3, it is implied although not clearly stated, that it is sufficient to check for cracking under quasi-permanent loading. In this example, the frequent loading combination will be taken, as explained in Chapter 1.

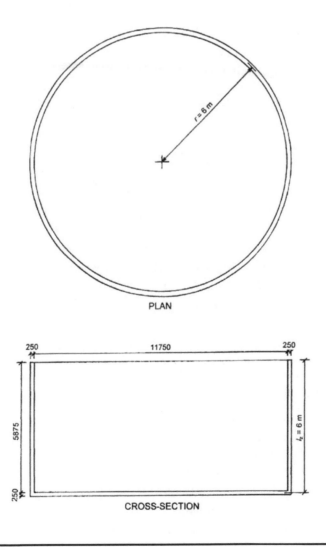

PLAN

CROSS-SECTION

Example 6

Calculation Sheet 1

Reference	CALCULATIONS	OUTPUT
	TANK WITH WATER AT AMBIENT TEMPERATURE	

Loading

The internal liquid level will be taken to the top of the wall assuming that the liquid outlets are blocked. Vertical loading due to water in the tank can be taken as a uniform load over the entire area of the floor, modified by an upward line load due to the displacement of water by concrete in the perimeter wall load. Uniform loading due to the weight of the water and the floor slab is transferred directly to the ground, with no structural effect on the floor or the wall.

Design ultimate values of $\gamma_G = 1.35$ for concrete, and $\gamma_Q = 1.2$ for water, apply with the maximum water depth taken as 5.875 m. Design ultimate perimeter load:

$$Q = (1.35 \times 25 - 1.2 \times 9.81) \times 5.875 \times 0.25 = 32.3 \text{ kN/m}$$

For serviceability (cracking), ratios of service load/ultimate load are:

Concrete: $1.0/1.35 = 0.74$ Liquid (frequent load): $(1.0 \times 0.9)/1.2 = 0.75$

Analysis

The floor, subjected to vertical loading from the perimeter wall, will be considered initially as a circular slab fixed at the edge and bearing on an elastic soil. The wall, subjected to hydrostatic loading, will be considered initially as fixed at the bottom. The out-of-balance moment at the junction of the wall and the slab will then be distributed according to the stiffness of the members and the resulting effects on the wall and the slab determined.

The fixed-edge moment and rotational stiffness of the wall can be determined from the data in Table C13, according to the value of the term $l_z^2/2rh$.

With $h = 0.25$ m, $l_z = 6$ m and $r = 6$ m, $l_z^2/2rh = 6.0^2/(2 \times 6.0 \times 0.25) = 12$

The fixed-edge moment and rotational stiffness of the slab can be determined from the data in Table C14, according to the radius of relative stiffness given by

$$r_k = [E_c h^3/12(1-v^2)k_s]^{0.25} \text{ where } v \text{ is Poisson's ratio}$$

With $E_c = 32$ GN/m^2, $h = 0.25$ m, $k_s = 18$ kN/m^3 and $v = 0.2$,

$$r_k = [32 \times 10^6 \times 0.25^3/(11.52 \times 18 \times 10^3)]^{1/4} = 1.25 \text{ m} \quad r/r_k = 6.0/1.25 = 4.8$$

The rotational stiffness values and moment distribution factors are as follows:

Wall: $K_w = \alpha_w E_c h^3/l_z = 1.108 \times 32 \times 10^6 \times 0.25^3/6.0 = 0.092 \times 10^6$ kN m/m

Slab: $K_s = \alpha_s E_c h^3/r = 0.272 \times 32 \times 10^6 \times 0.25^3/6.0 = 0.023 \times 10^6$ kN m/m

Distribution factors: $D_w = 0.092/(0.092 + 0.023) = 0.800$, $D_s = 0.200$

Fixed-edge moments at junction of wall and slab are as follows:

Wall: $M_w = \alpha_{w1}(1.2\gamma)l_z^3 = 0.0104 \times (1.2 \times 9.81) \times 6.0^3 = 26.5$ kN m/m

Slab: $M_s = \alpha_{s2} Q r = 0.146 \times 32.3 \times 6.0 = 28.3$ kN m/m

Resulting moment at bottom of wall, after releasing fixed-end moments, is

$$M = M_w - 0.8(M_w + M_s) = 26.5 - 43.8 = -17.3 \text{ kN m/m}$$

It can be seen that the moment at the bottom of the wall is of the opposite sign to the fixed-edge moment (i.e. the joint rotation exceeds that for a hinged condition).

The circumferential tensions and vertical moments, at various levels in the wall can now be obtained by combining the results for load cases (1) and (5) in Tables C10 and C12, respectively.

The resulting circumferential force n and vertical moment m in the wall are given by the following equations, in which $0.8(M_w + M_s) = 43.8$ kN m/m:

$$n = \alpha_{n1}(1.2\gamma)l_z r + \alpha_{n5} \times 0.8(M_w + M_s) r/l_z^2$$

$$= (1.2 \times 9.81 \times 6.0 \times 6.0)\alpha_{n1} + (43.8 \times 6.0/6.0^2)\alpha_{n5} = 424\alpha_{n1} + 7.3\alpha_{n5}$$

$$m = \alpha_{m1}(1.2\gamma)l_z^3 + \alpha_{m5} \times 0.8(M_w + M_s)$$

$$= (1.2 \times 9.81 \times 6.0^3)\alpha_{m1} + 43.8\alpha_{m5} = 2543\alpha_{m1} + 43.8\alpha_{m5}$$

OUTPUT:

Fixed-edge moments

$M_w = 26.5$

$M_s = 28.3$

Example 6

Calculation Sheet 2

Reference	CALCULATIONS	OUTPUT

Values of n and m for the bottom half of the wall, where z is the depth from the top of the wall, and $l_z^2/2rh = 12$, are shown in the following tables:

Circumferential Tension in Wall (kN/m)					
Level z/l_z	Load case (1)		Load case (5)		Force n
	α_{n1}	$424\alpha_{n1}$	α_{n5}	$7.3\alpha_{n5}$	
0.5	0.543	230.2	–0.2	–1.5	228.7
0.6	0.628	266.3	3.52	25.7	292.0
0.7	0.633	268.4	11.3	82.5	350.9
0.8	0.494	209.5	21.8	159.2	368.7
0.9	0.211	89.5	25.7	187.6	277.1
1.0	0	0	0	0	0

Vertical Moment in Wall (kN m/m)					
Level z/l_z	Load case (1)		Load case (5)		Moment m
	α_{m1}	$2543\alpha_{m1}$	α_{m5}	$43.8\alpha_{m5}$	
0.5	0.0003	0.8	–0.040	–1.8	–1.0
0.6	0.0013	3.3	–0.064	–2.8	0.5
0.7	0.0023	5.8	–0.049	–2.1	3.7
0.8	0.0026	6.6	0.081	3.6	10.2
0.9	–0.0005	–1.3	0.424	18.6	17.3
1.0	–0.0104	–26.5	1.0	43.8	17.3

The radial shear force at the bottom of the wall is given by the equation:

$$v = \alpha_{v1}(1.2\gamma)l_z^2 + \alpha_{v5} \times 0.8(M_w + M_s)/l_z$$

$$= (1.2 \times 9.81 \times 6.0^2)\,\alpha_{v1} + (43.8/6.0)\,\alpha_{v5} = 424\alpha_{v1} + 7.3\alpha_{v5}$$

Resulting value of v, with $\alpha_{v1} = 0.145$ and $\alpha_{v5} = -6.38$, is

$$v = 424 \times 0.145 - 7.3 \times 6.38 = 14.9 \text{ kN/m}$$

Radial and tangential moments at particular perimeters in the slab, can be obtained by combining the results for load cases (1) and (2) in Table C14.

The resulting moments, m_r and m_t, are given by the following equations, in which $0.2(M_w + M_s) = 11.0$ kN m/m and $Qr = 32.3 \times 6.0 = 193.8$ kN/m:

$$m_r = \alpha_{r1} \times 0.2(M_w + M_s) + \alpha_{r2}\,Qr \qquad m_t = \alpha_{t1} \times 0.2(M_w + M_s) + \alpha_{t2}\,Qr$$

Perimeter r_x/r	Radial Moment (kN m/m)			Tangential Moment (kN m/m)		
	α_{r1}	α_{r2}	m_r	α_{t1}	α_{t2}	m_t
1.0	1.0	–0.146	–17.3	0.509	–0.029	0
0.8	0.789	–0.018	5.2	0.332	0.015	6.6
0.6	0.385	0.028	13.0	0.129	0.028	6.9
0.4	0.089	0.032	7.2	–0.011	0.027	5.1
0.2	–0.061	0.025	4.2	–0.083	0.022	3.4
0	–0.104	0.021	2.9	–0.104	0.021	2.9

The maximum shear force at the edge of the slab is given by $Q = 32.3$ kN/m

Durability

BS 8500

External and internal surfaces of the tank are likely to be exposed to cyclic wet and dry conditions, class XC4, and moderate water saturation without de-icing agents, class XF1 (*Reynolds,* Table 4.5). Concrete of minimum strength class C28/35 will be specified, with covers $c_{min} = 30$ mm and $c_{nom} = 40$ mm.

Concrete strength class C28/35 with 40 mm cover to both faces

Example 6   Calculation Sheet 3

Reference	CALCULATIONS	OUTPUT
	WALL	
	When the tank is full, the wall is subjected to pure tension in the circumferential direction and, ignoring the axial load, pure bending in the vertical direction.	
	Design for circumferential tension and vertical flexure	
	Maximum circumferential tension, at depth $z = 0.8l_z$, is $N_{Ed} = 368.7$ kN/m.	
	$A_s = N_{Ed}/0.87f_{yk} = 368.7 \times 10^3/(0.87 \times 500) = 848$ mm²/m H12-250 (EF)	
	Maximum vertical moment, at bottom of wall, is $M_{Ed} = 17.3$ kN m/m	
	Allowing for 40 mm cover and 12 mm diameter horizontal bars in the outer layers, for the vertical bars, $d = 250 - (40 + 12 + 12/2) = 190$ mm say.	
	$M/bd^2f_{ck} = 17.3 \times 10^6/(1000 \times 190^2 \times 28) = 0.017$ $z/d = 0.95$ (max)	
	$A_s = 17.3 \times 10^6/(0.87 \times 500 \times 0.95 \times 190) = 221$ mm²/m H12-300 say	
	Design for radial shear	
	Maximum radial shear force at bottom of wall is $V_{Ed} = 14.9$ kN/m	
6.2.2 (1)	$v_{Ed} = V_{Ed}/b_w d = 14.9 \times 10^3/(1000 \times 190) = 0.08$ MPa ($< v_{min}$)	
	Cracking due to loading	
	Minimum area of reinforcement required in tension zone for crack control:	
	$A_{s,min} = k_c k f_{ct,eff} A_{ct}/\sigma_s$ where	
	$k = 1.0$ for $h \leq 300$ mm, $f_{ct,eff} = f_{ctm} = 0.3f_{ck}^{(2/3)} = 2.8$ MPa for general design purposes, and $\sigma_s \leq f_{yk} = 500$ MPa.	
	In circumferential direction (pure tension), $k_c = 1.0$, $A_{ct} = bh$	
	$A_{s,min} = 1.0 \times 1.0 \times 2.8 \times 1000 \times 250/500 = 1400$ mm²/m H12-150 (EF)	
	In vertical direction (pure bending), $k_c = 0.4$, $A_{ct} = bh/2$	
	$A_{s,min} = 0.4 \times 1.0 \times 2.8 \times 1000 \times 125/500 = 280$ mm²/m H12-300 say	
BS EN 1992-3 7.3.1	For cracks that can be expected to pass through the full thickness of the section, the crack width limit is given by	
	$w_{k,lim} = 0.225(1 - z_w/45h) \leq 0.2$ mm where z_w is depth of water at section	
	Hence, for $h = 0.25$ m: $w_{k,lim} = 0.225(1 - z_w/11.25) \leq 0.2$ mm from which	
	$w_{k,lim} = 0.2$ mm for $z_w \leq 1.25$ m decreasing linearly to 0.11 mm at $z_w = 5.75$ m	
BS EN 1992-3 7.3.3	For sections in tension, the crack width criterion can be met by limiting the bar spacing to values derived from the charts given in *Reynolds,* Table 4.25. The reinforcement stress under the characteristic load is given by $\sigma_s = 0.75 \times (N_{Ed}/A_s)$.	
	The following table shows the approximate crack width limit at each level, the reinforcement stress for an assumed bar spacing, and the limiting bar spacing.	
		Level z/l_z	Crack Width Limit (mm)	N_{Ed} (kN)	Assumed Bars (EF)	Stress σ_s (MPa)	Maximum Bar Spacing (mm)		Horizontal bars (EF) as follows:
		0.5	0.16	228.7	H12-150	114	300		H12-150 from top of wall to depth of 3.0 m
		0.6	0.15	292.0	H12-125	121	250		H12-100 for depth of more than 3.0 m
		0.7	0.14	350.9	H12-100	117	200		
		0.8	0.13	368.7	H12-100	123	150		
		0.9	0.12	277.1	H12-125	115	175		
BS EN 1992-3 7.3.1	For sections where the depth of the compression zone is at least equal to the lesser of $0.2h$ or 50 mm in all conditions, the requirements of Eurocode 2: Part 1 apply.	
	If vertical bars are made H16-300, $100A_s/bd = 100 \times 670/(1000 \times 190) = 0.352$	
	From Table A6, $x/d = 0.276$ (for $\alpha_e = 15$) and $x = 0.276 \times 190 = 52$ mm	

Example 6

Calculation Sheet 4

Reference	CALCULATIONS	OUTPUT
7.3.3 (2) Table 7.3	The crack width criterion can be met by limiting the bar spacing to values given in *Reynolds,* Table 4.24. The reinforcement stress is given approximately by $\quad \sigma_s = 0.75 \times (A_{s,req}/A_{s,prov}) = 327 \times (A_{s,req}/A_{s,prov})$ Hence, for the section at the bottom of the wall, reinforced with H16-300: $\quad \sigma_s = 327 \times 221/670 = 108$ MPa From the chart in *Reynolds,* Table 4.25, a crack width <0.3 mm is indicated.	Vertical bars: H16-300 (EF)
	Cracking due to restrained early thermal contraction	
7.3.2 (2) 7.3.4	Minimum area of horizontal reinforcement, with $f_{ct,eff} = 1.8$ MPa for cracking at age of 3 days, $k_c = 1.0$ for tension and $k = 1.0$ for $h \leq 300$ mm, is given by $\quad A_{s,min} = k_c k f_{ct,eff} A_{ct}/f_{yk} = 1.0 \times 1.0 \times 1.8 \times 1000 \times 250/500 = 900$ mm^2/m With $c = 40$ mm, $k_1 = 0.8$ for high bond bars, $k_2 = 1.0$ for tension, $h_{c,ef}$ as the lesser of $2.5(h-d)$ and $h/2$, and H12-150 (EF) as minimum reinforcement: $\quad s_{r,max} = 3.4c + 0.425 k_1 k_2 (A_{c,eff}/A_s)\,\varphi$ $\quad\quad\quad = 3.4 \times 40 + 0.425 \times 0.8 \times 1.0 \times (2.5 \times 46 \times 1000/754) \times 12 = 758$ mm	
PD 6687 2.16	With $R = 0.8$ for wall on a thick base, $\Delta T = 25°$C for 350 kg/m^3 Portland cement concrete and 250 mm thick wall (*Reynolds,* Table 2.18), and $\alpha = 12 \times 10^{-6}$ per $°$C: $\quad w_k = (0.8R\alpha\Delta T) \times s_{r,max} = 0.8 \times 0.8 \times 12 \times 10^{-6} \times 25 \times 758 = 0.15$ mm Similarly, with H12-125 (EF): $\quad s_{r,max} = 632$ mm and $w_k = 0.12$ mm Thus, H12-150 (EF) is sufficient for values of $z_w \leq 11.25(1 - 0.15/0.225) = 3.75$ m, and H12-125 (EF) is sufficient for $3.75 < z_w \leq 5.25$ m.	
	Lap requirements for circumferential bars	
Figure 8.2 8.4.3 8.7.3 (1)	From *Reynolds,* Table 4.30, assuming poor bond conditions, $l_{b,rqd} = 54\phi$. From *Reynolds,* Table 4.31, assuming laps are staggered by at least $0.65l_0$ in alternate rows, design lap length: $\quad l_0 = \alpha_6 \times (54\varphi) = 1.4 \times 54 \times 12 = 1000$ mm say.	For horizontal bars, provide 1000 mm laps with laps staggered in alternate rows
	FLOOR In the radial direction, the slab is subjected to bending and shear, in combination with direct tension equal in magnitude to the shear force at the bottom of the wall. **Design for combined flexure and tension** The maximum moment occurs at the edge of the floor, where $M_{Ed} = 17.3$ kN m/m and $N_{Ed} = 14.9$ kN/m. Allowing for 50 mm cover (bottom) and 12 mm diameter radial bars in the outer layer, $d = 250 - (50 + 12/2) = 190$ mm say $\quad M_1 = M_{Ed} - N_{Ed}(d - 0.5h) = 17.3 - 14.9 \times (0.190 - 0.125) = 16.3$ kN m/m $\quad M_1/bd^2 f_{ck} = 16.3 \times 10^6/(1000 \times 190^2 \times 28) = 0.016 \quad z/d = 0.95$ (max) $\quad A_s = M_1/(0.87f_{yk}z) + N/(0.87f_{yk})$ $\quad\quad = 16.3 \times 10^6/(0.87 \times 500 \times 0.95 \times 190) + 14.9 \times 10^3/(0.87 \times 500)$ $\quad\quad = 208 + 34 = 242$ mm^2/m $\quad\quad\quad\quad$ H12-300 say For tension at the top surface, the maximum moment occurs at $r_x/r = 0.6$, where $M_{Ed} = 13.0$ kN m/m. Allowing for 40 mm cover and 12 mm bars, $d_{min} = 190$ mm. $\quad M_1 = 13.0 - 14.9 \times 0.065 = 12.0$ kN m/m $\quad A_s = 12.0 \times 10^6/(0.87 \times 500 \times 0.95 \times 190) + 34 = 187$ mm^2/m $\quad$ H12-300 say **Design for shear** Maximum shear force at the edge of the slab is $V_{Ed} = 32.3$ kN/m	
6.2.2 (1)	$\quad v_{Ed} = V_{Ed}/b_w d = 32.3 \times 10^3/(1000 \times 190) = 0.17$ MPa ($< v_{min}$)	

Example 6   **Calculation Sheet 5**

Reference	CALCULATIONS	OUTPUT
	Cracking due to loading	
	For the section at the edge of the floor, assuming the ratio M_1/σ_s for SLS is the same as for ULS, increasing the reinforcement to H16-300, and allowing for the reduction in σ_s due to $A_{s,prov} > A_{s,req}$:	
	$100 M_1 / b d^2 \sigma_s = 100 \times 16.3 \times 10^6/(1000 \times 190^2 \times 435 \times 242/670) = 0.287$	
	From Table A6, $x/d = 0.264$ (for $\alpha_e = 15$) and $x = 0.262 \times 190 = 50$ mm	
	The reinforcement stress is given approximately by	
7.3.3 (2)	$\sigma_s = 363 \times (A_{s,req}/A_{s,prov}) = 363 \times (242/670) = 132$ MPa	Provide H16-300
Table 7.3	From the chart in *Reynolds,* Table 4.25, a crack width <0.2 mm is indicated.	top and bottom
	Cracking due to restrained early thermal contraction	
	From the calculations for the wall, minimum area of reinforcement:	
	$A_{s,min} = 900$ mm^2/m H16-300 (EF) say	
	Anchorage and lap requirements	
Figure 8.2	For tension due to loading, with H16-300, good bond conditions and allowing for $A_{s,prov} > A_{s,req}$, from *Reynolds,* Table 4.30:	
8.4.3 (2)	$l_{b,rqd} = 38\varphi \times A_{s,req} / A_{s,prov} = 38 \times 16 \times 242/670 = 220$ mm	
	For tension due to restrained early thermal contraction, with H16-300 (EF):	
8.4.2 (2)	$\sigma_s = 500 \times 900/1340 = 336$ MPa $f_b = 2.25 f_{ct,eff} = 2.25 \times 1.8 = 4.05$ MPa	
8.4.3 (2)	$l_{b,rqd} = (\varphi/4) \times (\sigma_s /f_b) = 16/4 \times 336/4.05 = 332$ mm	
	Allowing for all bars being lapped at the same section, required lap length	
8.7.3 (1)	$l_0 = 1.5 \times 332 = 600$ mm say	Provide 600 mm laps

Bar Marks	Commentary on Bar Arrangement (Drawings 1 and 2)
01, 02	Bars (shape code 21) in a radial arrangement. Bars 01 to lap with vertical bars in wall. Projection above floor to provide a lap length above kicker = $1.5 \times 38 \times 16 \times 221/670 + 100 = 500$ mm say. Cover = 50 mm (outer face) and 40 mm (inner face), to allow for circumferential bars in wall being in layer 1 (outer face) and layer 2 (inner face). Bars 02 to lap with grid of bars in floor. Cover = 50 mm (bottom) and 40 mm (top).
03, 07	Bars supplied straight and bent to follow curvature of wall during fixing. Lap length for bars 07 = 1000 mm (see calculation sheet 4). Lap length for bars 03 = $1.5 \times 38 \times 12 = 700$ mm.
04, 05	Straight bars arranged to form a rectangular grid. Lengths of bars 05a to 05p vary to suit dimensions of floor.
06	Straight bars bearing on kicker and curtailed 100 mm below top of wall.
08	Bars (shape code 21) lapping 300 mm say with bars 06.

Example 6: Reinforcement in Tank Floor

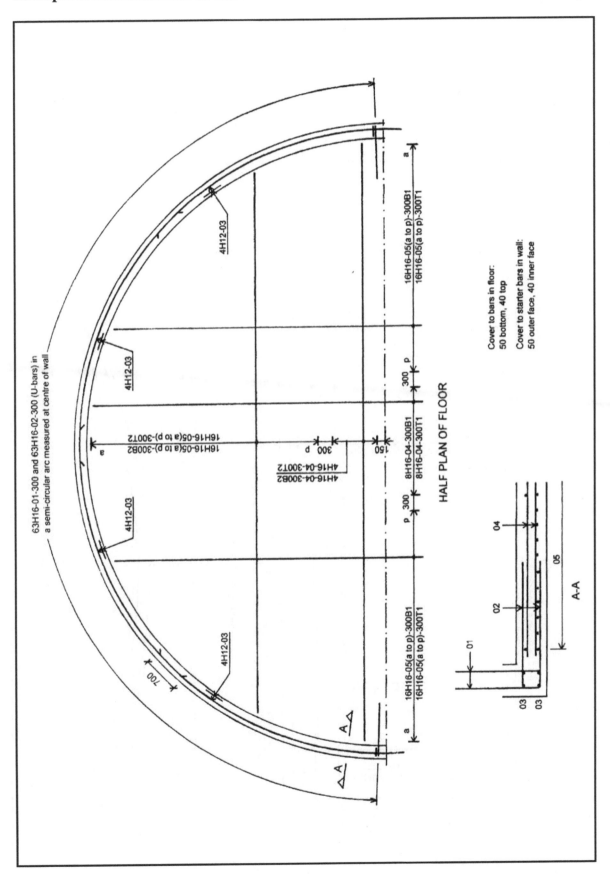

HALF PLAN OF FLOOR

63H16-01-300 and 63H16-02-300 (U-bars) in
a semi-circular arc measured at centre of wall

4H12-03

4H12-03

4H12-03

4H12-03

700

16H16-05(a to p)-300B1
16H16-05(a to p)-300T1

a

16H16-05(a to p)-300B1
16H16-05(a to p)-300T1

a

16H16-05(a to p)-300B2
16H16-05(a to p)-300T2

a

4H16-04-300B2
4H16-04-300T2

8H16-04-300B1
8H16-04-300T1

300 p

p 300

300 p

150

Cover to bars in floor:
50 bottom, 40 top

Cover to starter bars in wall:
50 outer face, 40 inner face

A-A

01

02

03

03

04

05

A

A

Example 6: Reinforcement in Tank Wall   **Drawing 2**

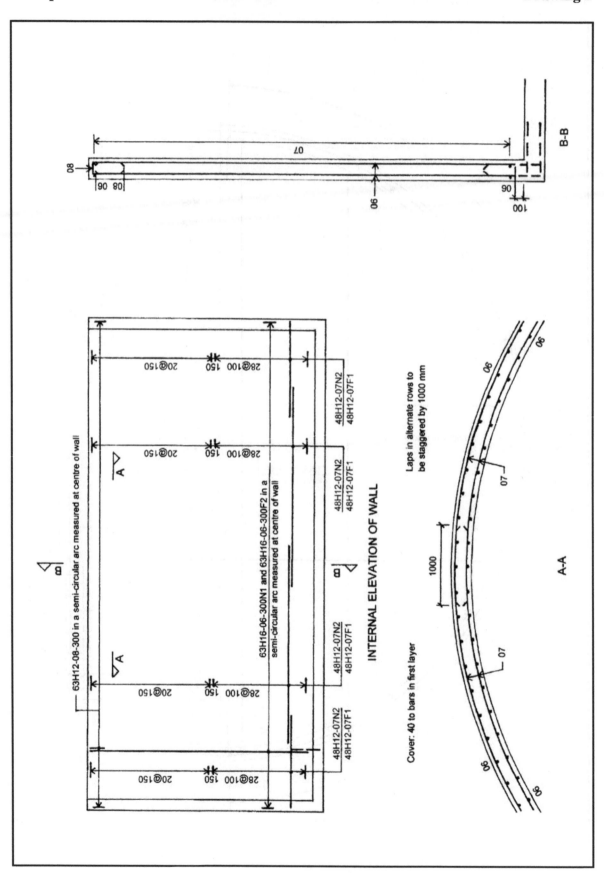

Example 6　　　　　　　　　　　　　　　　　　　**Calculation Sheet 6**

Reference	CALCULATIONS	OUTPUT
	TANK SUBJECTED TO HYDROSTATIC AND THERMAL ACTIONS	

For the purpose of this example, it is assumed that steady-state thermal conditions apply at the surfaces of the tank, with the temperature varying linearly through the thickness of the concrete section. This is not strictly correct for the outer surface, where differences can exist in the temperatures of the exposed and the buried parts of the tank, and the temperature of the exposed part of the tank can vary around the circumference due to incident solar radiation.

Suppose that the concrete section is subjected to a temperature variation from T_1 on the cold face to T_2 on the warm face, where $T_2 > T_1$. The temperature effect can be divided into two components: an average temperature rise $T_A = (T_2 - T_1)/2$, and a differential temperature change $T_D = \pm(T_2 - T_1)/2$.

Analysis for average temperature rise T_A

The radial deformation of the wall due to an average temperature rise is equal to that for a uniform pressure $p = \alpha T_A E_c h/r$. For a wall fixed at the bottom edge, the circumferential forces, vertical moments and radial shears can be determined from the coefficients in Table C11, where:

$$n = (1 - \alpha_{n3})\, pr \qquad\qquad m = \alpha_{m3} p l_z^2 \qquad\qquad v = \alpha_{v3} p l_z$$

Analysis for differential temperature change T_D

If the wall is considered initially as a cylinder with the top and bottom edges fixed, then the differential temperature change will cause a system of bending moments given by the following equation to occur vertically and circumferentially.

$$M = \alpha T_D E_c h^2/6(1 - \upsilon) \qquad\qquad \text{for } \upsilon = 0.2,\ M = \alpha T_D E_c h^2/4.8$$

Similarly, if the floor is considered as a circular plate fixed at the edge, a system of bending moments given by the same equation will occur radially and tangentially.

Design conditions

(a) Ambient temperature 0°C and temperature of water 40°C

Assuming $T_1 = 0°C$ and $T_2 = 40°C$ for both the wall and the floor, the average temperature rise T_A will cause a uniform expansion of the tank with no structural effects. It remains to consider the effect of the differential temperature change T_D.

Since the wall and floor are subjected to the same differential temperature change, and are of the same thickness, the resulting moments at the junction are of equal value. Since the top edge of the wall should be free, an equal and opposite moment needs to be applied at the top of the wall to restore the edge condition.

(b) Ambient temperature 20°C and temperature of water 0°C

In this case, it will be assumed that $T_2 = 20°C$ for the wall and 0°C for the floor. Thus, for the floor, both T_A and T_D are zero. For the wall, the bending moments at the bottom due to both T_A and T_D need to be distributed according to the stiffness of the wall and the floor. The bending moment at the top of the wall due to T_D needs to be released as indicated in (a).

Analysis for condition (a)

For the wall, due to the differential temperature change, the horizontal moment m_h, the vertical moment m_v and the circumferential force n, are given by the following equations, where α_{m6} and α_{n6} are found in Table C12, and $M_D = \alpha T_D E_c h^2/4.8$.

$$m_h = M_D \qquad\qquad m_v = (1 - \alpha_{m6})M_D \qquad\qquad n = -\alpha_{n6} M_D r/l_z^2$$

With $\alpha = 12 \times 10^{-6}$ per°C, $T_D = \pm 40/2 = \pm 20°C$, $E_c = 32$ GPa and $h = 0.25$ m:

$$M_D = 12 \times 10^{-6} \times 20 \times 32 \times 10^6 \times 0.25^2/4.8 = 100 \text{ kN m/m} \qquad m_h = 100 \text{ kN m/m}$$

$$m_v = 100(1 - \alpha_{m6}) \text{ kN m/m} \qquad n = -(100 \times 6.0/6.0^2)\,\alpha_{n6} = -(100/6)\,\alpha_{n6} \text{ kN/m}$$

Values of m_v and n for the wall, where z is the depth from the top of the wall, and $l_z^2/2rh = 12$, are shown in the following table.

Example 6 **Calculation Sheet 7**

Reference	CALCULATIONS	OUTPUT

Level z/l_z	Vertical Moment (kN m/m)		Circumferential Force (kN/m)	
	α_{m6}	m_v	α_{n6}	n
0	1.0	0	−81.46	1358
0.1	0.739	26.1	−8.93	149
0.2	0.348	65.2	15.16	−253
0.4	−0.022	102.2	8.80	−147
0.6	−0.031	103.1	0.23	−4
0.8	−0.004	100.4	−0.67	11
1.0	0	100	0	0

For the floor, the radial moment m_r and the tangential moment m_t are both equal to $M_D = 100$ kNm/m.

Analysis for condition (b)

Owing to the average temperature rise and the differential temperature change in the wall, the fixed-edge moments at the bottom of the wall are given by the following equations, where α_{m3} is found in Table C11, and $p = \alpha T_A E_c h/r$.

$$M_A = -\alpha_{m3} p l_z^2 \qquad\qquad M_D = \alpha T_D E_c h^2/4.8$$

With $T_A = 20/2 = 10° C$, $T_D = \pm 10° C$, $l_z = 6.0$ m, $r = 6.0$ m and $\alpha_{m3} = -0.0123$:

$$p = 12 \times 10^{-6} \times 10 \times 32 \times 10^6 \times 0.25/6.0 = 160 \text{ kN/m}^2$$

$$M_A = -0.0123 \times 160 \times 6.0^2 = -70.8 \text{ kNm/m}$$

$$M_D = -12 \times 10^{-6} \times 10 \times 32 \times 10^6 \times 0.25^2/4.8 = -50 \text{ kNm/m}$$

Resulting moment at bottom of wall, after releasing fixed-end moments, is

$$M = -0.2(M_A + M_D) = -24.2 \text{ kNm/m}$$

The horizontal moment m_h, vertical moment m_v and circumferential force n, are given by the following equations, where α_{m3} and α_{n3} are found in Table C11, and α_{m5}, α_{m6}, α_{n5} and α_{n6} are found in Table C12.

$$m_h = M_D = -50 \text{ kNm/m} \qquad m_v = \alpha_{m3} p l_z^2 - \alpha_{m5} \times 0.8(M_A + M_D) + (1 - \alpha_{m6})M_D$$

$$n = (\alpha_{n3} - 1)pr + [\alpha_{n5} \times 0.8(M_A + M_D) - \alpha_{n6}M_D]r/l_z^2$$

Values of m_v and n for the wall, where z is the depth from the top of the wall, and $l_z^2/2rh = 12$, are shown in the following table.

Level z/l_z	Vertical Moment (kNm/m)				Circumferential Force (kN/m)			
	α_{m3}	α_{m5}	α_{m6}	m_v	α_{n3}	α_{n5}	α_{n6}	n
0	0	0	1.0	0	0.994	0.32	−81.5	−680
0.1	0	0	0.739	−13.1	0.997	−0.05	−8.9	−78
0.2	0	0	0.348	−32.6	1.003	−0.46	15.2	122
0.4	0	−0.016	−0.022	−52.7	1.031	−1.15	8.8	85
0.6	0.0014	−0.064	−0.031	−49.7	1.022	3.52	0.2	80
0.8	0.0022	0.081	−0.044	−31.7	0.652	21.8	−0.7	11
1.0	−0.0123	1.0	0	−24.2	0	0	0	−960

The radial shear force at the bottom of the wall is given by the equation:

$$v = \alpha_{v3} p l_z + \alpha_{v5} \times 0.8(M_A + M_D)/l_z$$

$$= 0.158 \times 160 \times 6.0 - 6.38 \times 0.8 \times 120.8/6.0 = 48.9 \text{ kN/m}$$

For the floor slab, the edge moment is $M = 24.2$ kNm/m, and radial and tangential moments can be obtained from the coefficients given for load case 1 in Table C14, as shown on calculation sheet 2.

The thermal action effects have been determined on the basis of an uncracked section, and the calculated values can be multiplied by a factor that takes account of the local reduction in stiffness resulting from cracking in each direction.

Example 6 Calculation Sheet 8

Reference	CALCULATIONS	OUTPUT

Values of the reduction factor given in the New Zealand Standard NZS 3106 are shown in the following chart, where the factor K_T is equal to the second moment of area of the cracked section, including tension stiffening in the concrete, divided by the second moment of area of the uncracked section. The presence of axial forces has been ignored in deriving the second moment of area of the cracked section, and the multiplier will be further reduced by axial tension.

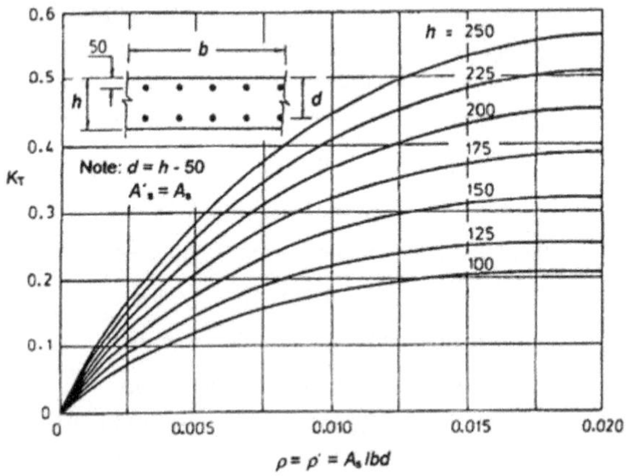

Reduction factor for effects of thermal action on cracked section

Action effects for serviceability (cracking): frequent load combination

BS EN 1991-4 Table NA.A.5

Taking $y_1 = 0.5$ for thermal actions (leading) and $y_2 = 0.9$ (instead of 0.3) for liquid loads, the forces and moments in the following tables have been obtained by multiplying those derived on sheet 2 by 0.9/1.2 = 0.75, and those derived on sheet 7 by $0.5K_T = 0.3$, conservatively taking $K_T = 0.6$.

Circumferential Forces (kN/m) in Tank Wall for Frequent Load Combination

Level z/l_z	ULS Hydrostatic Load × 0.75	Thermal Action × 0.3 (a)	Combined Actions (a)	Thermal Action × 0.3 (b)	Combined Actions (b)
0	0	408	408	–204	–204
0.1	31	45	76	–24	7
0.2	62	–76	–14	37	99
0.4	131	–44	87	26	157
0.6	219	–1	218	24	243
0.8	277	3	280	3	280
1.0	0	0	0	–288	–288

Note: Positive values indicate tension

Vertical Moments (kNm/m) in Tank Wall for Frequent Load Combination

Level z/l_z	ULS Hydrostatic Load × 0.75	Thermal Action × 0.3 (a)	Combined Actions (a)	Thermal Action × 0.3 (b)	Combined Actions (b)
0	0	0	0	0	0
0.1	0	7.8	7.8	–3.9	–3.9
0.2	0	19.6	19.6	–9.8	–9.8
0.4	0	30.7	30.7	–15.8	–15.8
0.6	0.4	30.9	31.3	–14.9	–14.5
0.8	7.7	30.1	37.8	–9.5	–1.8
1.0	13.0	30.0	43.0	–7.3	5.7

Note: Positive values indicate tension on the outside face.

Horizontal moments in tank wall due to 0.3 × thermal action are:

(a) $m_h = 0.3 \times 100 = 30$ kNm/m (b) $m_h = -0.3 \times 50 = -15$ kNm/m

Example 6   **Calculation Sheet 9**

Reference	CALCULATIONS	OUTPUT
	WALL	
	The elastic effects of thermal actions do not normally need to be considered at the ULS, since the effects reduce with increasing inelastic strain.	
	Cracking due to loading	
2.3.1.2 (2)	In the horizontal direction, the section is subjected to a bending moment and a tensile force. In the following equations, $e = M/N$, where the values of M and N are those appropriate to the characteristic load combination, and σ_s is the maximum stress applicable to the bar spacing used. Solutions can be obtained by assuming a value for σ_s, calculating A_s, and choosing bars to satisfy the spacing limitation.	
	For $e < (d - 0.5\,h)$, reinforcement is needed on both faces, and is given by	
	$A_{s1} = (0.5N/\sigma_s)\,[1 + e/(d - 0.5\,h)]$ $A_{s1} = (0.5N/\sigma_s)\,[1 - e/(d - 0.5\,h)]$	
	For $e \geq (d - 0.5\,h)$, reinforcement is needed on one face only, and is given by	
	$A_s = M_1/(\sigma_s z) + N/\sigma_s$ where $M_1 = M - N(d - 0.5\,h)$	
	With the bars in layer 1 for the outer ring, and layer 2 for the inner ring:	
	$d - 0.5\,h = 200 - 125 = 75$ mm (outer ring), and 60 mm (inner ring)	
	Condition (a): $m_h = 30$ kN m/m at all levels. Reinforcement is needed only on the outer face of the wall for values of $N \leq M/(d - 0.5\,h) = 30/0.075 = 400$ kN/m.	
7.3.3 (2) Table 7.3	If a crack width limit of 0.2 mm is assumed for the outer face, from *Reynolds*, Table 4.24, the maximum values of σ_s are 200 MPa for a bar spacing of 150 mm, 220 MPa for a bar spacing of 125 mm, and 240 MPa for a bar spacing of 100 mm.	
	Since the value of N reduces rapidly from a maximum of 408 kN/m at the top of the wall to 76 kN/m at a depth of $0.1l_z$, it is reasonable to take an average value of $N = 242$ kN/m for the top 600 mm of the wall. Hence, the maximum value of N is reached at a depth of $0.8l_z$, where $N = 280$ kN/m.	
	$M_1 = 30 - 280 \times 0.075 = 9$ kNm/m Assuming $\sigma_s = 220$ MPa,	
	$100M_1/bd^2\sigma_s = 100 \times 9 \times 10^6/(1000 \times 200^2 \times 220) = 0.102$	
	From Table A9, $x/d = 0.165$, $z/d = 1 - (1/3)(x/d) = 0.945$	
	$A_s = 9 \times 10^6/(220 \times 0.945 \times 200) + 280 \times 10^3/220 = 1490$ mm²/m (H16-125)	
	Condition (b): $m_h = -15$ kNm/m at all levels. Reinforcement is needed only on the inner face of the wall for values of $N \leq M/(d - 0.5\,h) = 15/0.060 = 250$ kN/m.	
BS EN 1992-3 7.3.3	The circumferential forces are compressive at the top and bottom of the wall, and tensile within the middle four-fifths of the wall height. The maximum tensile force is reached at a depth of $0.8l_z$, where $N = 280$ kN/m and the crack width limit for the inner face is 0.13 mm. By interpolation from the chart in *Reynolds*, Table 4.25, maximum values of σ_s are 130 MPa for a bar spacing of 125 mm, and 140 MPa for a bar spacing of 100 mm. With $e = M/N = (15/280) \times 1000 = 54$ mm,	
	$A_{s1} = (0.5N/\sigma_s)\,[1 + e/(d - 0.5\,h)]$	
	$= 0.5 \times 280 \times 10^3/140 \times (1 + 54/60) = 1900$ mm²/m (H16-100)	
	At a depth of $0.6l_z$, the crack width limit is 0.15 mm, and $N = 243$ kN/m (< 250). If the chart in *Reynolds*, Table 4.25 is used, maximum values of σ_s are 150 MPa for a bar spacing of 125 mm, and 160 MPa for a bar spacing of 100 mm.	
	$M_1 = 15 - 243 \times 0.060 = 0.4$ kNm/m Assuming $\sigma_s = 150$ MPa,	
	$A_s = 0.4 \times 10^6/(150 \times 0.95 \times 190) + 243 \times 10^3/150 = 1635$ mm²/m	Horizontal bars (EF) as follows:
	It can be seen that H16-125 is almost sufficient at this depth and, for simplicity, the following arrangement of bars will be used: H16-125 (EF) for top half of wall, and H16-100 (EF) for bottom half of wall.	H16-125 from top of wall to depth of 3.0 m
	In the vertical direction, the section is subjected to pure bending, and a crack width limit of 0.2 mm will be taken for each face.	H16-100 for depth of more than 3.0 m

Example 6

Calculation Sheet 10

Reference	CALCULATIONS	OUTPUT
	For condition (a), the maximum moment occurs at the bottom of the wall, where $M = 43.0$ kNm/m. At the outer face, with bars in layer 2, $d = 185$ mm. Assuming $\sigma_s = 200$ MPa, $\quad 100M/bd^2\sigma_s = 100 \times 43.0 \times 10^6/(1000 \times 185^2 \times 200) = 0.628$ $\quad$ From Table A6, $x/d = 0.369$, $z/d = 1 - (1/3)(x/d) = 0.877$ $\quad A_s = 43.0 \times 10^6/(200 \times 0.877 \times 185) = 1325$ mm^2/m (H16-150) For condition (b), the maximum vertical moment occurs at depth $z = 0.4l_z$, where $M = -15.8$ kNm/m. At the inner face, with bars in layer 1, $d = 200$ mm. Assuming $\sigma_s = 200$ MPa, $\quad 100M/bd^2\sigma_s = 100 \times 15.8 \times 10^6/(1000 \times 200^2 \times 200) = 0.198$ $\quad$ From Table A9, $x/d = 0.224$, $z/d = 1 - (1/3)(x/d) = 0.925$ $\quad A_s = 15.8 \times 10^6/(200 \times 0.925 \times 200) = 427$ mm^2/m (H12-150)	Vertical bars: H16-150 (outer face) H12-150 (inner face)
	FLOOR For condition (a), with the characteristic load combination, $M = 43.0$ kNm/m at the edge of the floor and the coexisting radial force is $N = 14.9 \times 0.75 = 11.2$ kN/m. $\quad M_1 = 43.0 - 11.2 \times 0.065 = 42.3$ kNm/m $\qquad$ Assuming $\sigma_s = 200$ MPa, $\quad 100M_1/bd^2\sigma_s = 100 \times 42.3 \times 10^6/(1000 \times 190^2 \times 200) = 0.586$ $\quad$ From Table A9, $x/d = 0.358$, $z/d = 1 - (1/3)(x/d) = 0.881$ $\quad A_s = 42.3 \times 10^6/(200 \times 0.881 \times 190) + 11.2 \times 10^3/200 = 1320$ mm^2/m (H16-150) Since the moments due to thermal action are dominant, H16-150 will be provided in both directions at the bottom surface throughout the floor. For condition (b), $N = 14.9 \times 0.75 + 48.9 \times 0.3 = 25.9$ kN/m. For tension at the top surface of the floor, the maximum moments occur at $r_x/r = 0.6$, where the radial moment is $M = 13.0 \times 0.75 + 0.385 \times (24.2 \times 0.3) = 12.6$ kNm/m $\quad M_1 = 12.6 - 25.9 \times 0.065 = 10.9$ kNm/m $\qquad$ Assuming $\sigma_s = 200$ MPa, $\quad 100M_1/bd^2\sigma_s = 100 \times 10.9 \times 10^6/(1000 \times 190^2 \times 200) = 0.151$ $\quad$ From Table A9, $x/d = 0.197$, $z/d = 1 - (1/3)(x/d) = 0.934$ $\quad A_s = 10.9 \times 10^6/(200 \times 0.934 \times 190) + 25.9 \times 10^3/200 = 4377$ mm^2/m (H12-150)	Bars in each direction: H16-150 (bottom) H12-150 (top)
	ARRANGEMENT OF REINFORCEMENT AT BOTTOM OF WALL 	

Appendix A: General Information

Table A1 Design Formulae for Rectangular Beams

The following equations, based on the simplest stress–strain relationships specified for concrete and reinforcement, apply for values of $f_{ck} \leq 50$ MPa and $f_{yk} = 500$ MPa. The condition $z/d \leq 0.95$, although not in Eurocode 2, is normal in UK practice.

Singly reinforced sections

For values of $K \leq K'$, compression reinforcement is not required and the following equations apply:

$A_s = M/(0.87 f_{yk} z)$ where $z/d = (0.5 + \sqrt{0.25 - 0.882K}) \leq 0.95$. Hence, $z/d = 0.95$ for $K \leq 0.054$.

Values of $A_s f_{yk}/(bd f_{ck})$, $x/d = 2.5(1 - z/d)$ and z/d, according to the value of K, may be obtained from the table below.

Doubly reinforced sections

For values of $K > K'$, compression reinforcement is required and the following equations apply:

$A_{s2} = (K - K')(bd^2 f_{ck})/[0.87 f_{yk}(d - d_2)]$ and $A_{s1} = A_{s2} + K'bd^2 f_{ck}/(0.87 f_{yk} z)$ where $z/d = (0.5 + \sqrt{0.25 - 0.882K'})$

For $d_2/x > 0.375$, A_{s2} should be replaced by $1.6(1 - d_2/x)A_{s2}$ in the above equations.

Values of K and K'

In the above equations: $K = M/(bd^2 f_{ck})$ and for linear elastic analysis with no redistribution of moment, $K' = 0.210$.

For values of $\delta < 1.0$, $K' = 0.6\delta - 0.18\delta^2 - 0.21$, where $\delta =$ redistributed moment/elastic moment at section considered.

$\dfrac{M}{bd^2 f_{ck}}$	$\dfrac{A_s f_{yk}}{bd f_{ck}}$	$\dfrac{x}{d}$	$\dfrac{z}{d}$	$\dfrac{M}{bd^2 f_{ck}}$	$\dfrac{A_s f_{yk}}{bd f_{ck}}$	$\dfrac{x}{d}$	$\dfrac{z}{d}$	$\dfrac{M}{bd^2 f_{ck}}$	$\dfrac{A_s f_{yk}}{bd f_{ck}}$	$\dfrac{x}{d}$	$\dfrac{z}{d}$
				0.110	0.142	0.272	0.891	0.170	0.240	0.459	0.816
				0.112	0.145	0.278	0.889	0.172	0.243	0.466	0.814
≤0.054	1.21K	0.125	0.950	0.114	0.148	0.284	0.887	0.174	0.247	0.473	0.811
0.056	0.068	0.130	0.948	0.116	0.151	0.289	0.884	0.176	0.251	0.480	0.808
0.058	0.071	0.135	0.946	0.118	0.154	0.295	0.882	0.178	0.255	0.488	0.805
0.060	0.073	0.140	0.944	0.120	0.157	0.301	0.880	0.180	0.258	0.495	0.802
0.062	0.076	0.145	0.942	0.122	0.160	0.307	0.877	0.182	0.262	0.502	0.799
0.064	0.079	0.150	0.940	0.124	0.163	0.313	0.875	0.184	0.266	0.510	0.796
0.066	0.081	0.155	0.938	0.126	0.166	0.319	0.873	0.186	0.270	0.517	0.793
0.068	0.084	0.160	0.936	0.128	0.169	0.324	0.870	0.188	0.274	0.525	0.790
0.070	0.086	0.165	0.934	0.130	0.173	0.330	0.868	0.190	0.278	0.532	0.787
0.072	0.089	0.170	0.932	0.132	0.176	0.336	0.865	0.192	0.282	0.540	0.784
0.074	0.092	0.176	0.930	0.134	0.179	0.343	0.863	0.194	0.286	0.548	0.781
0.076	0.094	0.181	0.928	0.136	0.182	0.349	0.861	0.196	0.290	0.556	0.778
0.078	0.097	0.186	0.926	0.138	0.185	0.355	0.858	0.198	0.294	0.564	0.775
0.080	0.100	0.191	0.924	0.140	0.188	0.361	0.856	0.200	0.298	0.572	0.771
0.082	0.103	0.196	0.921	0.142	0.192	0.367	0.853	0.202	0.303	0.580	0.768
0.084	0.105	0.202	0.919	0.144	0.195	0.373	0.851	0.204	0.307	0.588	0.765
0.086	0.108	0.207	0.917	0.146	0.198	0.380	0.848	0.206	0.311	0.597	0.761
0.088	0.111	0.212	0.915	0.148	0.201	0.386	0.846	0.208	0.316	0.605	0.758
0.090	0.114	0.217	0.913	0.150	0.205	0.393	0.843	0.210	0.320	0.614	0.754
0.092	0.117	0.223	0.911	0.152	0.208	0.399	0.840				
0.094	0.119	0.228	0.909	0.154	0.212	0.405	0.838				
0.096	0.122	0.234	0.907	0.156	0.215	0.412	0.835				
0.098	0.125	0.239	0.904	0.158	0.218	0.419	0.833				
0.100	0.128	0.245	0.902	0.160	0.222	0.425	0.830				
0.102	0.131	0.250	0.900	0.162	0.226	0.432	0.827				
0.104	0.133	0.256	0.898	0.164	0.229	0.439	0.824				
0.106	0.136	0.261	0.896	0.166	0.233	0.446	0.822				
0.108	0.139	0.267	0.893	0.168	0.236	0.452	0.819				

Limiting K values according to % moment redistribution for singly-reinforced sections							
% Redistribution	0	5	10	15	20	25	30
$K = M/(bd^2 f_{ck})$	0.207	0.194	0.181	0.167	0.152	0.136	0.120

Table A2 Design Chart for Rectangular Columns – 1

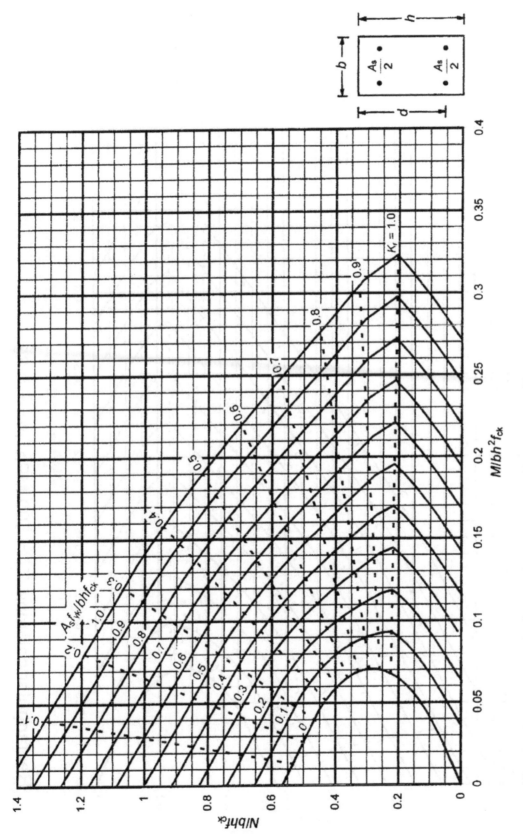

Rectangular columns (f_{yk} = 500 MPa, d/h = 0.8)

Table A3 Design Chart for Rectangular Columns – 2

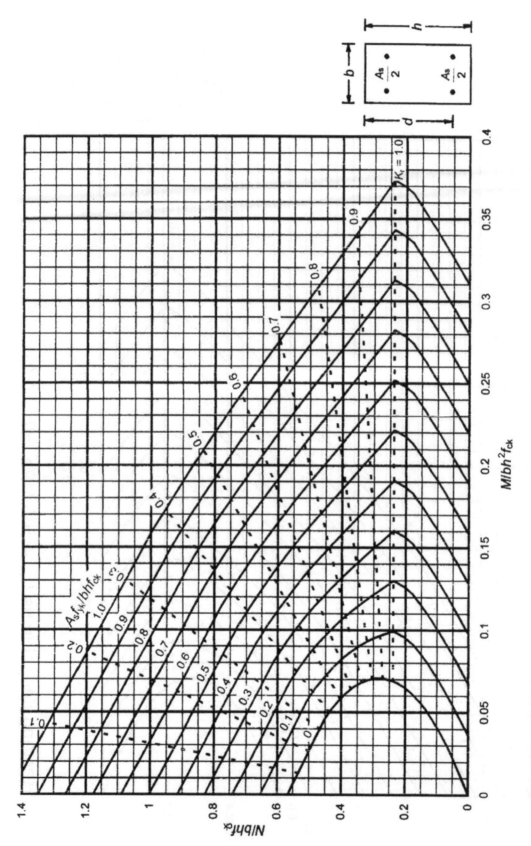

Rectangular columns (f_{yk} = 500 MPa, d/h = 0.85)

Table A4 Design Chart for Circular Columns – 1

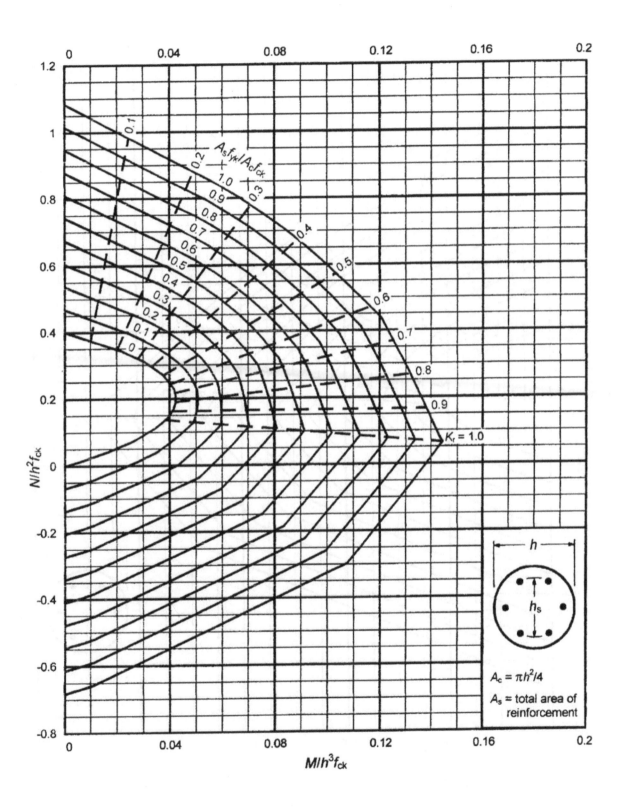

Circular columns (f_{yk} = 500 MPa, h_s/h = 0.6)

Table A5 Design Chart for Circular Columns – 2

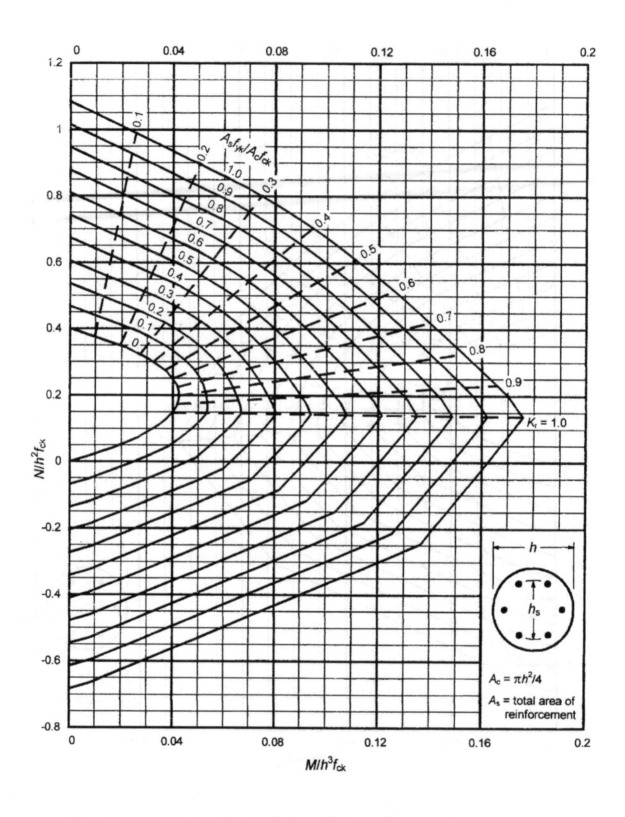

Circular columns (f_{yk} = 500 MPa, h_s/h = 0.7)

Table A6 Elastic Properties of Cracked Rectangular Sections in Flexure

$\dfrac{100A_s}{bd}$	$\dfrac{100M}{\sigma_s bd^2}$	$\dfrac{x}{d}$	$\dfrac{100A_s}{bd}$	$\dfrac{100M}{\sigma_s bd^2}$	$\dfrac{x}{d}$	$\dfrac{100A_s}{bd}$	$\dfrac{100M}{\sigma_s bd^2}$	$\dfrac{x}{d}$
0.10	0.095	0.159	0.80	0.697	0.384	1.50	1.259	0.483
0.12	0.113	0.173	0.82	0.714	0.388	1.52	1.274	0.485
0.14	0.131	0.185	0.84	0.730	0.392	1.54	1.290	0.487
0.16	0.150	0.196	0.86	0.747	0.395	1.56	1.306	0.489
0.18	0.168	0.207	0.88	0.763	0.398	1.58	1.321	0.491
0.20	0.186	0.217	0.90	0.779	0.402	1.60	1.337	0.493
0.22	0.203	0.226	0.92	0.796	0.405	1.62	1.353	0.495
0.24	0.221	0.235	0.94	0.812	0.408	1.64	1.368	0.497
0.26	0.239	0.243	0.96	0.828	0.412	1.66	1.384	0.499
0.28	0.257	0.251	0.98	0.845	0.415	1.68	1.399	0.501
0.30	0.274	0.258	1.00	0.861	0.418	1.70	1.415	0.503
0.32	0.292	0.266	1.02	0.877	0.421	1.72	1.430	0.505
0.34	0.309	0.272	1.04	0.893	0.424	1.74	1.446	0.507
0.36	0.327	0.279	1.06	0.909	0.427	1.76	1.461	0.509
0.38	0.344	0.286	1.08	0.925	0.430	1.78	1.477	0.511
0.40	0.361	0.292	1.10	0.941	0.433	1.80	1.492	0.513
0.42	0.378	0.298	1.12	0.957	0.436	1.82	1.508	0.515
0.44	0.396	0.303	1.14	0.973	0.438	1.84	1.523	0.517
0.46	0.413	0.309	1.16	0.989	0.441	1.86	1.539	0.518
0.48	0.430	0.314	1.18	1.005	0.444	1.88	1.554	0.520
0.50	0.447	0.319	1.20	1.021	0.446	1.90	1.569	0.522
0.52	0.464	0.325	1.22	1.037	0.449	1.92	1.585	0.524
0.54	0.481	0.330	1.24	1.053	0.452	1.94	1.600	0.526
0.56	0.498	0.334	1.26	1.069	0.454	1.96	1.616	0.527
0.58	0.514	0.339	1.28	1.085	0.457	1.98	1.631	0.529
0.60	0.531	0.344	1.30	1.101	0.459	2.00	1.646	0.531
0.62	0.548	0.348	1.32	1.117	0.462			
0.64	0.565	0.353	1.34	1.133	0.464			
0.66	0.581	0.357	1.36	1.149	0.467			
0.68	0.598	0.361	1.38	1.164	0.469			
0.70	0.615	0.365	1.40	1.180	0.471			
0.72	0.631	0.369	1.42	1.196	0.474			
0.74	0.648	0.373	1.44	1.212	0.476			
0.76	0.665	0.377	1.46	1.227	0.478			
0.78	0.681	0.381	1.48	1.243	0.480			

Elastic properties of singly reinforced rectangular sections subjected to bending (or combined bending and tension)

The values in the above table are derived from the following equations:

$$\frac{x}{d} = \sqrt{\left(\frac{\alpha_e A_s}{bd}\right)^2 + \frac{2\alpha_e A_s}{bd}} - \frac{\alpha_e A_s}{bd} \qquad \frac{100M}{\sigma_s bd^2} = \frac{100A_s}{bd}\left(1 - \frac{x}{3d}\right) \qquad \alpha_e = \frac{2E_s}{E_c} = 15$$

For a section subjected to combined bending and tension, where $M/N > (d - 0.5h)$, the value of M in these equations should be replaced by $M_1 = M - N(d - 0.5h)$, and the total area of reinforcement is given by $A_s + N/\sigma_s$. For a section containing a given area of reinforcement, analysis involves an iterative process to determine the values of x and σ_s.

A_s is the area of tension reinforcement to resist M or M_1

M is the bending moment due to design service loading

N is the direct tension due to design service loading

E_c is the modulus of elasticity of concrete

E_s is the modulus of elasticity of reinforcement

b is the breadth of section

d is the effective depth of tension reinforcement

h is the overall depth of section

x is the neutral axis depth

σ_s is the stress in tension reinforcement

For a section where $M/N < (d - 0.5h)$, tension reinforcement is required on both faces where the areas, for a particular reinforcement stress, are given by the following equations:

$$A_{s1} = \frac{0.5N}{\sigma_s}\left(1 + \frac{M/N}{d - 0.5h}\right) \text{ on the face in tension due to } M \qquad A_{s2} = \frac{0.5N}{\sigma_s}\left(1 - \frac{M/N}{d - 0.5h}\right) \text{ on the other face}$$

Table A7 Early Thermal Cracking in End Restrained Panels

For a concrete element restrained at its ends, the mean tensile strain contributing to cracking, with $k_c = 1.0$ (pure tension), may be calculated from the expression $(\varepsilon_{sm} - \varepsilon_{cm}) = 0.5\,k f_{ct,eff}(1 + \alpha_e A_s/A_{ct})/(A_s/A_{ct})E_s$.

With $k_1 = 0.8$ (high bond bars) and $k_2 = 1.0$ (pure tension), the maximum crack spacing $s_{r,max} = 3.4[c + 0.1(A_{c,eff}/A_s)\varphi]$.

With $\alpha_e = 6$, the design crack width $w_k = 1.7k(1 + 6A_s/A_{ct})[c + 0.1(A_{c,eff}/A_s)\varphi]\,f_{ct,eff}/(A_s/A_{ct})E_s$.

Maximum values of $f_{ct,eff}$ for $w_k = 0.2$ mm are given below. For other values of w_k, multiply values of $f_{ct,eff}$ by $5w_k$.

	Thickness of section (mm)	Bar size (EF)	Maximum values of $f_{ct,eff}$ (MPa) according to bar spacing (mm) for $w_k = 0.2$ mm								
			300	250	225	200	175	150	125	100	75
Cover = 40 mm	200	H12								1.70	2.72
		H16					1.41	1.83	2.47	3.53	5.45
		H20		1.38	1.65	2.02	2.52	3.23	4.30	6.00	8.98
	250	H16							1.76	2.55	4.05
		H20				1.38	1.73	2.25	3.03	4.30	6.59
		H25		1.69	2.03	2.47	3.07	3.93	5.20	7.21	10.7
	300	H16							1.49	2.15	3.40
		H20					1.46	1.90	2.57	3.66	5.65
		H25		1.38	1.65	2.02	2.52	3.23	4.31	6.02	9.03
	350	H16								1.93	3.06
		H20						1.71	2.31	3.30	5.12
		H25			1.48	1.81	2.27	2.92	3.90	5.48	8.26
	400	H16								1.77	2.81
		H20						1.56	2.12	3.04	4.72
		H25			1.36	1.66	2.08	2.69	3.60	5.07	7.68
Cover = 50 mm	200	H12								1.60	2.51
		H16					1.34	1.72	2.31	3.26	4.95
		H20		1.31	1.57	1.90	2.36	3.01	3.96	5.47	8.06
	250	H16							1.61	2.31	3.59
		H20				1.31	1.64	2.11	2.82	3.97	5.98
		H25		1.61	1.92	2.33	2.88	3.66	4.80	6.58	9.59
	300	H16							1.22	1.76	2.78
		H20						1.57	2.11	3.01	4.63
		H25			1.42	1.73	2.15	2.76	3.66	5.09	7.58
	350	H16								1.58	2.50
		H20						1.41	1.90	2.72	4.20
		H25				1.51	1.88	2.42	3.22	4.52	6.79
	400	H20							1.74	2.50	3.87
		H25				1.38	1.73	2.22	2.97	4.18	6.31
		H32		1.73	2.06	2.50	3.10	3.93	5.16	7.09	10.3
Cover = 60 mm	200	H12								1.50	2.33
		H16						1.63	2.16	3.02	4.54
		H20			1.49	1.80	2.22	2.81	3.68	5.03	7.31
	250	H16							1.52	2.16	3.32
		H20					1.56	2.00	2.65	3.68	5.48
		H25		1.54	1.83	2.20	2.71	3.42	4.45	6.05	8.70
	300	H16								1.63	2.54
		H20						1.49	2.00	2.81	4.27
		H25			1.36	1.65	2.04	2.60	3.42	4.72	6.94
	350	H20							1.62	2.30	3.55
		H25					1.65	2.11	2.81	3.92	5.87
		H32		1.72	2.05	2.47	3.04	3.84	4.99	6.77	9.73
	400	H20							1.48	2.12	3.28
		H25					1.48	1.90	2.53	3.55	5.35
		H32		1.49	1.77	2.15	2.66	3.37	4.41	6.04	8.79

Table A8 Early Thermal Cracking in Edge Restrained Panels

For a long concrete panel restrained along an edge, the mean tensile strain contributing to cracking may be taken as $R_{ax}\,\varepsilon_{free}$.

With $k_1 = 0.8$ (high bond bars) and $k_2 = 1.0$ (pure tension), the maximum crack spacing $s_{r,max} = 3.4\,[c + 0.1(A_{c,eff}/A_s)\varphi]$.

The design crack width $w_k = 3.4\,[\,c + 0.1(A_{c,eff}/A_s)\varphi\,]\,R_{ax}\,\varepsilon_{free}$.

Maximum values of $R_{ax}\,\varepsilon_{free}$ for $w_k = 0.2$ mm are given below. For other values of w_k, multiply values of $R_{ax}\,\varepsilon_{free}$ by $5w_k$.

	Thickness of section (mm)	Bar size (EF)	Maximum values of $R_{ax}\,\varepsilon_{free}$ ($\times 10^{-6}$) according to bar spacing (mm) for $w_k = 0.2$ mm								
			300	250	225	200	175	150	125	100	75
Cover = 40 mm	200	H10	139	164	180	200	224	255	295	351	434
		H12	164	192	211	233	260	295	341	402	492
		H16	211	246	268	295	328	369	422	492	590
		H20	254	295	321	351	388	434	492	567	670
	250	H12	145	170	187	207	232	264	305	363	447
		H16	180	211	231	254	284	321	369	434	527
		H20	211	246	268	295	328	369	422	492	590
	300	H25	244	284	309	338	375	419	476	550	652
		H32	284	328	356	388	428	476	536	614	719

Note: Values of $R_{ax}\,\varepsilon_{free}$ given for section thickness $h = 250$ mm apply for $h \geq 230$ mm (H12), $h \geq 240$ mm (H16) and $h \geq 250$ mm (H20). Values of $R_{ax}\,\varepsilon_{free}$ given for $h = 300$ mm apply for $h \geq 262.5$ mm (H25) and $h \geq 280$ mm (H32).

	Thickness of section (mm)	Bar size (EF)	300	250	225	200	175	150	125	100	75
Cover = 50 mm	200	H10	136	159	175	193	216	244	281	331	404
		H12	160	186	204	224	249	281	322	377	454
		H16	203	236	257	281	311	347	393	454	536
		H20	244	281	304	332	364	404	454	517	602
	250	H12	131	154	169	186	208	236	272	322	393
		H16	169	197	215	236	262	295	337	393	472
		H20	204	236	257	281	311	347	393	454	536
	300	H12	119	139	153	169	190	215	249	296	364
		H16	148	174	190	209	233	263	303	356	431
		H20	175	204	222	244	271	304	347	404	484
	350	H25	204	236	257	281	311	347	393	454	536
		H32	238	275	297	324	357	396	445	508	592

Note: Values of $R_{ax}\,\varepsilon_{free}$ given for section thickness $h = 300$ mm apply for $h \geq 280$ mm (H12), $h \geq 290$ mm (H16) and $h \geq 300$ mm (H20). Values of $R_{ax}\,\varepsilon_{free}$ given for $h = 350$ mm apply for $h \geq 312.5$ mm (H25) and $h \geq 330$ mm (H32).

	Thickness of section (mm)	Bar size (EF)	300	250	225	200	175	150	125	100	75
Cover = 60 mm	200	H10	133	155	170	187	208	234	268	314	378
		H12	155	181	197	216	239	268	305	354	421
		H16	197	227	246	268	295	328	369	421	491
		H20	234	268	289	314	343	378	421	476	546
	250	H12	128	150	164	181	201	227	260	305	369
		H16	164	190	207	227	251	281	319	369	437
		H20	197	227	246	268	295	328	369	421	491
	300	H12	109	128	141	155	174	197	227	268	328
		H16	141	164	179	197	219	246	281	328	393
		H20	170	197	214	234	259	289	328	378	447
	350	H12	100	118	130	143	160	182	211	250	307
		H16	126	148	161	178	198	224	257	301	364
		H20	149	174	189	208	231	259	295	343	410
		H25	180	208	226	247	272	304	343	394	464
		H32	219	251	271	295	323	358	400	454	524
	400	H25	174	202	220	240	265	296	335	386	455
		H32	205	236	255	278	306	339	381	434	504

Note: Values of $R_{ax}\,\varepsilon_{free}$ given for section thickness $h = 350$ mm apply for $h \geq 330$ mm (H12), $h \geq 340$ mm (H16) and $h \geq 350$ mm (H20). Values of $R_{ax}\,\varepsilon_{free}$ given for $h = 400$ mm apply for $h \geq 362.5$ mm (H25) and $h \geq 380$ mm (H32).

Table A9 Cross-Sectional Areas of Reinforcing Bars and Fabric

	Number of bars	Cross-sectional area of number of bars (mm²) for size of bars (mm)								
		6	8	10	12	16	20	25	32	40
Bars in specified numbers	1	28	50	78	113	201	314	491	804	1257
	2	57	101	157	226	402	628	982	1608	2513
	3	85	151	236	339	603	942	1473	2413	3770
	4	113	201	314	452	804	1257	1963	3217	5027
	5	141	251	393	565	1005	1571	2454	4021	6283
	6	170	302	471	679	1206	1885	2945	4825	7540
	7	198	352	550	792	1407	2199	3436	5630	8796
	8	226	402	628	905	1608	2513	3927	6434	10050
	9	254	452	707	1018	1810	2827	4418	7238	11310
	10	283	503	785	1131	2011	3142	4909	8042	12570
	11	311	553	864	1244	2212	3456	5400	8847	12570
	12	339	603	942	1357	2413	3770	5890	9651	15080

	Spacing of bars (mm)	Cross-sectional area of bars per unit width (mm²/m) for size of bars (mm)								
		6	8	10	12	16	20	25	32	40
Bars at specified spacing	75	377	670	1047	1508	2681	4189	6545	10720	–
	100	283	503	785	1131	2011	3142	4909	8042	12570
	125	226	402	628	905	1608	2513	3927	6434	10053
	150	188	335	524	754	1340	2094	3272	5362	8378
	175	162	287	449	646	1149	1795	2805	4596	7181
	200	141	251	393	565	1005	1571	2454	4021	6283
	225	–	223	349	503	894	1396	2182	3574	5585
	250	–	201	314	452	804	1257	1963	3217	5027
	300	–	168	262	377	670	1047	1636	2681	4189
	400	–	–	196	283	503	785	1227	2011	3142
	500	–	–	–	226	402	628	982	1608	2513
	600	–	–	–	–	335	524	818	1340	2094

*6 mm is a non-preferred size.

		Standard fabric types to BS 4483						
	Fabric reference	Longitudinal bars			Cross bars			Mass per unit area
		Nominal bar size mm	Pitch of bars (mm)	Area of bars per unit width (mm²/m)	Nominal bar size (mm)	Pitch of bars (mm)	Area of bars per unit width (mm²/m)	(kg/m²)
Standard fabrics	A393	10	200	393	10	200	393	6.16
	A252	8	200	252	8	200	252	3.95
	A193	7	200	193	7	200	193	3.02
	A142	6	200	142	6	200	142	2.22
	B1131	12	100	1131	8	200	252	10.90
	B785	10	100	785	8	200	252	8.14
	B503	8	100	503	8	200	252	5.93
	B385	7	100	383	7	200	193	4.53
	B283	6	100	283	7	200	193	3.73
	C785	10	100	785	6	400	71	6.72
	C636	9	100	636	6	400	71	5.55
	C503	8	100	503	6	400	71	4.51
	C385	7	100	385	6	400	71	3.58
	C283	6	100	283	6	400	71	2.78
	D98	5	200	98	5	200	98	1.54
	D49	2.5	100	49	2.5	100	49	0.77

Notes: Bars used for fabric are in accordance with BS 4449 except for D98 and D49, where wire to BS 4442 may be used. Stock sheet size is 4.8 m (longitudinal bars) × 2.4 m (cross bars).

Appendix B: Beam on Elastic Foundation

Table B1 Load Cases and Modulus of Subgrade Reaction

The information given in Tables B1 to B10 is derived from the formulae developed by Hetenyi for a beam of finite length on an elastic foundation. Values are given for five load cases as shown below, where the symbols used are as follows:

B	Width of beam (m)		a	Distance from end of beam to application of load (m)
F	Total load on beam (kN)		c	Half length of a distributed load (m)
L	Length of beam (m)		k_s	Modulus of subgrade reaction (kN/m^3)
M	Moment at distance x from end of beam (kNm)		q	Bearing pressure (kN/m^2)
M_0	Moment applied at both ends of beam (kNm)		x	Distance from end of beam to position considered (m)
V	Shear force at distance x from end of beam (kN)		θ_0	Slope at end of beam (rad)

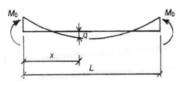

Concentrated moments at both ends
(Tables B2 and B3)

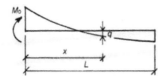

Concentrated moment at LH end
(Tables B2 and B3)

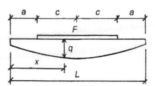

Centrally placed distributed load
(Tables B2 and B3)

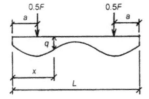

Two symmetrically placed loads
(Tables B2, B5 and B6)

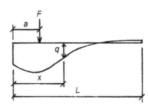

Concentrated load at any point
(Tables B2, and B7 to B10)

In Tables B2 to B10, the factor

$$\lambda L = (Bk_sL^4/4E_cI)^{1/4}$$

$$= (3k_sL^4/E_ch^3)^{1/4}$$

where

E_c is modulus of elasticity of concrete (kN/m^2)

h is overall depth of beam (m)

The information given in Table B11 is derived from the formulae developed by Hetenyi for a concentrated load on a slab of infinite dimensions on an elastic foundation, with Poisson's ratio taken as 0.2.

In principle, the value of k_s used in design should be related to the range of influence of the load, but it is normal practice to base k_s on a loaded area of diameter 750 mm. To this end, it is strongly recommended that the value of k_s is determined from a BS plate–loading test, using a 750 mm diameter plate and a fixed settlement of 1.25 mm. If a smaller plate is used, or a value of k_s appropriate to a particular area is required, the following approximate relationship may be assumed:

$k_s = 0.5(1 + 0.3/D)^2 k_{0.75}$, where D is the diameter of the loaded area, and $k_{0.75}$ is a value for $D = 0.75$ m.

This gives values of $k_s/k_{0.75}$ for particular values of D as follows:

D (m)	0.3	0.5	0.75	1.5	3.0	∞
$k_s/k_{0.75}$	2.0	1.28	1.0	0.72	0.6	0.5

In the absence of more accurate information, the values given below (refer. *Bowles J E, Foundation Analysis and Design*) may be used as a guide:

Soil Type	Values of k_s (MN/m^3)	
	Lower	Upper
Loose sand	5	16
Medium dense sand	10	80
Dense sand	64	128
Clayey medium dense sand	32	80
Silty medium dense sand	24	48
Clayey soil:		
$\quad c_u \leq 100$ kN/m^2	12	24
$\quad 100 < c_u \leq 200$ kN/m^2	24	48
$\quad c_u > 400$ kN/m^2	–	>48
(c_u is undrained shear strength)		

Table B2 End Slope Coefficients for Different Load Cases

END SLOPE COEFFICIENTS FOR CONCENTRATED MOMENTS AT ONE OR BOTH ENDS

λL	$\theta_0/(M_0/k_sBL^3)$ for Clockwise Moment M_0 at LH End ($x/L=0$)		$\theta_0/(M_0/k_sBL^3)$ for Anti-Clockwise Moment M_0 at RH End ($x/L=1.0$)		$\theta_0/(M_0/k_sBL^3)$ for Moments M_0 at Both Ends	
	$x/L=0$	1.0	0	1.0	0	1.0
1.0	13.48	11.49	−11.49	−13.48	1.99	−1.99
2.0	34.44	4.96	−4.96	−34.44	29.48	−29.48
3.0	108.4	−9.2	9.2	−108.4	117.6	−117.6
4.0	256.5	−13.3	13.3	−256.5	269.8	−269.8
5.0	500.1	−4.6	4.6	−500.1	504.7	−504.7
6.0	864	2.9	−2.9	−864	861.1	−861.1
7.0	1372	3.5	−3.5	−1372	1368	−1368
8.0	2048	1.2	−1.2	−2048	2047	−2047

END SLOPE COEFFICIENTS FOR SYMMETRICALLY PLACED LOADS

λL	$\theta_0/(F/k_sBL^2)$ for Two Symmetrically Placed Loads $0.5F$ at Distance a from Each End					$\theta_0/(F/k_sBL^2)$ for a Centrally Placed Distributed Load F of Length $2c$			
	$a/L=0$	0.1	0.2	0.3	0.4	$c/L=0.1$	0.2	0.3	0.4
1.0	−0.166	−0.076	−0.007	0.043	0.073	0.079	0.070	0.053	0.030
2.0	−2.396	−1.082	−0.081	0.620	1.034	1.125	0.987	0.756	0.428
3.0	−8.750	−3.672	−0.092	2.216	3.487	3.757	3.343	2.615	1.520
4.0	−16.91	−5.922	0.658	4.072	5.552	5.818	5.374	4.440	2.752
5.0	−25.66	−6.453	2.674	5.642	5.954	5.869	5.887	5.428	3.769
6.0	−36.10	−5.393	5.841	6.932	5.057	4.400	5.217	5.771	4.691
7.0	−48.88	−2.997	9.589	7.769	3.517	2.334	3.950	5.715	5.591
8.0	−63.92	0.616	13.20	7.865	1.879	0.464	2.542	5.373	6.433

Negative/positive signs indicate anticlockwise/clockwise at LH end, and clockwise/anticlockwise at RH end.

END SLOPE COEFFICIENTS FOR A CONCENTRATED LOAD AT ANY POINT

λL	$\theta_0/(F/k_sBL^2)$ for Load F at Distance a from LH End ($a/L=0$)											
	$a/L=0$		0.1		0.2		0.3		0.4		0.5	
	$x/L=0$	1.0	0	1.0	0	1.0	0	1.0	0	1.0	0	1.0
1.0	−6.208	5.877	−4.880	4.728	−3.591	3.578	−2.337	2.423	−1.114	1.260	0.083	0.083
2.0	−9.073	4.280	−5.947	3.783	−3.432	3.270	−1.466	2.706	0.036	2.031	1.170	1.170
3.0	−18.01	0.507	−8.765	1.421	−2.501	2.318	1.297	3.136	3.236	3.739	3.891	3.891
4.0	−32.05	−1.776	−11.40	−0.445	0.358	0.958	5.609	2.535	6.819	4.285	5.946	5.946
5.0	−50.02	−1.293	−12.08	−0.825	5.546	−0.198	10.36	0.922	9.015	2.894	5.814	5.814
6.0	−72.00	−0.200	−10.30	−0.486	12.36	−0.674	14.30	−0.432	9.230	0.885	4.053	4.053
7.0	−98.00	0.235	−5.870	−0.125	19.71	−0.530	16.42	−0.882	7.611	−0.577	1.728	1.728
8.0	−128.0	0.170	1.188	0.044	26.59	−0.195	16.41	−0.676	4.905	−1.147	−0.243	−0.243

λL	$\theta_0/(F/k_sBL^2)$ for Load F at Distance a from LH End ($a/L=0$)											
	$a/L=0.5$		0.6		0.7		0.8		0.9		1.0	
	$x/L=0$	1.0	0	1.0	0	1.0	0	1.0	0	1.0	0	1.0
1.0	0.083	0.083	1.260	−1.114	2.423	−2.337	3.578	−3.591	4.728	−4.880	5.877	−6.208
2.0	1.170	1.170	2.031	0.036	2.706	−1.466	3.270	−3.432	3.783	−5.947	4.280	−9.073
3.0	3.891	3.891	3.739	3.236	3.136	1.297	2.318	−2.501	1.421	−8.765	0.507	−18.01
4.0	5.946	5.946	4.285	6.819	2.535	5.609	0.958	0.358	−0.445	−11.40	−1.776	−32.05
5.0	5.814	5.814	2.894	9.015	0.922	10.36	−0.198	5.546	−0.825	−12.08	−1.293	−50.02
6.0	4.053	4.053	0.885	9.230	−0.432	14.30	−0.674	12.36	−0.486	−10.30	−0.200	−72.00
7.0	1.728	1.728	−0.577	7.611	−0.882	16.42	−0.530	19.71	−0.125	−5.870	0.235	−98.00
8.0	−0.243	−0.243	−1.147	4.905	−0.676	16.41	−0.195	26.59	0.044	1.188	0.170	−128.0

For definitions of symbols, see Table B1.

Table B3 Bearing, Bending and Shear Coefficients for Moments at One or Both Ends

BEARING, BENDING AND SHEAR COEFFICIENTS FOR CONCENTRATED MOMENT M_0 AT LH END ($x/L = 0$)

x/L	$q/(M_0/BL^2)$ for Values of λL				$M/(M_0)$ for Values of λL				$V/(M_0/L)$ for Values of λL			
	1.0	2.0	3.0	4.0	1.0	2.0	3.0	4.0	1.0	2.0	3.0	4.0
0	−6.208	−9.073	−18.01	−32.05	1.0	1.0	1.0	1.0	0	0	0	0
0.1	−4.880	−5.947	−8.765	−11.40	0.971	0.960	0.927	0.878	−0.554	−0.746	−1.312	−2.091
0.2	−3.591	−3.432	−2.501	0.357	0.894	0.860	0.763	0.635	−0.977	−1.210	−1.853	−2.578
0.3	−2.337	−1.466	1.297	5.609	0.780	0.726	0.573	0.390	−1.274	−1.451	−1.895	−2.236
0.4	−1.115	0.036	3.236	6.819	0.643	0.576	0.394	0.198	−1.446	−1.519	−1.655	−1.590
0.5	0.083	1.170	3.891	5.946	0.495	0.426	0.246	0.072	−1.497	−1.456	−1.290	−0.941
0.6	1.260	2.031	3.739	4.285	0.348	0.288	0.136	0.005	−1.430	−1.294	−0.904	−0.426
0.7	2.423	2.706	3.136	2.535	0.213	0.170	0.064	−0.019	−1.246	−1.056	−0.557	−0.086
0.8	3.578	3.271	2.318	0.958	0.102	0.079	0.022	−0.018	−0.946	−0.756	−0.284	0.087
0.9	4.728	3.783	1.421	−0.445	0.028	0.021	0.004	−0.007	−0.530	−0.403	−0.096	0.111
1.0	5.877	4.281	0.507	−1.776	0	0	0	0	0	0	0	0

x/L	$q/(M_0/BL^2)$ for Values of λL				$M/(M_0)$ for Values of λL				$V/(M_0/L)$ for Values of λL			
	5.0	6.0	7.0	8.0	5.0	6.0	7.0	8.0	5.0	6.0	7.0	8.0
0	−50.02	−72.00	−98.00	−128.0	1.0	1.0	1.0	1.0	0	0	0	0
0.1	−12.08	−10.30	−5.870	1.188	0.823	0.763	0.700	0.635	−2.909	−3.719	−4.479	−5.157
0.2	5.546	12.36	19.71	26.59	0.508	0.390	0.285	0.196	−3.097	−3.369	−3.402	−3.229
0.3	10.36	14.30	16.42	16.41	0.238	0.124	0.044	−0.006	−2.226	−1.932	−1.480	−0.981
0.4	9.015	9.229	7.611	4.905	0.067	−0.006	−0.037	−0.043	−1.227	−0.735	−0.285	0.038
0.5	5.814	4.053	1.728	−0.243	−0.016	−0.042	−0.039	−0.026	−0.482	−0.084	0.148	0.222
0.6	2.894	0.885	−0.577	−1.146	−0.040	−0.037	−0.021	−0.008	−0.053	0.145	0.182	0.131
0.7	0.922	−0.432	−0.882	−0.676	−0.035	−0.020	−0.006	0.001	0.130	0.155	0.099	0.036
0.8	−0.198	−0.674	−0.530	−0.195	−0.020	−0.008	0	0.002	0.160	0.094	0.027	−0.005
0.9	−0.825	−0.486	−0.125	0.044	−0.006	−0.002	0	0.001	0.106	0.035	−0.006	−0.011
1.0	−1.293	−0.200	0.235	0.170	0	0	0	0	0	0	0	0

BEARING, BENDING AND SHEAR COEFFICIENTS FOR CONCENTRATED MOMENTS M_0 AT BOTH ENDS

x/L	$q/(M_0/BL^2)$ for Values of λL				$M/(M_0)$ for Values of λL				$V/(M_0/L)$ for Values of λL			
	1.0	2.0	3.0	4.0	1.0	2.0	3.0	4.0	1.0	2.0	3.0	4.0
0	−0.331	−4.793	−17.50	−33.83	1.0	1.0	1.0	1.0	0	0	0	0
0.1	−0.152	−2.164	−7.344	−11.85	0.999	0.981	0.931	0.872	−0.024	−0.343	−1.216	−2.202
0.2	−0.013	−0.162	−0.184	1.315	0.996	0.939	0.786	0.617	−0.032	−0.454	−1.569	−2.664
0.3	0.086	1.240	4.433	8.143	0.993	0.896	0.637	0.371	−0.028	−0.395	−1.337	−2.150
0.4	0.146	2.067	6.975	11.10	0.991	0.864	0.530	0.203	−0.016	−0.225	−0.751	−1.164
0.5	0.166	2.340	7.782	11.89	0.990	0.853	0.492	0.144	0	0	0	0

x/L	$q/(M_0/BL^2)$ for Values of λL				$M/(M_0)$ for Values of λL				$V/(M_0/L)$ for Values of λL			
	5.0	6.0	7.0	8.0	5.0	6.0	7.0	8.0	5.0	6.0	7.0	8.0
0	−51.31	−72.20	−97.77	−127.8	1.0	1.0	1.0	1.0	0	0	0	0
0.1	−12.91	−10.79	−5.995	1.232	0.817	0.761	0.700	0.636	−3.015	−3.753	−4.473	−5.146
0.2	5.348	11.68	19.18	26.39	0.489	0.382	0.285	0.198	−3.257	−3.463	−3.429	−3.224
0.3	11.28	13.86	15.54	15.73	0.203	0.103	0.038	−0.005	−2.355	−2.087	−1.579	−1.017
0.4	11.91	10.12	7.034	3.758	0.026	−0.042	−0.057	−0.051	−1.174	−0.880	−0.467	−0.093
0.5	11.63	8.106	3.455	−0.486	−0.032	−0.084	−0.078	−0.052	0	0	0	0

Negative signs indicate upward pressure for q, hogging moment for M and downward force for V.

For definitions of symbols, see Table B1.

Table B4 Bearing, Bending and Shear Coefficients for Centrally Placed Distributed Load

BEARING, BENDING AND SHEAR COEFFICIENTS FOR A CENTRALLY PLACED DISTRIBUTED LOAD

λL	x/L	$q/(F/BL)$ for Values of c/L				$M/(FL)$ for Values of c/L				V/F for Values of c/L			
		0.1	0.2	0.3	0.4	0.1	0.2	0.3	0.4	0.1	0.2	0.3	0.4
1.0	0	0.982	0.986	0.989	0.994	0	0	0	0	0	0	0	0
	0.1	0.990	0.992	0.994	0.997	0.005	0.005	0.005	0.005	0.099	0.099	0.099	0.100
	0.2	0.998	0.998	0.999	1.000	0.020	0.020	0.020	0.014	0.198	0.199	0.199	0.075
	0.3	1.005	1.004	1.003	1.002	0.045	0.045	0.037	0.020	0.298	0.299	0.132	0.050
	0.4	1.010	1.008	1.006	1.003	0.080	0.067	0.047	0.024	0.399	0.149	0.066	0.025
	0.5	1.012	1.010	1.007	1.004	0.100	0.075	0.050	0.025	0	0	0	0
2.0	0	0.747	0.781	0.837	0.911	0	0	0	0	0	0	0	0
	0.1	0.859	0.880	0.912	0.953	0.004	0.004	0.005	0.005	0.081	0.083	0.088	0.093
	0.2	0.968	0.975	0.984	0.993	0.017	0.017	0.018	0.013	0.172	0.176	0.182	0.066
	0.3	1.067	1.059	1.045	1.025	0.039	0.040	0.033	0.018	0.274	0.278	0.117	0.042
	0.4	1.140	1.118	1.085	1.045	0.072	0.060	0.041	0.021	0.385	0.137	0.057	0.020
	0.5	1.168	1.139	1.099	1.052	0.091	0.067	0.044	0.022	0	0	0	0
3.0	0	0.136	0.248	0.434	0.688	0	0	0	0	0	0	0	0
	0.1	0.511	0.582	0.695	0.839	0.001	0.002	0.003	0.004	0.033	0.042	0.057	0.076
	0.2	0.881	0.909	0.945	0.977	0.008	0.010	0.013	0.010	0.102	0.116	0.139	0.042
	0.3	1.224	1.202	1.156	1.086	0.024	0.027	0.023	0.013	0.207	0.222	0.077	0.020
	0.4	1.488	1.412	1.295	1.154	0.051	0.043	0.028	0.015	0.346	0.105	0.035	0.008
	0.5	1.593	1.487	1.343	1.177	0.068	0.048	0.030	0.015	0	0	0	0
4.0	0	−0.419	−0.254	0.033	0.452	0	0	0	0	0	0	0	0
	0.1	0.196	0.284	0.477	0.725	−0.001	0	0.001	0.003	−0.011	0.002	0.026	0.059
	0.2	0.780	0.824	0.907	0.970	0	0.003	0.007	0.007	0.038	0.057	0.095	0.019
	0.3	1.359	1.335	1.269	1.151	0.009	0.014	0.014	0.007	0.145	0.165	0.037	0
	0.4	1.841	1.714	1.503	1.258	0.029	0.025	0.015	0.007	0.305	0.067	0.012	−0.003
	0.5	2.067	1.852	1.583	1.294	0.044	0.028	0.016	0.007	0	0	0	0
5.0	0	−0.628	−0.485	−0.197	0.289	0	0	0	0	0	0	0	0
	0.1	−0.036	0.108	0.347	0.662	−0.002	−0.001	0	0.002	−0.028	−0.014	0.008	0.048
	0.2	0.611	0.734	0.886	0.983	−0.004	−0.002	0.004	0.005	0	0.028	0.069	0.008
	0.3	1.361	1.390	1.340	1.197	0	0.006	0.008	0.005	0.099	0.134	0.014	−0.008
	0.4	2.106	1.910	1.622	1.310	0.016	0.014	0.008	0.004	0.272	0.049	0	−0.008
	0.5	2.447	2.103	1.716	1.345	0.030	0.016	0.008	0.003	0	0	0	0
6.0	0	−0.581	−0.513	−0.302	0.171	0	0	0	0	0	0	0	0
	0.1	−0.131	0.018	0.279	0.634	−0.002	−0.002	0	0.002	−0.036	−0.025	−0.001	0.040
	0.2	0.438	0.635	0.876	1.013	−0.005	−0.003	0.002	0.004	−0.020	0.008	0.057	−0.002
	0.3	1.275	1.387	1.384	1.231	−0.003	0.003	0.005	0.003	0.066	0.109	0.003	−0.011
	0.4	2.280	2.035	1.679	1.324	0.010	0.008	0.004	0.001	0.247	0.034	−0.006	−0.008
	0.5	2.785	2.277	1.771	1.347	0.022	0.010	0.004	0.001	0	0	0	0
7.0	0	−0.417	−0.440	−0.343	0.073	0	0	0	0	0	0	0	0
	0.1	−0.169	−0.031	0.237	0.625	−0.002	−0.002	−0.001	0.001	−0.029	−0.024	−0.005	0.035
	0.2	0.256	0.528	0.872	1.053	−0.005	−0.003	0.002	0.003	−0.025	0.002	0.050	−0.006
	0.3	1.130	1.357	1.417	1.258	−0.004	0.002	0.004	0.002	0.045	0.096	−0.002	−0.011
	0.4	2.415	2.127	1.705	1.318	0.005	0.005	0.003	0	0.223	0.023	−0.008	−0.007
	0.5	3.116	2.413	1.787	1.327	0.016	0.006	0.002	0	0	0	0	0
8.0	0	−0.236	−0.333	−0.348	−0.011	0	0	0	0	0	0	0	0
	0.1	−0.174	−0.060	0.206	0.624	−0.001	−0.001	−0.001	0.001	−0.021	−0.020	−0.007	0.031
	0.2	0.091	0.423	0.868	1.094	−0.004	−0.003	0.001	0.003	−0.025	−0.002	0.047	−0.009
	0.3	0.959	1.321	1.448	1.278	−0.004	0.001	0.003	0.002	0.028	0.086	−0.004	−0.010
	0.4	2.520	2.201	1.717	1.306	0.003	0.003	0.002	0	0.202	0.014	−0.008	−0.005
	0.5	3.438	2.519	1.782	1.300	0.013	0.004	0.001	0	0	0	0	0

Negative signs indicate upward pressure for q, hogging moment for M and downward force for V.

For definitions of symbols, see Table B1.

Table B5 Bearing, Bending and Shear Coefficients for Two Symmetrical Loads –1

BEARING, BENDING AND SHEAR COEFFICIENTS FOR TWO SYMMETRICALLY PLACED LOADS

λL	x/L	$a/L = 0$			$a/L = 0.1$			$a/L = 0.2$		
		$q/(F/BL)$	$M/(FL)$	V/F	$q/(F/BL)$	$M/(FL)$	V/F	$q/(F/BL)$	$M/(FL)$	V/F
1.0	0	1.033	0	0/–0.5	1.017	0	0	1.003	0	0
	0.1	1.017	–0.045	–0.398	1.009	0.005	0.101/–0.399	1.002	0.005	0.100
	0.2	1.002	–0.080	–0.297	1.002	–0.030	–0.298	1.001	0.020	0.200/–0.300
	0.3	0.991	–0.104	–0.197	0.995	–0.055	0.198	1.000	–0.005	–0.200
	0.4	0.984	–0.119	–0.098	0.991	–0.070	–0.099	0.998	–0.020	–0.100
	0.5	0.982	–0.124	0	0.990	–0.075	0	0.998	–0.025	0
2.0	0	1.475	0	0/–0.5	1.241	0	0	1.033	0	0
	0.1	1.241	–0.043	–0.364	1.132	0.006	0.119/–0.381	1.025	0.005	0.103
	0.2	1.033	–0.074	–0.251	1.025	–0.027	–0.274	1.012	0.021	0.205/–0.295
	0.3	0.872	–0.094	–0.156	0.933	–0.049	–0.176	0.992	–0.004	–0.195
	0.4	0.770	–0.105	–0.075	0.873	–0.062	–0.086	0.973	–0.019	–0.097
	0.5	0.735	–0.109	0	0.852	–0.066	0	0.966	–0.024	0
3.0	0	2.681	0	0/–0.5	1.829	0	0	1.096	0	0
	0.1	1.829	–0.038	–0.275	1.460	0.009	0.164/–0.336	1.086	0.006	0.109
	0.2	1.096	–0.058	–0.130	1.086	–0.018	–0.208	1.054	0.022	0.216/–0.284
	0.3	0.547	–0.066	–0.050	0.768	–0.034	–0.117	0.977	–0.002	–0.182
	0.4	0.211	–0.069	–0.014	0.559	–0.043	–0.051	0.901	–0.015	–0.088
	0.5	0.098	–0.070	0	0.974	–0.045	0	0.871	–0.019	0
4.0	0	4.018	0	0/–0.5	2.397	0	0	1.089	0	0
	0.1	2.397	–0.033	–0.181	1.795	0.011	0.210/–0.290	1.150	0.006	0.112
	0.2	1.089	–0.041	–0.010	1.150	–0.010	–0.143	1.145	0.023	0.228/–0.272
	0.3	0.194	–0.038	0.051	0.600	–0.020	–0.057	0.984	0.001	–0.165
	0.4	–0.311	–0.033	0.042	0.245	–0.023	–0.017	0.806	–0.011	–0.076
	0.5	–0.472	–0.031	0	0.124	–0.024	0	0.733	–0.015	0
5.0	0	5.085	0	0/–0.5	2.681	0	0	0.938	0	0
	0.1	2.681	–0.029	–0.116	2.009	0.012	0.235/–0.265	1.196	0.005	0.107
	0.2	0.938	–0.030	0.059	1.196	–0.006	–0.104	1.299	0.022	0.234/–0.266
	0.3	–0.070	–0.022	0.097	0.494	–0.011	–0.022	1.032	0.002	–0.147
	0.4	–0.540	–0.013	0.063	0.060	–0.012	0.004	0.701	–0.009	–0.061
	0.5	–0.671	–0.010	0	–0.084	0.012	0	0.565	–0.012	0
6.0	0	6.037	0	0/–0.5	2.744	0	0	0.661	0	0
	0.1	2.744	–0.026	–0.068	2.148	0.013	0.246/–0.254	1.228	0.004	0.095
	0.2	0.661	–0.023	0.091	1.228	–0.003	–0.084	1.521	0.021	0.237/–0.263
	0.3	–0.271	–0.013	0.103	0.425	–0.007	–0.004	1.113	0.002	–0.127
	0.4	–0.551	–0.004	0.058	–0.035	–0.006	0.013	0.581	–0.007	–0.044
	0.5	–0.594	–0.001	0	–0.178	–0.005	0	0.365	–0.009	0
7.0	0	7.001	0	0/–0.5	2.672	0	0	0.314	0	0
	0.1	2.672	–0.023	–0.029	2.267	0.013	0.250/–0.250	1.253	0.003	0.079
	0.2	0.314	–0.017	0.103	1.253	–0.002	–0.071	1.788	0.019	0.238/–0.262
	0.3	–0.423	–0.007	0.088	0.365	–0.005	0.006	1.202	0.001	–0.106
	0.4	–0.452	–0.001	0.042	–0.085	–0.004	0.017	0.449	–0.005	–0.026
	0.5	–0.396	0.002	0	–0.209	–0.003	0	0.157	–0.006	0
8.0	0	7.994	0	0/–0.5	2.507	0	0	–0.034	0	0
	0.1	2.507	–0.020	0.005	2.397	0.013	0.250/–0.250	1.271	0.002	0.063
	0.2	–0.034	–0.013	0.105	1.271	–0.002	–0.062	2.070	0.017	0.240/–0.260
	0.3	–0.512	–0.004	0.067	0.299	–0.004	0.012	1.271	0	–0.085
	0.4	–0.320	0.001	0.024	–0.113	–0.002	0.018	0.319	–0.004	–0.009
	0.5	–0.192	0.002	0	–0.202	–0.001	0	–0.024	–0.004	0

Negative signs indicate upward pressure for q, hogging moment for M and downward force for V.

For values of $a/L = 0.3$ to 0.5, see Table B6.

For definitions of symbols, see Table B1.

Table B6 Bearing, Bending and Shear Coefficients for Two Symmetrical Loads – 2

BEARING, BENDING AND SHEAR COEFFICIENTS FOR TWO SYMMETRICALLY PLACED LOADS

λL	x/L	$a/L = 0.3$			$a/L = 0..4$			$a/L = 0.5$		
		$q/(F/BL)$	$M/(FL)$	V/F	$q/(F/BL)$	$M/(FL)$	V/F	$q/(F/BL)$	$M/(FL)$	V/F
1.0	0	0.991	0	0	0.984	0	0	0.982	0	0
	0.1	0.995	0.005	0.099	0.991	0.005	0.099	0.990	0.005	0.099
	0.2	1.000	0.020	0.199	0.998	0.020	0.198	0.998	0.020	0.198
	0.3	1.003	0.045	0.299/–0.201	1.004	0.045	0.298	1.005	0.045	0.298
	0.4	1.004	0.030	–0.101	1.009	0.080	0.399/–0.101	1.010	0.080	0.399
	0.5	1.005	0.025	0	1.010	0.075	0	1.013	0.125	0.5/–0.5
2.0	0	0.872	0	0	0.770	0	0	0.735	0	0
	0.1	0.933	0.005	0.090	0.873	0.004	0.082	0.852	0.004	0.079
	0.2	0.992	0.018	0.187	0.973	0.017	0.175	0.966	0.016	0.170
	0.3	1.038	0.042	0.288/–0.212	1.062	0.039	0.277	1.069	0.038	0.272
	0.4	1.062	0.026	–0.107	1.125	0.072	0.386/–0.114	1.147	0.071	0.383
	0.5	1.069	0.021	0	1.147	0.067	0	1.179	0.115	0.5/–0.5
3.0	0	0.547	0	0	0.211	0	0	0.098	0	0
	0.1	0.768	0.003	0.066	0.559	0.002	0.039	0.487	0.001	0.029
	0.2	0.977	0.014	0.153	0.901	0.009	0.112	0.871	0.007	0.097
	0.3	1.138	0.034	0.260/–0.240	1.211	0.025	0.218	1.229	0.022	0.203
	0.4	1.211	0.016	–0.122	1.437	0.053	0.351/–0.149	1.514	0.049	0.341
	0.5	1.229	0.010	0	1.514	0.046	0	1.636	0.091	0.5/–0.5
4.0	0	0.194	0	0	–0.311	0	0	–0.472	0	0
	0.1	0.600	0.002	0.040	0.245	–0.001	–0.003	0.124	–0.001	–0.018
	0.2	0.984	0.009	0.119	0.806	0.001	0.049	0.733	–0.002	0.025
	0.3	1.264	0.027	0.233/–0.267	1.347	0.011	0.157	1.351	0.006	0.130
	0.4	1.347	0.007	–0.135	1.763	0.034	0.315/–0.185	1.899	0.027	0.293
	0.5	1.351	0	0	1.899	0.025	0	2.160	0.066	0.5/–0.5
5.0	0	–0.070	0	0	–0.540	0	0	–0.671	0	0
	0.1	0.494	0.001	0.021	0.060	–0.002	–0.024	–0.084	–0.003	–0.038
	0.2	1.032	0.006	0.098	0.701	–0.003	0.013	0.565	–0.006	–0.015
	0.3	1.397	0.022	0.222/–0.278	1.397	0.003	0.118	1.340	–0.003	0.080
	0.4	1.397	0.001	–0.136	1.984	0.023	0.290/–0.210	2.162	0.013	0.255
	0.5	1.340	–0.006	0	2.162	0.013	0	2.611	0.051	0.5/–0.5
6.0	0	–0.271	0	0	–0.551	0	0	–0.594	0	0
	0.1	0.425	0	0.008	–0.035	–0.002	–0.030	–0.178	–0.002	–0.039
	0.2	1.113	0.004	0.085	0.581	–0.004	–0.004	0.365	–0.006	–0.032
	0.3	1.553	0.019	0.223/–0.277	1.383	0	0.093	1.216	–0.006	0.045
	0.4	1.383	–0.001	–0.128	2.140	0.017	0.272/–0.228	2.340	0.006	0.222
	0.5	1.216	–0.008	0	2.340	0.006	0	3.048	0.042	0.5/–0.5
7.0	0	–0.423	0	0	–0.452	0	0	–0.396	0	0
	0.1	0.365	–0.001	–0.003	–0.085	–0.002	–0.027	–0.209	–0.002	–0.031
	0.2	1.202	0.002	0.075	0.449	–0.004	–0.012	0.157	–0.005	–0.036
	0.3	1.742	0.017	0.229/–0.271	1.335	–0.002	0.075	1.019	–0.007	0.018
	0.4	1.335	–0.002	–0.113	2.272	0.014	0.259/–0.241	2.466	0.002	0.185
	0.5	1.019	–0.008	0	2.466	0.002	0	3.514	0.036	0.5/–0.5
8.0	0	–0.512	0	0	–0.320	0	0	–0.192	0	0
	0.1	0.299	–0.001	–0.011	–0.113	–0.001	–0.022	–0.203	–0.001	–0.020
	0.2	1.271	0.001	0.066	0.319	–0.004	–0.015	–0.024	–0.004	–0.035
	0.3	1.961	0.015	0.236/–0.264	1.277	–0.002	0.060	0.789	–0.007	–0.004
	0.4	1.277	–0.002	–0.096	2.400	0.012	0.248/–0252	2.545	0	0.156
	0.5	0.789	–0.007	0	2.545	0	0	4.002	0.031	0.5/–0.5

Negative signs indicate upward pressure for q, hogging moment for M and downward force for V.

Values for $a/L = 0.5$ apply to a single load F at the centre of the beam.

For definitions of symbols, see Table B1.

Table B7 Bearing, Bending and Shear Coefficients for a Load at any Point – 1

BEARING, BENDING AND SHEAR COEFFICIENTS FOR A CONCENTRATED LOAD AT ANY POINT

λL	x/L	$a/L = 0$			$a/L = 0.1$			$a/L = 0.2$		
		$q/(F/BL)$	$M/(FL)$	V/F	$q/(F/BL)$	$M/(FL)$	V/F	$q/(F/BL)$	$M/(FL)$	V/F
1.0	0	4.038	0	0/–1.0	3.418	0	0	2.800	0	0
	0.1	3.418	–0.081	–0.627	2.930	0.016	0.317/–0.683	2.441	0.014	0.262
	0.2	2.800	–0.128	–0.316	2.441	–0.038	–0.414	2.082	0.051	0.488/–0.512
	0.3	2.188	–0.146	–0.067	1.955	–0.068	–0.194	1.720	0.010	–0.322
	0.4	1.582	–0.143	0.122	1.471	–0.079	–0.023	1.359	–0.014	–0.168
	0.5	0.982	–0.124	0.250	0.990	–0.075	0.100	0.998	–0.025	–0.050
	0.6	0.386	–0.095	0.318	0.512	–0.060	0.175	0.638	–0.026	0.032
	0.7	–0.206	–0.062	0.327	0.036	–0.041	0.202	0.278	–0.020	0.078
	0.8	–0.796	–0.032	0.277	–0.438	–0.021	0.182	–0.080	–0.011	0.088
	0.9	–1.384	–0.009	0.168	–0.911	–0.006	0.115	–0.438	–0.004	0.062
	1.0	–1.972	0	0	–1.384	0	0	–0.796	0	0
2.0	0	4.550	0	0/–1.0	3.653	0	0	2.803	0	0
	0.1	3.653	–0.079	–0.590	3.057	0.017	0.335/–0.665	2.459	0.014	0.263
	0.2	2.803	–0.121	–0.268	2.459	–0.035	–0.389	2.105	0.052	0.492/–0.508
	0.3	2.030	–0.135	–0.027	1.883	–0.063	–0.172	1.728	0.011	–0.317
	0.4	1.342	–0.129	0.141	1.345	–0.071	–0.011	1.343	–0.013	–0.163
	0.5	0.735	–0.109	0.244	0.852	–0.066	0.099	0.966	–0.023	–0.048
	0.6	0.198	–0.082	0.290	0.401	–0.053	0.161	0.603	–0.024	0.031
	0.7	–0.287	–0.053	0.286	–0.016	–0.036	0.180	0.255	–0.019	0.073
	0.8	–0.738	–0.026	0.234	–0.410	–0.018	0.158	–0.081	–0.011	0.082
	0.9	–1.171	–0.007	0.139	–0.793	–0.005	0.098	–0.410	–0.003	0.058
	1.0	–1.600	0	0	–1.171	0	0	–0.738	0	0
3.0	0	6.040	0	0/–1.0	4.285	0	0	2.755	0	0
	0.1	4.285	–0.073	–0.485	3.403	0.020	0.385/–0.615	2.501	0.014	0.263
	0.2	2.755	–0.103	–0.135	2.501	–0.026	–0.320	2.197	0.052	0.499/–0.501
	0.3	1.550	–0.105	0.077	1.677	–0.047	–0.113	1.774	0.012	–0.302
	0.4	0.678	–0.091	0.186	1.001	–0.051	0.020	1.308	–0.010	–0.148
	0.5	0.098	–0.070	0.222	0.487	–0.045	0.093	0.871	–0.019	–0.039
	0.6	–0.256	–0.048	0.213	0.118	–0.034	0.122	0.494	–0.020	0.028
	0.7	–0.455	–0.028	0.176	–0.141	–0.022	0.120	0.179	–0.015	0.061
	0.8	–0.563	–0.013	0.125	–0.330	–0.011	0.096	–0.089	–0.008	0.066
	0.9	–0.627	–0.003	0.065	–0.484	–0.003	0.056	–0.033	–0.003	0.045
	1.0	–0.678	0	0	–0.627	0	0	–0.563	0	0
4.0	0	8.006	0	0/–1.0	4.941	0	0	2.501	0	0
	0.1	4.941	–0.065	–0.356	3.781	0.023	0.437/–0.563	2.526	0.013	0.252
	0.2	2.501	–0.081	0.010	2.526	–0.017	–0.248	2.401	0.050	0.500/–0.500
	0.3	0.864	–0.070	0.171	1.420	–0.031	–0.053	1.911	0.012	–0.282
	0.4	–0.063	–0.051	0.206	0.619	–0.031	0.046	1.287	–0.008	–0.122
	0.5	–0.472	–0.031	0.176	0.124	–0.024	0.081	0.733	–0.015	–0.022
	0.6	–0.559	–0.016	0.122	–0.129	–0.016	0.079	0.325	–0.014	0.030
	0.7	–0.476	–0.007	0.070	–0.221	–0.009	0.061	0.056	–0.010	0.048
	0.8	–0.322	–0.002	0.029	–0.225	–0.004	0.038	–0.111	–0.005	0.044
	0.9	–0.147	0	0.006	–0.191	–0.001	0.017	–0.225	–0.002	0.027
	1.0	0.031	0	0	–0.147	0	0	–0.322	0	0

Negative signs indicate upward pressure for q, hogging moment for M and downward force for V.

For $\lambda L = 5.0$ to 8.0, see Table B8; for values of $a/L = 0.3$ to 0.5, see Table B9.

For definitions of symbols, see Table B1.

Table B8 Bearing, Bending and Shear Coefficients for a Load at any Point – 2

BEARING, BENDING AND SHEAR COEFFICIENTS FOR A CONCENTRATED LOAD AT ANY POINT

λL	x/L	$a/L = 0$			$a/L = 0.1$			$a/L = 0.2$		
		$q/(F/BL)$	$M/(FL)$	V/F	$q/(F/BL)$	$M/(FL)$	V/F	$q/(F/BL)$	$M/(FL)$	V/F
5.0	0	10.00	0	0/–1.0	5.324	0	0	1.987	0	0
	0.1	5.324	–0.058	–0.241	4.063	0.025	0.471/–0.529	2.520	0.011	0.226
	0.2	1.987	–0.062	0.111	2.520	–0.011	–0.199	2.729	0.047	0.494/–0.506
	0.3	0.154	–0.045	0.207	1.187	–0.020	–0.018	2.132	0.009	–0.258
	0.4	–0.571	–0.025	0.179	0.335	–0.018	0.055	1.267	–0.008	–0.088
	0.5	–0.671	–0.010	0.113	–0.084	–0.012	0.064	0.565	–0.012	0.001
	0.6	–0.509	–0.002	0.053	–0.215	–0.006	0.048	0.135	–0.009	0.034
	0.7	–0.295	0.002	0.013	–0.200	–0.002	0.026	–0.068	–0.006	0.036
	0.8	–0.110	0.002	–0.007	–0.129	0	0.009	–0.130	–0.003	0.025
	0.9	0.037	0.001	–0.010	–0.046	0	0.001	–0.129	–0.001	0.012
	1.0	0.168	0	0	0.037	0	0	–0.110	0	0
6.0	0	12.00	0	0/–1.0	5.436	0	0	1.310	0	0
	0.1	5.436	–0.052	0.143	4.293	0.026	0.490/–0.510	2.512	0.009	0.192
	0.2	1.310	–0.047	–0.172	2.512	–0.007	–0.168	3.160	0.042	0.485/–0.515
	0.3	–0.451	–0.027	–0.199	0.982	–0.014	0.002	2.366	0.006	–0.230
	0.4	–0.804	–0.010	–0.128	0.133	–0.011	0.052	1.191	–0.008	–0.053
	0.5	–0.594	–0.001	–0.056	–0.178	–0.006	0.047	0.365	–0.009	0.021
	0.6	–0.298	0.002	–0.012	–0.203	–0.002	0.026	–0.031	–0.006	0.035
	0.7	–0.091	0.002	0.007	–0.132	0	0.009	–0.140	–0.003	0.025
	0.8	0.011	0.001	0.010	–0.056	0	0	–0.118	–0.001	0.011
	0.9	0.052	0.001	0.006	0.003	0	–0.003	–0.056	0	0.002
	1.0	0.074	0	0	0.052	0	0	0.011	0	0
7.0	0	14.00	0	0/–1.0	5.317	0	0	0.587	0	0
	0.1	5.317	–0.046	0.060	4.523	0.026	0.498/–0.502	2.519	0.006	0.157
	0.2	0.587	–0.035	–0.201	2.518	–0.005	–0.145	3.654	0.037	0.480/–0.520
	0.3	–0.866	–0.015	–0.168	0.798	–0.010	0.014	2.549	0.002	–0.197
	0.4	–0.802	–0.003	–0.078	–0.015	–0.007	0.046	1.035	–0.008	–0.021
	0.5	–0.396	0.002	–0.018	–0.209	–0.003	0.032	0.157	–0.006	0.032
	0.6	–0.102	0.002	0.006	–0.156	0	0.013	–0.138	–0.003	0.030
	0.7	0.021	0.001	0.009	–0.068	0.001	0.002	–0.146	–0.001	0.014
	0.8	0.042	0.001	0.005	–0.013	0	–0.002	–0.077	0	0.003
	0.9	0.026	0	0.002	0.012	0	–0.002	–0.013	0	0.002
	1.0	0.003	0	0	0.026	0	0	0.042	0	0
8.0	0	16.00	0	0	5.009	0	0	–0.095	0	0
	0.1	5.009	–0.040	–0.009	4.785	0.025	0.500/–0.500	2.535	0.004	0.123
	0.2	–0.095	–0.025	–0.208	2.535	–0.004	–0.125	4.173	0.033	0.478//–0.522
	0.3	–1.070	–0.008	–0.128	0.621	–0.008	0.023	2.651	0	–0.163
	0.4	–0.651	0	–0.038	–0.123	–0.004	0.040	0.818	–0.007	0.003
	0.5	–0.192	0.002	0.002	–0.203	–0.001	0.020	–0.024	–0.004	0.035
	0.6	0.012	0.001	0.009	–0.102	0	0.005	–0.181	–0.001	0.021
	0.7	0.047	0	0.005	–0.023	0.001	–0.001	–0.110	0	0.006
	0.8	0.028	0	0.001	0.007	0	–0.002	–0.033	0	–0.001
	0.9	0.006	0	0	0.010	0	–0.001	0.007	0	–0.002
	1.0	–0.012	0	0	0.006	0	0	0.028	0	0

Negative signs indicate upward pressure for q, hogging moment for M and downward force for V.

For $\lambda L = 1.0$ to 4.0, see Table B7; for values of $a/L = 0.3$ to 0.5, see Table B10.

For definitions of symbols, see Table B1.

Table B9 Bearing, Bending and Shear Coefficients for a Load at any Point – 3

BEARING, BENDING AND SHEAR COEFFICIENTS FOR A CONCENTRATED LOAD AT ANY POINT

λL	x/L	$a/L = 0.3$			$a/L = 0.4$			$a/L = 0.5$		
		$q/(F/BL)$	$M/(FL)$	V/F	$q/(F/BL)$	$M/(FL)$	V/F	$q/(F/BL)$	$M/(FL)$	V/F
1.0	0	2.189	0	0	1.582	0	0	0.981	0	0
	0.1	1.955	0.011	0.207	1.471	0.008	0.153	0.990	0.005	0.099
	0.2	1.720	0.041	0.391	1.359	0.030	0.294	0.998	0.020	0.198
	0.3	1.485	0.088	0.551/−0.449	1.246	0.066	0.424	1.005	0.045	0.298
	0.4	1.246	0.050	−0.312	1.130	0.115	0.543/−0.457	1.010	0.080	0.399
	0.5	1.005	0.025	−0.200	1.010	0.075	−0.350	1.012	0.124	0.500/−0.500
	0.6	0.763	0.010	−0.111	0.888	0.044	−0.255	1.010	0.080	−0.399
	0.7	0.521	0.002	−0.047	0.763	0.023	−0.172	1.005	0.045	−0.298
	0.8	0.278	−0.001	−0.007	0.638	0.010	−0.102	0.998	0.020	−0.198
	0.9	0.036	−0.001	0.009	0.512	0.002	−0.045	0.990	0.005	−0.099
	1.0	−0.206	0	0	0.386	0	0	0.981	0	0
2.0	0	2.030	0	0	1.342	0	0	0.735	0	0
	0.1	1.883	0.010	0.196	1.345	0.007	0.134	0.852	0.004	0.079
	0.2	1.728	0.039	0.376	1.343	0.027	0.269	0.966	0.016	0.170
	0.3	1.548	0.085	0.540/−0.460	1.324	0.061	0.402	1.069	0.038	0.272
	0.4	1.324	0.046	−0.316	1.265	0.107	0.532/−0.468	1.147	0.071	0.383
	0.5	1.069	0.021	−0.196	1.147	0.067	−0.347	1.179	0.115	0.500/−0.500
	0.6	0.801	0.006	−0.102	0.986	0.038	−0.240	1.147	0.071	−0.383
	0.7	0.528	−0.001	−0.036	0.801	0.018	−0.150	1.069	0.038	−0.272
	0.8	0.255	−0.002	0.003	0.603	0.007	−0.080	0.966	0.016	−0.170
	0.9	−0.016	−0.001	0.015	0.401	0.001	−0.030	0.852	0.004	−0.079
	1.0	−0.287	0	0	0.198	0	0	0.735	0	0
3.0	0	1.550	0	0	0.678	0	0	0.098	0	0
	0.1	1.677	0.008	0.161	1.001	0.004	0.084	0.487	0.001	0.029
	0.2	1.774	0.033	0.334	1.308	0.018	0.200	0.871	0.007	0.097
	0.3	1.761	0.075	0.513/−0.487	1.554	0.045	0.344	1.229	0.022	0.203
	0.4	1.554	0.035	−0.320	1.650	0.087	0.505/−0.495	1.514	0.049	0.341
	0.5	1.230	0.010	−0.181	1.514	0.046	−0.335	1.636	0.091	0.500/−0.500
	0.6	0.869	−0.003	−0.076	1.225	0.020	−0.197	1.514	0.049	−0.341
	0.7	0.515	−0.006	−0.007	0.869	0.006	−0.092	1.230	0.022	−0.203
	0.8	0.179	−0.005	0.028	0.494	0	−0.024	0.871	0.007	−0.097
	0.9	−0.141	−0.002	0.030	0.118	−0.001	0.007	0.487	0.001	−0.029
	1.0	−0.455	0	0	−0.256	0	0	0.098	0	0
4.0	0	0.864	0	0	−0.063	0	0	−0.472	0	0
	0.1	1.420	0.005	0.114	0.619	0.001	0.028	0.124	−0.001	−0.017
	0.2	1.911	0.025	0.282	1.287	0.008	0.123	0.733	−0.002	0.025
	0.3	2.133	0.063	0.488/−0.512	1.864	0.028	0.282	1.351	0.006	0.130
	0.4	1.864	0.022	−0.309	2.142	0.066	0.486/−0.514	1.899	0.027	0.293
	0.5	1.351	0	−0.147	1.899	0.025	−0.308	2.160	0.066	0.500/−0.500
	0.6	0.830	−0.009	−0.039	1.385	0.003	−0.143	1.899	0.027	−0.293
	0.7	0.396	−0.010	0.022	0.830	−0.006	−0.032	1.351	0.006	−0.130
	0.8	0.056	−0.006	0.043	0.325	−0.006	0.025	0.733	−0.002	−0.025
	0.9	−0.221	−0.002	0.035	−0.129	−0.002	0.034	0.124	−0.001	0.017
	1.0	−0.476	0	0	−0.559	0	0	−0.472	0	0

Negative signs indicate upward pressure for q, hogging moment for M and downward force for V.

For $\lambda L = 5.0$ to 8.0, see Table B10; for values of $a/L = 0$ to 0.2, see Table B7.

For definitions of symbols, see Table B1.

Table B10 Bearing, Bending and Shear Coefficients for a Load at any Point – 4

BEARING, BENDING AND SHEAR COEFFICIENTS FOR A CONCENTRATED LOAD AT ANY POINT

λL	x/L	$a/L = 0.3$			$a/L = 0.4$			$a/L = 0.5$		
		$q/(F/BL)$	$M/(FL)$	V/F	$q/(F/BL)$	$M/(FL)$	V/F	$q/(F/BL)$	$M/(FL)$	V/F
5.0	0	0.154	0	0	−0.571	0	0	−0.671	0	0
	0.1	1.187	0.003	0.067	0.335	−0.002	−0.012	−0.084	−0.003	−0.038
	0.2	2.132	0.017	0.235	1.267	0.001	0.068	0.565	−0.006	−0.015
	0.3	2.614	0.052	0.479/−0.521	2.156	0.015	0.240	1.340	−0.003	0.080
	0.4	2.156	0.013	−0.277	2.618	0.051	0.485/−0.515	2.162	0.013	0.255
	0.5	1.340	−0.006	−0.102	2.162	0.013	−0.270	2.611	0.051	0.500/−0.500
	0.6	0.638	−0.011	−0.005	1.351	−0.005	−0.094	2.162	0.013	−0.255
	0.7	0.181	−0.009	0.035	0.638	−0.009	0.004	1.340	−0.003	−0.080
	0.8	−0.068	−0.005	0.039	0.135	−0.006	0.041	0.565	−0.006	0.015
	0.9	−0.200	−0.001	0.025	−0.215	−0.002	0.037	−0.084	−0.003	0.038
	1.0	−0.295	0	0	−0.509	0	0	−0.671	0	0
6.0	0	−0.451	0	0	−0.804	0	0	−0.594	0	0
	0.1	0.982	0	0.027	0.133	−0.003	−0.034	−0.178	−0.002	−0.039
	0.2	2.366	0.010	0.196	1.191	−0.004	0.031	0.365	−0.006	−0.032
	0.3	3.127	0.043	0.480/−0.520	2.381	0.008	0.210	1.216	−0.006	0.045
	0.4	2.381	0.006	−0.235	3.080	0.042	0.492/−0.508	2.340	0.006	0.222
	0.5	1.216	−0.008	−0.056	2.340	0.006	−0.228	3.048	0.042	0.500/−0.500
	0.6	0.384	−0.009	0.020	1.199	−0.007	−0.052	2.340	0.006	−0.222
	0.7	−0.021	−0.006	0.035	0.384	−0.008	0.023	1.216	−0.006	−0.045
	0.8	−0.140	−0.003	0.026	−0.031	−0.005	0.038	0.365	−0.006	0.032
	0.9	−0.132	−0.001	0.011	−0.203	−0.001	0.025	−0.178	−0.002	0.039
	1.0	−0.091	0	0	−0.298	0	0	−0.594	0	0
7.0	0	−0.866	0	0	−0.802	0	0	−0.396	0	0
	0.1	0.798	−0.002	−0.004	−0.015	−0.003	−0.042	−0.209	−0.002	−0.031
	0.2	2.549	0.005	0.164	1.035	−0.006	0.006	0.157	−0.005	−0.036
	0.3	3.625	0.037	0.486/−0.514	2.520	0.003	0.181	1.019	−0.007	0.018
	0.4	2.520	0.002	−0.194	3.545	0.036	0.497/−0.503	2.466	0.002	0.189
	0.5	1.019	−0.008	−0.021	2.466	0.002	−0.190	3.513	0.036	0.500/−0.500
	0.6	0.151	−0.006	0.032	1.000	−0.007	−0.020	2.466	0.002	−0.189
	0.7	−0.141	−0.003	0.029	0.151	−0.006	0.032	1.019	−0.007	−0.018
	0.8	−0.146	−0.001	0.013	−0.138	−0.003	0.029	0.157	−0.005	0.036
	0.9	−0.−68	0	0.002	−0.156	−0.001	0.013	−0.209	−0.002	0.031
	1.0	0.021	0	0	−0.102	0	0	−0.396	0	0
8.0	0	−1.071	0	0	−0.651	0	0	−0.192	0	0
	0.1	0.621	−0.003	−0.024	−0.123	−0.003	−0.040	−0.203	−0.001	−0.020
	0.2	2.651	0.002	0.137	0.818	−0.006	−0.011	−0.024	−0.004	−0.035
	0.3	4.102	0.032	0.492/−0.508	2.583	0	0.153	0.789	−0.007	−0.004
	0.4	2.583	−0.001	−0.158	4.020	0.031	0.499/−0.501	2.545	0	0.156
	0.5	0.789	−0.007	0.004	2.545	−0.001	−0.156	4.002	0.031	0.500/−0.500
	0.6	−0.023	−0.004	0.034	0.781	−0.007	−0.003	2.545	0	−0.156
	0.7	−0.181	−0.001	0.020	−0.029	−0.004	−0.033	0.789	−0.007	0.004
	0.8	−0.110	0	0.005	−0.181	−0.001	−0.020	−0.024	−0.004	0.035
	0.9	−0.023	0	−0.001	−0.102	0	−0.005	−0.203	−0.001	0.020
	1.0	0.047	0	0	0.012	0	0	−0.192	0	0

Negative signs indicate upward pressure for q, hogging moment for M and downward force for V.

For $\lambda L = 1.0$ to 4.0, see Table B9; for values of $a/L = 0$ to 0.2, see Table B8.

For definitions of symbols, see Table B1.

Table B11 Coefficients for a Concentrated Load on a Slab on an Elastic Foundation

BEARING, BENDING AND SHEAR COEFFICIENTS FOR A CONCENTRATED LOAD ON AN INFINITE SLAB

r_x/r_k	$q/(F/r_k^2)$	m_r/F	m_t/F	$v/(F/r_k)$	r_x/r_k	$q/(F/r_k^2)$	m_r/F	m_t/F	$v/(F/r_k)$
0	0.125	∞	∞	$-\infty$					
0.1	0.124	0.200	0.263	−1.585	2.1	0.029	−0.021	0.011	−0.013
0.2	0.121	0.134	0.197	−0.784	2.2	0.026	−0.020	0.009	−0.010
0.3	0.117	0.096	0.159	−0.513	2.3	0.023	−0.020	0.008	−0.007
0.4	0.112	0.071	0.132	−0.375	2.4	0.020	−0.019	0.007	−0.005
0.5	0.107	0.052	0.112	−0.290	2.5	0.018	−0.019	0.006	−0.003
0.6	0.102	0.037	0.096	−0.232	2.6	0.016	−0.018	0.005	−0.001
0.7	0.096	0.025	0.083	−0.190	2.7	0.014	−0.017	0.004	0.001
0.8	0.090	0.015	0.072	−0.157	2.8	0.012	−0.017	0.003	0.002
0.9	0.085	0.007	0.062	−0.132	2.9	0.010	−0.016	0.002	0.003
1.0	0.079	0.001	0.054	−0.111	3.0	0.008	−0.015	0.002	0.004
1.1	0.073	−0.005	0.047	−0.093	3.1	0.007	−0.014	0.002	0.004
1.2	0.068	−0.009	0.041	−0.079	3.2	0.006	−0.013	0.001	0.005
1.3	0.063	−0.012	0.036	−0.067	3.3	0.005	−0.012	0.001	0.005
1.4	0.058	−0.015	0.031	−0.056	3.4	0.004	−0.011	0.001	0.005
1.5	0.053	−0.017	0.027	−0.047	3.5	0.003	−0.010	0.001	0.005
1.6	0.048	−0.018	0.023	−0.039	3.6	0.002	−0.010	0	0.005
1.7	0.044	−0.019	0.020	−0.032	3.7	0.001	−0.009	0	0.005
1.8	0.040	−0.020	0.017	−0.027	3.8	0.005	−0.008	0	0.005
1.9	0.036	−0.021	0.015	−0.022	3.9	0	−0.007	0	0.005
2.0	0.032	−0.021	0.013	−0.017	4.0	0	−0.007	0	0.005

1. Moments in terms of rectangular coordinates, where θ is the angle between the radius and the x axis, are given by
$$m_x = m_r \cos^2\theta + m_t \sin^2\theta \quad \text{and} \quad m_y = m_r \sin^2\theta + m_t \cos^2\theta$$
2. When the range of influence of one load overlaps with that of another load, the principle of superposition can be applied.
3. When the edge of the slab is within the range of influence of the load, edge moments and shears of opposite sign to those corresponding to the value of r_x/r_k should be applied. This approach has been used to obtain the bending coefficients for a concentrated load at the centre of a circular slab of radius r (see below), where the corrections due to the release of the edge moments and shears were derived from the coefficients in Table C12.

BENDING COEFFICIENTS FOR A CONCENTRATED LOAD AT THE CENTRE OF A CIRCULAR SLAB

r/r_k	m_r/F for Values of r_x/r						m_t/F for Values of r_x/r					
	0.1	0.2	0.4	0.6	0.8	1.0	0.1	0.2	0.4	0.6	0.8	1.0
0.5	0.326	0.259	0.179	0.118	0.060	0	0.392	0.324	0.251	0.202	0.158	0.117
1	0.242	0.176	0.108	0.065	0.030	0	0.306	0.240	0.172	0.132	0.102	0.075
1.5	0.192	0.127	0.067	0.034	0.014	0	0.257	0.190	0.126	0.092	0.068	0.052
2	0.155	0.092	0.037	0.014	0.005	0	0.218	0.153	0.094	0.063	0.045	0.035
2.5	0.123	0.063	0.014	−0.001	−0.002	0	0.186	0.123	0.066	0.040	0.028	0.022
3	0.100	0.042	−0.002	−0.010	−0.006	0	0.163	0.100	0.046	0.024	0.016	0.012
3.5	0.084	0.027	−0.012	−0.015	−0.009	0	0.146	0.084	0.033	0.014	0.008	0.007
4	0.071	0.015	−0.018	−0.019	−0.013	0	0.132	0.072	0.024	0.009	0.004	0.003

Symbols

E Modulus of elasticity of concrete (kN/m^2)
F Concentrated load on slab (kN)
h Overall thickness of slab (m)
k_s Modulus of subgrade reaction (kN/m^3)
m_r Radial moment at distance r_x from load (kNm/m)
m_t Tangential moment at distance r_x from load (kNm/m)

q Bearing pressure (kN/m^2)
r Radius of circular slab (m)
r_k Radius of relative stiffness (m)
r_x Radial distance from load to position considered (m)
v Shear force at distance r_x from load (kN/m)
υ Poisson's ratio (taken as 0.2 for uncracked concrete)

The radius of relative stiffness is given by the equation: $r_k = [E_c h^3/\{12(1-\upsilon^2)k_s\}]^{1/4}$

Appendix C: Rectangular and Cylindrical Tanks

Notes

1. The coefficients given in Tables C2 to C9 have been taken from *Rectangular Concrete Tanks: Revised Fifth Edition*, published by the Portland Cement Association 1998. The publication provides a complete map of moment values at intervals of one-tenth of the height and length of each panel. Values were derived by finite element analysis with a Poisson's ratio of 0.2.
2. The coefficients given in Tables C10 to C13 have been taken from *Circular Concrete Tanks without Prestressing*, published by the Portland Cement Association. The coefficients given in Table C14 have been determined from those given in the paper by Lightfoot and Michael, *The Analysis of Ground-Supported Open Circular Concrete Tanks*, published in the *Civil Engineering and Public Works Review*, September 1965.

Table C1 Rectangular Tanks: General Data

The values given in Tables C2 and C4 enable the maximum values of bending moments and shearing forces on vertical and horizontal strips of unit width to be determined, for panels of different aspect ratios and edge conditions. For each moment shown, a corresponding moment of one-fifth the value occurs in the perpendicular direction. Maximum values for negative moments at the bottom edge, and shear forces at the bottom and top edges, occur midway along the panel. Maximum values for negative moments and shear forces at the side edges, and positive moments in the spans, occur at a height z above the bottom of the panel where values of z/l_z are given in Tables C3 and C5. The moments obtained for an individual panel apply directly to a square tank. For rectangular tanks, a further distribution of the unequal negative moments at the sides is needed and the resulting coefficients are given in Tables C6 to C9.

The values given in the tables below enable the panel stiffness on vertical and horizontal strips of unit width to be determined for panels of different aspect ratios and edge conditions. These values have been derived, at positions midway along the edge, by dividing the fixed edge moment (FEM) by the rotation that occurs in the hinged condition. The rotation has been determined from the deflection equations given in *Theory of Plates and Shells* by Timoshenko and Woinowski-Krieger.

Rectangular Panel with Top Edge Hinged or Free, and Other Edges Fixed											
Restraint at Top Edge		Stiffness Coefficient α_{kz} for Values of l_x/l_z									
		0.5	0.75	1.0	1.25	1.5	2.0	2.5	3.0	4.0	∞
Top hinged	α_{kz}	16	11	8.3	6.5	5.4	4.2	3.6	3.3	3.1	3.0
Top free	α_{kz}	16	11	8.3	6.5	5.2	3.7	2.8	2.0	1.2	0

Value of panel stiffness per unit width, at middle of panel length, is given by the following relationship, where l_x is the panel length and l_z the panel height. Flexural rigidity $D = E_c h^3/12(1 - \upsilon^2)$, where E_c is the modulus of elasticity of concrete, h the section thickness and υ is Poisson's ratio (taken as 0.2 for uncracked concrete).

$$\text{Stiffness for span } l_z: \quad K_z = \alpha_{kz} D/l_z$$

Rectangular Panel with All Edges Fixed											
Direction of Span		Stiffness Coefficient α_k for Values of l_x/l_z									
		0.5	0.75	1.0	1.25	1.5	2.0	2.5	3.0	4.0	∞
Span l_x	α_{kx}	5.0	4.3	3.8	3.4	3.1	3.0	2.9	2.8	2.8	2.8
Span l_z	α_{kz}	6.0	4.4	3.8	3.4	3.0	2.5	2.2	2.1	2.0	2.0

Values of panel stiffness per unit width, at middle of panel length or height, are given by the following relationships, where l_x is the panel length, and l_z the panel height (or width). Flexural rigidity $D = E_c h^3/12(1 - \upsilon^2)$, where E_c is the modulus of elasticity of concrete, h the section thickness and υ is Poisson's ratio (taken as 0.2 for uncracked concrete).

$$\text{Stiffness for span } l_x: \quad K_x = \alpha_{kx} D/l_z \qquad \text{Stiffness for span } l_z: \quad K_z = \alpha_{kz} D/l_z$$

Table C2 Rectangular Panels: Triangular Load – 1

Rectangular Panels with Provision for Torsion at Corners										
Type of Panel with Moments and Shears Considered		Coefficients for Values of l_x/l_z								
		0.5	0.75	1.0	1.25	1.5	2.0	2.5	3.0	4.0
1. Top hinged, bottom fixed										
Negative moment at side edge	α_{mx}	0.012	0.022	0.029	0.033	0.036	0.037	0.037	0.037	0.037
Positive moment for span l_x	α_{mx}	0.006	0.010	0.012	0.013	0.012	0.010	0.009	0.009	0.009
Shear force at side edge (maximum)	α_{vx}	0.17	0.22	0.24	0.25	0.26	0.27	0.26	0.26	0.26
Shear force at side edge (mid-height)	α_{vx}	0.13	0.19	0.23	0.25	0.26	0.26	0.26	0.26	0.26
Negative moment at bottom edge	α_{mz}	0.011	0.023	0.035	0.045	0.053	0.062	0.065	0.066	0.067
Positive moment for span l_z	α_{mz}	0.003	0.007	0.011	0.016	0.021	0.026	0.028	0.029	0.029
Shear force at bottom edge	α_{vz}	0.20	0.26	0.32	0.36	0.38	0.40	0.40	0.40	0.40
Shear force at top edge	α_{vz}	0.03	0.05	0.07	0.09	0.11	0.11	0.11	0.11	0.10
2. Top free, bottom fixed										
Negative moment at side edge	α_{mx}	0.012	0.022	0.030	0.037	0.044	0.066	0.082	0.091	0.099
Positive moment for span l_x	α_{mx}	0.002	0.010	0.013	0.016	0.021	0.028	0.028	0.024	0.017
Shear force at side edge (maximum)	α_{vx}	0.17	0.22	0.24	0.25	0.26	0.27	0.33	0.37	0.38
Shear force at side edge (mid-height)	α_{vx}	0.13	0.19	0.23	0.25	0.26	0.26	0.25	0.24	0.23
Negative moment at bottom edge	α_{mz}	0.011	0.023	0.035	0.048	0.061	0.086	0.109	0.127	0.149
Positive moment for span l_z	α_{mz}	0.003	0.007	0.010	0.013	0.015	0.016	0.014	0.011	0.007
Shear force at bottom edge	α_{vz}	0.19	0.26	0.32	0.36	0.40	0.45	0.48	0.50	0.50
3. Top hinged, bottom hinged										
Negative moment at side edge	α_{mx}	0.014	0.026	0.038	0.047	0.054	0.061	0.063	0.064	0.064
Positive moment for span l_x	α_{mx}	0.007	0.012	0.017	0.019	0.021	0.020	0.018	0.017	0.017
Shear force at side edge (maximum)	α_{vx}	0.20	0.26	0.32	0.35	0.38	0.40	0.41	0.41	0.41
Shear force at side edge (mid-height)	α_{vx}	0.13	0.20	0.26	0.30	0.33	0.36	0.37	0.37	0.37
Positive moment for span l_z	α_{mz}	0.004	0.009	0.015	0.023	0.031	0.045	0.054	0.059	0.063
Shear force at bottom edge	α_{vz}	0.11	0.16	0.20	0.23	0.26	0.30	0.32	0.33	0.33
Shear force at top edge	α_{vz}	0.01	0.03	0.05	0.07	0.10	0.13	0.15	0.16	0.17
4. Top free, bottom hinged										
Negative moment at side edge	α_{mx}	0.014	0.026	0.038	0.050	0.063	0.098	0.150	0.205	0.317
Positive moment for span l_x	α_{mx}	0.007	0.012	0.017	0.022	0.028	0.046	0.062	0.074	0.089
Shear force at side edge (maximum)	α_{vx}	0.20	0.26	0.31	0.35	0.37	0.41	0.58	0.76	1.14
Shear force at side edge (mid-height)	α_{vx}	0.13	0.19	0.25	0.30	0.34	0.39	0.43	0.45	0.51
Positive moment for span l_z	α_{mz}	0.004	0.009	0.014	0.021	0.027	0.037	0.045	0.051	0.058
Shear force at bottom edge	α_{vz}	0.11	0.15	0.19	0.23	0.26	0.31	0.33	0.36	0.39

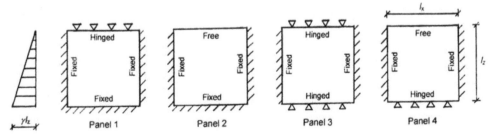

Note: Maximum values of moment per unit width and shear force per unit width are given by the following relationships, where l_x is the panel length, l_z the panel height and γ the unit weight of liquid. For details of the positions at which the maximum values occur, see Table C3.

$$\text{Horizontal span: } m_x = \alpha_{mx}\gamma l_z^3, \; v_x = \alpha_{vx}\gamma l_z^2 \qquad \text{Vertical span: } m_z = \alpha_{mz}\gamma l_z^3, \; v_x = \alpha_{vz}\gamma l_z^2$$

Table C3 Rectangular Panels: Triangular Load – 2

Type of Panel	Coefficient	Height at which Maximum Values of Coefficients Occur in Table C2								
		Value of z/l_z for Values of l_x/l_z								
		0.5	0.75	1.0	1.25	1.5	2.0	2.5	3.0	4.0
1	α_{mx} (negative) and α_{vx}	0.3	0.4	0.5	0.5	0.5	0.5	0.5	0.5	0.5
	α_{mx} (positive)	0.3	0.4	0.5	0.5	0.5	0.5	0.5	0.5	0.5
	α_{mz} (positive)	0.3	0.3	0.4	0.5	0.5	0.5	0.5	0.5	0.5
2	α_{mx} (negative) and α_{vx}	0.3	0.4	0.5	0.6	0.8	0.9	0.9	0.9	1.0
	α_{mx} (positive)	0.4	0.5	0.6	0.8	1.0	1.0	1.0	1.0	1.0
	α_{mz} (positive)	0.3	0.3	0.4	0.4	0.5	0.6	0.6	0.7	0.7
3	α_{mx} (negative) and α_{vx}	0.3	0.3	0.4	0.4	0.4	0.4	0.4	0.4	0.4
	α_{mx} (positive)	0.3	0.3	0.4	0.4	0.4	0.4	0.4	0.4	0.4
	α_{mz} (positive)	0.2	0.2	0.3	0.3	0.3	0.4	0.4	0.4	0.4
4	α_{mx} (negative) and α_{vx}	0.2	0.3	0.4	0.5	0.5	0.9	0.9	0.9	0.9
	α_{mx} (positive)	0.3	0.4	0.5	0.6	0.6	1.0	1.0	1.0	1.0
	α_{mz} (positive)	0.1	0.2	0.2	0.3	0.3	0.4	0.4	0.4	0.4

Edge Conditions	Rectangular Panel with Triangular Load								
	Deflection Coefficient α_d for Values of l_x/l_z								
	0.5	0.75	1.0	1.25	1.5	2.0	2.5	3.0	4.0
Top edge hinged, other edges fixed	0.0001	0.0003	0.0007	0.0012	0.0016	0.0020	0.0022	0.0023	0.0024
Top and bottom edges hinged	0.0001	0.0004	0.0010	0.0018	0.0027	0.0042	0.0052	0.0058	0.0063
Top edge free, other edges fixed	0.0001	0.0003	0.0008	0.0016	0.0031	0.0077	0.0132	0.0184	0.0258
Top edge free, bottom edge hinged	0.0001	0.0004	0.0010	0.0021	0.0039	0.0122	0.0269	0.0487	0.1132

The maximum deflection is given by the following relationship, where l_x is the panel length, l_z the panel height and γ the unit weight of the liquid. Flexural rigidity $D = E_c h^3/12(1 - \upsilon^2)$, where E_c is the modulus of elasticity of concrete, h the section thickness and υ is Poisson's ratio (taken as 0.2 for uncracked concrete).

$$\text{Deflection:} \quad a = \alpha_d \gamma l_z^5/D$$

Table C4 Rectangular Panels: Uniform Load – 1

Rectangular Panels with Provision for Torsion at Corners										
Type of Panel with Moments and Shears Considered		Coefficients for Values of l_x/l_z								
		0.5	0.75	1.0	1.25	1.5	2.0	2.5	3.0	4.0
1. Top hinged, bottom fixed										
Negative moment at side edge	α_{mx}	0.021	0.042	0.061	0.072	0.080	0.081	0.081	0.081	0.081
Positive moment for span l_x	α_{mx}	0.010	0.020	0.027	0.029	0.025	0.023	0.021	0.021	0.020
Shear force at side edge (maximum)	α_{vx}	0.26	0.39	0.48	0.53	0.56	0.56	0.56	0.56	0.55
Shear force at side edge (mid-height)	α_{vx}	0.26	0.38	0.47	0.52	0.54	0.54	0.54	0.54	0.53
Negative moment at bottom edge	α_{mz}	0.014	0.032	0.055	0.077	0.107	0.115	0.122	0.124	0.125
Positive moment for span l_z	α_{mz}	0.004	0.011	0.022	0.036	0.055	0.061	0.067	0.069	0.070
Shear force at bottom edge	α_{vz}	0.22	0.34	0.45	0.53	0.58	0.62	0.63	0.62	0.62
Shear force at top edge	α_{vz}	0.18	0.25	0.32	0.35	0.39	0.40	0.39	0.39	0.39
2. Top free, bottom fixed										
Negative moment at side edge	α_{mx}	0.021	0.049	0.087	0.133	0.181	0.275	0.334	0.379	0.404
Positive moment for span l_x	α_{mx}	0.011	0.025	0.043	0.063	0.081	0.102	0.102	0.089	0.064
Shear force at side edge (maximum)	α_{vx}	0.25	0.40	0.58	0.77	0.95	1.26	1.47	1.59	1.68
Shear force at side edge (mid-height)	α_{vx}	0.25	0.37	0.46	0.51	0.54	0.53	0.50	0.47	0.45
Negative moment at bottom edge	α_{mz}	0.014	0.032	0.056	0.087	0.124	0.206	0.286	0.351	0.433
Positive moment for span l_z	α_{mz}	0.004	0.008	0.014	0.020	0.025	0.028	0.024	0.018	0.009
Shear force at bottom edge	α_{vz}	0.22	0.33	0.45	0.56	0.66	0.85	0.95	1.01	1.03
3. Top hinged, bottom hinged										
Negative moment at side edge	α_{mx}	0.021	0.045	0.070	0.090	0.105	0.119	0.123	0.125	0.125
Positive moment for span l_x	α_{mx}	0.010	0.022	0.032	0.038	0.041	0.039	0.035	0.034	0.034
Shear force at side edge (maximum)	α_{vx}	0.26	0.40	0.52	0.60	0.67	0.72	0.74	0.74	0.74
Shear force at side edge (mid-height)	α_{vx}	0.26	0.40	0.52	0.60	0.67	0.72	0.74	0.74	0.74
Positive moment for span l_z	α_{mz}	0.004	0.010	0.022	0.038	0.055	0.085	0.103	0.114	0.122
Shear force at bottom edge	α_{vz}	0.12	0.18	0.24	0.31	0.36	0.43	0.47	0.48	0.50
Shear force at top edge	α_{vz}	0.12	0.18	0.24	0.31	0.36	0.43	0.47	0.48	0.50
4. Top free, bottom hinged										
Negative moment at side edge	α_{mx}	0.021	0.049	0.088	0.139	0.200	0.340	0.496	0.660	0.995
Positive moment for span l_x	α_{mx}	0.011	0.025	0.044	0.068	0.093	0.144	0.188	0.188	0.266
Shear force at side edge (maximum)	α_{vx}	0.26	0.40	0.58	0.78	1.00	1.45	2.10	2.61	3.74
Shear force at side edge (mid-height)	α_{vx}	0.25	0.38	0.50	0.60	0.69	0.83	0.94	1.02	1.17
Positive moment for span l_z	α_{mz}	0.004	0.010	0.017	0.026	0.036	0.055	0.072	0.087	0.106
Shear force at bottom edge	α_{vz}	0.12	0.18	0.24	0.30	0.35	0.45	0.52	0.61	0.68

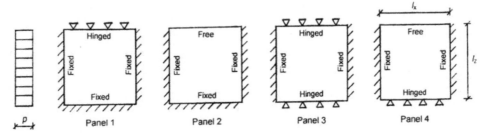

Note: Maximum values of moment per unit width and shear force per unit width are given by the following relationships, where l_x is the panel length, l_z the panel height and p the unit pressure. For details of the positions at which the maximum values occur, see Table C5.

Horizontal span: $m_x = \alpha_{mx}pl_z^2$, $v_x = \alpha_{vx}pl_z$ Vertical span: $m_z = \alpha_{mz}pl_z^2$, $v_x = \alpha_{vz}pl_z$

Table C5 Rectangular Panels: Uniform Load – 2

		Height at which Maximum Values of Coefficients Occur in Table C4								
Type of Panel	Coefficient	Value of z/l_z for Values of l_x/l_z								
		0.5	0.75	1.0	1.25	1.5	2.0	2.5	3.0	4.0
1	α_{mx} (negative) and α_{vx}	0.6	0.6	0.6	0.6	0.6	0.6	0.6	0.6	0.6
	α_{mx} (positive)	0.6	0.6	0.6	0.6	0.6	0.6	0.6	0.6	0.6
	α_{mz} (positive)	0.8	0.7	0.6	0.6	0.6	0.6	0.6	0.6	0.6
2	α_{mx} (negative) and α_{vx}	0.7	0.9	0.9	0.9	0.9	1.0	1.0	1.0	1.0
	α_{mx} (positive)	1.0	1.0	1.0	1.0	1.0	1.0	1.0	1.0	1.0
	α_{mz} (positive)	0.3	0.4	0.5	0.6	0.6	0.7	0.8	0.8	0.9
3	α_{mx} (negative) and α_{vx}	0.5	0.5	0.5	0.5	0.5	0.5	0.5	0.5	0.5
	α_{mx} (positive)	0.5	0.5	0.5	0.5	0.5	0.5	0.5	0.5	0.5
	α_{mz} (positive)	0.5	0.5	0.5	0.5	0.5	0.5	0.5	0.5	0.5
4	α_{mx} (negative) and α_{vx}	0.8	0.9	0.9	0.9	0.9	0.9	0.9	0.9	0.9
	α_{mx} (positive)	1.0	1.0	1.0	1.0	1.0	1.0	1.0	1.0	1.0
	α_{mz} (positive)	0.2	0.2	0.3	0.4	0.4	0.5	0.5	0.5	0.5

Rectangular Panels with Uniform Load									
Edge Conditions	Deflection Coefficient α_d for Values of l_x/l_z								
	0.5	0.75	1.0	1.25	1.5	2.0	2.5	3.0	4.0
All edges fixed	0.0002	0.0006	0.0013	0.0018	0.0022	0.0026	0.0026	0.0026	0.0026
Top edge hinged, other edges fixed	0.0002	0.0007	0.0016	0.0026	0.0035	0.0046	0.0051	0.0053	0.0054
Top and bottom edges hinged	0.0002	0.0007	0.0019	0.0035	0.0053	0.0084	0.0105	0.0117	0.0127
All edges hinged	0.0006	0.0021	0.0041	0.0060	0.0077	0.0101	0.0115	0.0122	0.0128
Top edge free, other edges fixed	0.0002	0.0009	0.0028	0.0066	0.0124	0.0296	0.0500	0.0690	0.0965
Top edge free, bottom edge hinged	0.0002	0.0009	0.0028	0.0069	0.0139	0.0398	0.0845	0.1498	0.3434

Maximum deflection is given by the following relationship, where l_x is the panel length, l_z the panel height and p the unit pressure. Flexural rigidity $D = E_c h^3/12(1 - v^2)$, where E_c is the modulus of elasticity of concrete, h the section thickness, and v is Poisson's ratio (taken as 0.2 for uncracked concrete).

$$\text{Deflection:} \quad a = \alpha_d p l_z^4/D$$

Rectangular Panels with Uniform Load									
Edge Conditions and Moments	Moment Coefficient α_m for Values of l_x/l_z								
	1.0	1.25	1.5	2.0	2.5	3.0	4.0	∞	
All edges fixed									
Edge moment for span l_x α_{mx}	0.052	0.056	0.057	0.057	0.057	0.057	0.057	0.057	
Mid-span moment for span l_x α_{mx}	0.023	0.022	0.020	0.016	0.015	0.014	0.013	0.013	
Edge moment for span l_z α_{mz}	0.052	0.067	0.076	0.083	0.083	0.083	0.083	0.083	
Mid-span moment for span l_z α_{mz}	0.023	0.031	0.037	0.042	0.042	0.042	0.042	0.042	
All edges hinged									
Mid-span moment for span l_x α_{mx}	0.044	0.045	0.043	0.037	0.032	0.029	0.026	0.025	
Mid-span moment for span l_z α_{mz}	0.044	0.063	0.078	0.100	0.112	0.118	0.123	0.125	
Edge shear for span l_x α_{vx}	0.34	0.36	0.36	0.37	0.37	0.37	0.37	0.37	
Edge shear for span l_z α_{vz}	0.34	0.39	0.42	0.46	0.48	0.49	0.50	0.50	

Maximum values of moment per unit width and shear force per unit width are given by the following relationships, where l_x is the panel length, l_z the panel height (or width), and p the unit pressure.

$$\text{Span } l_x: \quad m_x = \alpha_{mx} p l_z^2, \ v_x = \alpha_{vx} p l_z \qquad \text{Span } l_z: \quad m_z = \alpha_{mz} p l_z^2, \ v_z = \alpha_{vz} p l_z$$

Table C6 Rectangular Tanks: Triangular Load – 1

Span Ratios and Moments Considered		(1) Top Hinged, Bottom Fixed					(2) Top Free, Bottom Fixed				
		Coefficients for Short Span Ratio l_y/l_z					Coefficients for Short Span Ratio l_y/l_z				
		0.5	1.0	1.5	2.0	3.0	0.5	1.0	1.5	2.0	3.0
Long span ratio $l_x/l_z = 4.0$											
Negative moments at corners	α_{mx}	0.022	0.032	0.036	0.037	0.037	0.057	0.056	0.069	0.081	0.095
Positive moment for span l_x	α_{mx}	0.009	0.009	0.009	0.009	0.009	0.016	0.017	0.017	0.017	0.017
Positive moment for span l_y	α_{mx}	0.003	0.012	0.012	0.010	0.009	0.001	0.007	0.017	0.027	0.024
Negative moments at bottom	$\alpha_{mz,x}$	0.067	0.067	0.067	0.067	0.067	0.152	0.152	0.151	0.150	0.149
	$\alpha_{mz,y}$	0.005	0.033	0.053	0.062	0.066	0	0.019	0.050	0.081	0.126
Positive moment for span $l_{z,x}$	$\alpha_{mz,x}$	0.029	0.029	0.029	0.029	0.029	0.007	0.007	0.006	0.006	0.007
Positive moment for span $l_{z,y}$	$\alpha_{mz,y}$	0.003	0.011	0.021	0.026	0.029	0.007	0.012	0.016	0.016	0.011
Long span ratio $l_x/l_z = 3.0$											
Negative moments at corners	α_{mx}	0.022	0.032	0.036	0.037		0.054	0.053	0.066	0.078	
Positive moment for span l_x	α_{mx}	0.009	0.009	0.009	0.009		0.022	0.022	0.023	0.024	
Positive moment for span l_y	α_{my}	0.003	0.012	0.012	0.010		0.001	0.007	0.017	0.027	
Negative moments at bottom	$\alpha_{mz,x}$	0.067	0.066	0.066	0.066		0.134	0.133	0.131	0.129	
	$\alpha_{mz,y}$	0.005	0.033	0.053	0.062		0	0.020	0.051	0.082	
Positive moment for span $l_{z,x}$	$\alpha_{mz,x}$	0.029	0.029	0.029	0.029		0.009	0.009	0.010	0.010	
Positive moment for span $l_{z,y}$	$\alpha_{mz,y}$	0.003	0.011	0.021	0.026		0.006	0.012	0.016	0.016	
Long span ratio $l_x/l_z = 2.0$											
Negative moments at corners	α_{mx}	0.022	0.032	0.036			0.041	0.042	0.054		
Positive moment for span l_x	α_{mx}	0.010	0.010	0.010			0.029	0.029	0.028		
Positive moment for span l_y	α_{my}	0.003	0.012	0.012			0.001	0.009	0.019		
Negative moments at bottom	$\alpha_{mz,x}$	0.063	0.063	0.062			0.097	0.095	0.090		
	$\alpha_{mz,y}$	0.005	0.033	0.053			0	0.023	0.056		
Positive moment for span $l_{z,x}$	$\alpha_{mz,x}$	0.027	0.026	0.026			0.015	0.015	0.016		
Positive moment for span $l_{z,y}$	$\alpha_{mz,y}$	0.003	0.011	0.021			0.006	0.011	0.016		
Long span ratio $l_x/l_z = 1.5$											
Negative moments at corners	α_{mx}	0.022	0.032				0.028	0.036			
Positive moment for span l_x	α_{mx}	0.012	0.012				0.025	0.024			
Positive moment for span l_y	α_{my}	0.003	0.012				0.001	0.010			
Negative moments at bottom	$\alpha_{mz,x}$	0.056	0.054				0.071	0.067			
	$\alpha_{mz,y}$	0.005	0.033				0.001	0.028			
Positive moment for span $l_{z,x}$	$\alpha_{mz,x}$	0.022	0.021				0.016	0.015			
Positive moment for span $l_{z,y}$	$\alpha_{mz,y}$	0.003	0.011				0.005	0.010			
Long span ratio $l_x/l_z = 1.0$											
Negative moments at corners	α_{mx}	0.020					0.021				
Positive moment for span l_x	α_{mx}	0.013					0.016				
Positive moment for span l_y	α_{my}	0.003					0.003				
Negative moments at bottom	$\alpha_{mz,x}$	0.039					0.041				
	$\alpha_{mz,y}$	0.006					0.005				
Positive moment for span $l_{z,x}$	$\alpha_{mz,x}$	0.013					0.011				
Positive moment for span $l_{z,y}$	$\alpha_{mz,y}$	0.003					0.003				

For details of tank dimensions and notes, see Table C7.

Table C7 Rectangular Tanks: Triangular Load – 2

Span Ratios and Moments Considered		(3) Top Hinged, Bottom Hinged					(4) Top Free, Bottom Hinged				
		Coefficients for Short Span Ratio l_y/l_z					Coefficients for Short Span Ratio l_y/l_z				
		0.5	1.0	1.5	2.0	3.0	0.5	1.0	1.5	2.0	3.0
Long span ratio $l_x/l_z = 4.0$											
Negative moments at corners	α_{mx}	0.037	0.050	0.059	0.062	0.064	0.216	0.187	0.191	0.209	0.261
Positive moment for span l_x	α_{mx}	0.017	0.017	0.017	0.017	0.017	0.091	0.092	0.092	0.091	0.090
Positive moment for span l_y	α_{my}	0	0.014	0.021	0.020	0.017	0	0	0.004	0.024	0.070
Positive moment for span $l_{z,x}$	$\alpha_{mz,x}$	0.063	0.063	0.063	0.063	0.063	0.059	0.059	0.059	0.059	0.058
Positive moment for span $l_{z,y}$	$\alpha_{mz,y}$	0.001	0.013	0.030	0.044	0.059	0	0.006	0.018	0.031	0.049
Long span ratio $l_x/l_z = 3.0$											
Negative moments at corners	α_{mx}	0.037	0.050	0.059	0.062		0.142	0.124	0.132	0.152	
Positive moment for span l_x	α_{mx}	0.017	0.017	0.017	0.017		0.080	0.081	0.080	0.078	
Positive moment for span l_y	α_{my}	0	0.015	0.021	0.020		0	0	0.012	0.034	
Positive moment for span $l_{z,x}$	$\alpha_{mz,x}$	0.060	0.059	0.059	0.059		0.053	0.053	0.053	0.052	
Positive moment for span $l_{z,y}$	$\alpha_{mz,y}$	0.001	0.013	0.030	0.044		0	0.008	0.021	0.034	
Long span ratio $l_x/l_z = 2.0$											
Negative moments at corners	α_{mx}	0.036	0.049	0.058			0.071	0.065	0.078		
Positive moment for span l_x	α_{mx}	0.019	0.019	0.020			0.055	0.055	0.051		
Positive moment for span l_y	α_{my}	0	0.015	0.021			0	0.005	0.022		
Positive moment for span $l_{z,x}$	$\alpha_{mz,x}$	0.048	0.046	0.045			0.040	0.040	0.038		
Positive moment for span $l_{z,y}$	$\alpha_{mz,y}$	0.001	0.013	0.030			0.001	0.011	0.025		
Long span ratio $l_x/l_z = 1.5$											
Negative moments at corners	α_{mx}	0.033	0.046				0.042	0.050			
Positive moment for span l_x	α_{mx}	0.021	0.021				0.037	0.035			
Positive moment for span l_y	α_{my}	0	0.015				0	0.011			
Positive moment for span $l_{z,x}$	$\alpha_{mz,x}$	0.035	0.032				0.030	0.029			
Positive moment for span $l_{z,y}$	$\alpha_{mz,y}$	0.002	0.013				0.001	0.013			
Long span ratio $l_x/l_z = 1.0$											
Negative moments at corners	α_{mx}	0.025					0.027				
Positive moment for span l_x	α_{mx}	0.018					0.021				
Positive moment for span l_y	α_{my}	0.002					0.001				
Positive moment for span $l_{z,x}$	$\alpha_{mz,x}$	0.018					0.017				
Positive moment for span $l_{z,y}$	$\alpha_{mz,y}$	0.002					0.002				

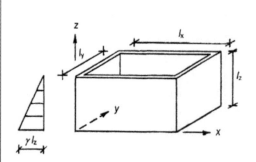

Dimensions of tank

Note: Maximum values of moment per unit width are given by the following relationships, where l_x, l_y and l_z are length, breadth and height, respectively, of the tank, and γ is unit weight of liquid.

Horizontal (long span): $m_x = \alpha_{mx}\gamma l_z^3$

Horizontal (short span): $m_y = \alpha_{my}\gamma l_z^3$

Vertical (long wall): $m_{z,x} = \alpha_{mz,x}\gamma l_z^3$

Vertical (short wall): $m_{z,y} = \alpha_{mz,y}\gamma l_z^3$

Maximum values of shear per unit width may be determined for each wall, according to the value of l_x/l_z or l_y/l_z, from Table C2.

Table C8 Rectangular Tanks: Uniform Load – 1

Span Ratios and Moments Considered		(1) Top Hinged, Bottom Fixed					(2) Top Free, Bottom Fixed				
		Coefficients for Short Span Ratio l_y/l_z					Coefficients for Short Span Ratio l_y/l_z				
		0.5	1.0	1.5	2.0	3.0	0.5	1.0	1.5	2.0	3.0
Long span ratio $l_x/l_z = 4.0$											
Negative moments at corners	α_{mx}	0.048	0.070	0.079	0.081	0.081	0.225	0.227	0.276	0.323	0.373
Positive moment for span l_x	α_{mx}	0.020	0.020	0.020	0.020	0.020	0.061	0.061	0.062	0.063	0.064
Positive moment for span l_y	α_{my}	0.002	0.026	0.027	0.023	0.021	0	0.008	0.065	0.100	0.089
Negative moments at bottom	$\alpha_{mz,x}$	0.125	0.125	0.125	0.125	0.125	0.445	0.444	0.440	0.437	0.433
	$\alpha_{mz,y}$	0	0.051	0.094	0.115	0.124	0	0	0.086	0.188	0.349
Positive moment for span $l_{z,x}$	$\alpha_{mz,x}$	0.070	0.070	0.070	0.070	0.070	0.008	0.008	0.008	0.008	0.009
Positive moment for span $l_{z,y}$	$\alpha_{mz,y}$	0.005	0.020	0.047	0.061	0.069	0.030	0.022	0.025	0.029	0.018
Long span ratio $l_x/l_z = 3.0$											
Negative moments at corners	α_{mx}	0.048	0.071	0.080	0.081		0.213	0.216	0.264	0.311	
Positive moment for span l_x	α_{mx}	0.022	0.021	0.021	0.021		0.081	0.082	0.085	0.087	
Positive moment for span l_y	α_{my}	0.002	0.026	0.027	0.023		0.006	0.008	0.066	0.100	
Negative moments at bottom	$\alpha_{mz,x}$	0.125	0.125	0.124	0.124		0.377	0.375	0.366	0.359	
	$\alpha_{mz,y}$	0	0.051	0.094	0.115		0	0.002	0.090	0.192	
Positive moment for span $l_{z,x}$	$\alpha_{mz,x}$	0.069	0.069	0.069	0.069		0.015	0.015	0.016	0.017	
Positive moment for span $l_{z,y}$	$\alpha_{mz,y}$	0.005	0.020	0.047	0.061		0.029	0.021	0.024	0.028	
Long span ratio $l_x/l_z = 2.0$											
Negative moments at corners	α_{mx}	0.048	0.070	0.079			0.167	0.176	0.223		
Positive moment for span l_x	α_{mx}	0.023	0.023	0.023			0.106	0.106	0.103		
Positive moment for span l_y	α_{my}	0.002	0.026	0.027			0.004	0.012	0.073		
Negative moments at bottom	$\alpha_{mz,x}$	0.117	0.115	0.115			0.245	0.238	0.221		
	$\alpha_{mz,y}$	0	0.051	0.094			0	0.017	0.106		
Positive moment for span $l_{z,x}$	$\alpha_{mz,x}$	0.063	0.062	0.061			0.027	0.027	0.027		
Positive moment for span $l_{z,y}$	$\alpha_{mz,y}$	0.005	0.020	0.047			0.022	0.017	0.023		
Long span ratio $l_x/l_z = 1.5$											
Negative moments at corners	α_{mx}	0.047	0.069				0.119	0.135			
Positive moment for span l_x	α_{mx}	0.026	0.027				0.095	0.090			
Positive moment for span l_y	α_{my}	0.002	0.026				0.003	0.021			
Negative moments at bottom	$\alpha_{mz,x}$	0.101	0.096				0.158	0.146			
	$\alpha_{mz,y}$	0	0.052				0	0.033			
Positive moment for span $l_{z,x}$	$\alpha_{mz,x}$	0.051	0.048				0.027	0.026			
Positive moment for span $l_{z,y}$	$\alpha_{mz,y}$	0.005	0.020				0.014	0.014			
Long span ratio $l_x/l_z = 1.0$											
Negative moments at corners	α_{mx}	0.040					0.062				
Positive moment for span l_x	α_{mx}	0.028					0.058				
Positive moment for span l_y	α_{my}	0.003					0.002				
Negative moments at bottom	$\alpha_{mz,x}$	0.064					0.075				
	$\alpha_{mz,y}$	0.004					0				
Positive moment for span $l_{z,x}$	$\alpha_{mz,x}$	0.028					0.017				
Positive moment for span $l_{z,y}$	$\alpha_{mz,y}$	0.004					0.006				

For details of tank dimensions and notes, see Table C9.

Table C9 Rectangular Tanks: Uniform Load – 2

Span Ratios and Moments Considered		(3) Top Hinged, Bottom Hinged					(4) Top Free, Bottom Hinged				
		Coefficients for Short Span Ratio l_y/l_z					Coefficients for Short Span Ratio l_y/l_z				
		0.5	1.0	1.5	2.0	3.0	0.5	1.0	1.5	2.0	3.0
Long span ratio $l_x/l_z = 4.0$											
Negative moments at corners	α_{mx}	0.070	0.095	0.114	0.122	0.125	0.675	0.598	0.615	0.672	0.827
Positive moment for span l_x	α_{mx}	0.033	0.034	0.034	0.034	0.034	0.274	0.276	0.276	0.274	0.270
Positive moment for span l_y	α_{my}	0	0.027	0.041	0.039	0.034	0	0	0	0.073	0.210
Positive moment for span $l_{z,x}$	$\alpha_{mz,x}$	0.123	0.122	0.122	0.122	0.122	0.110	0.111	0.111	0.110	0.108
Positive moment for span $l_{z,y}$	$\alpha_{mz,y}$	0	0.014	0.053	0.084	0.114	0	0	0.011	0.035	0.082
Long span ratio $l_x/l_z = 3.0$											
Negative moments at corners	α_{mx}	0.070	0.095	0.114	0.122		0.453	0.410	0.440	0.502	
Positive moment for span l_x	α_{mx}	0.033	0.034	0.034	0.034		0.240	0.243	0.240	0.235	
Positive moment for span l_y	α_{my}	0	0.027	0.041	0.039		0	0	0.015	0.109	
Positive moment for span $l_{z,x}$	$\alpha_{mz,x}$	0.116	0.115	0.114	0.114		0.094	0.095	0.094	0.092	
Positive moment for span $l_{z,y}$	$\alpha_{mz,y}$	0	0.015	0.053	0.084		0	0	0.019	0.043	
Long span ratio $l_x/l_z = 2.0$											
Negative moments at corners	α_{mx}	0.068	0.093	0.112			0.238	0.229	0.272		
Positive moment for span l_x	α_{mx}	0.038	0.039	0.039			0.171	0.171	0.160		
Positive moment for span l_y	α_{my}	0	0.028	0.041			0	0	0.066		
Positive moment for span $l_{z,x}$	$\alpha_{mz,x}$	0.091	0.088	0.085			0.064	0.064	0.060		
Positive moment for span $l_{z,y}$	$\alpha_{mz,y}$	0	0.015	0.054			0	0.008	0.030		
Long span ratio $l_x/l_z = 1.5$											
Negative moments at corners	α_{mx}	0.062	0.087				0.142	0.149			
Positive moment for span l_x	α_{mx}	0.042	0.041				0.120	0.114			
Positive moment for span l_y	α_{my}	0	0.029				0	0.014			
Positive moment for span $l_{z,x}$	$\alpha_{mz,x}$	0.064	0.059				0.044	0.042			
Positive moment for span $l_{z,y}$	$\alpha_{mz,y}$	0	0.017				0	0.012			
Long span ratio $l_x/l_z = 1.0$											
Negative moments at corners	α_{mx}	0.046					0.064				
Positive moment for span l_x	α_{mx}	0.036					0.062				
Positive moment for span l_y	α_{my}	0.001					0				
Positive moment for span $l_{z,x}$	$\alpha_{mz,x}$	0.029					0.022				
Positive moment for span $l_{z,y}$	$\alpha_{mz,y}$	0.002					0.002				

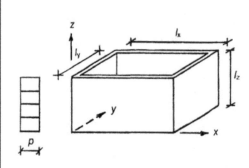

Dimensions of tank

Note: Maximum values of moment per unit width are given by the following relationships, where l_x, l_y and l_z are length, breadth and height, respectively, of the tank, and p is unit pressure.

Horizontal (long span): $m_x = \alpha_{mx} p\, l_z^2$

Horizontal (short span): $m_y = \alpha_{my} p\, l_z^2$

Vertical (long wall): $m_{z,x} = \alpha_{mz,x} p\, l_z^2$

Vertical (short wall): $m_{z,y} = \alpha_{mz,y} p\, l_z^2$

Maximum values of shear per unit width may be determined for each wall, according to value of l_x/l_z or l_y/l_z, from Table C4.

Table C10 Cylindrical Tanks: Elastic Analysis – 1

			Coefficients for Circumferential Forces, Vertical Moments and Radial Shears in Wall of Constant Thickness								
Load Case	α	z/l_z	Values of Coefficient α for Values of $l_z^2/2rh$								
			2	3	4	5	6	8	10	12	16
(1) Triangular load (fixed base, free top)	α_{n1}	0	0.234	0.134	0.067	0.025	0.018	−0.011	−0.011	−0.005	0.000
		0.1	0.251	0.203	0.164	0.137	0.119	0.104	0.098	0.097	0.099
		0.2	0.273	0.267	0.256	0.245	0.234	0.218	0.208	0.202	0.199
		0.3	0.285	0.322	0.339	0.346	0.344	0.335	0.323	0.312	0.304
		0.4	**0.285**	0.357	0.403	0.428	0.441	0.443	0.437	0.429	0.412
		0.5	0.274	**0.362**	**0.429**	**0.477**	0.504	0.534	0.542	0.543	0.531
		0.6	0.232	0.330	0.409	0.469	**0.514**	**0.575**	**0.608**	0.628	0.641
		0.7	0.172	0.262	0.334	0.398	0.447	0.530	0.589	**0.633**	**0.687**
		0.8	0.104	0.157	0.210	0.259	0.301	0.381	0.440	0.494	0.582
		0.9	0.031	0.052	0.073	0.092	0.112	0.151	0.179	0.211	0.265
	α_{m1}	0.1	0.0010	0.0006	0.0003	0.0002	0.0001	0.0000	0.0000	0.0000	0.0000
		0.2	0.0035	0.0024	0.0015	0.0008	0.0003	0.0001	0.0000	−0.0001	0.0000
		0.3	0.0068	0.0047	0.0028	0.0016	0.0008	0.0002	0.0001	0.0001	−0.0001
		0.4	0.0099	0.0071	0.0047	0.0029	0.0019	0.0008	0.0004	0.0002	−0.0002
		0.5	**0.0120**	0.0090	0.0066	0.0046	0.0032	0.0016	0.0007	0.0003	−0.0001
		0.6	0.0115	**0.0097**	**0.0077**	**0.0059**	0.0046	0.0028	0.0019	0.0013	0.0004
		0.7	0.0075	0.0077	0.0069	0.0059	**0.0051**	**0.0038**	**0.0029**	0.0023	0.0013
		0.8	−0.0021	0.0012	0.0023	0.0028	0.0029	0.0029	0.0028	**0.0026**	**0.0019**
		0.9	−0.0185	−0.0119	−0.0080	−0.0058	−0.0041	−0.0022	−0.0012	−0.0005	0.0001
		1.0	**−0.0436**	**−0.0333**	**−0.0268**	**−0.0222**	**−0.0187**	**−0.0146**	**−0.0122**	**−0.0104**	**−0.0079**
	α_{v1}	1.0	**0.299**	**0.262**	**0.236**	**0.213**	**0.197**	**0.174**	**0.158**	**0.145**	**0.127**
(2) Triangular load (hinged base, free top)	α_{n2}	0	0.205	0.074	0.017	−0.008	−0.011	−0.015	−0.008	−0.002	0.002
		0.1	0.260	0.179	0.137	0.114	0.103	0.096	0.095	0.097	0.100
		0.2	0.321	0.281	0.253	0.235	0.223	0.208	0.200	0.197	0.198
		0.3	0.373	0.375	0.367	0.356	0.343	0.324	0.311	0.302	0.299
		0.4	0.411	0.449	0.469	0.469	0.463	0.443	0.428	0.417	0.403
		0.5	**0.434**	0.506	0.545	0.562	0.566	0.564	0.552	0.541	0.521
		0.6	0.419	**0.519**	**0.579**	**0.617**	0.639	0.661	0.666	0.664	0.650
		0.7	0.369	0.479	0.553	0.606	**0.643**	**0.697**	**0.730**	**0.750**	0.764
		0.8	0.280	0.375	0.447	0.503	0.547	0.621	0.678	0.720	**0.776**
		0.9	0.151	0.210	0.256	0.294	0.327	0.386	0.433	0.477	0.536
	α_{m2}	0.1	0.0009	0.0004	0.0001	0.0000	0.0000	0.0000	0.0000	0.0000	0.0000
		0.2	0.0033	0.0018	0.0007	0.0001	0.0000	0.0000	0.0000	0.0000	0.0000
		0.3	0.0073	0.0040	0.0016	0.0006	0.0002	−0.0002	−0.0002	−0.0001	0.0000
		0.4	0.0114	0.0063	0.0033	0.0016	0.0008	0.0000	−0.0001	−0.0002	−0.0002
		0.5	0.0158	0.0092	0.0057	0.0034	0.0019	0.0007	0.0002	0.0000	−0.0004
		0.6	0.0199	0.0127	0.0083	0.0057	0.0039	0.0020	0.0011	0.0005	0.0008
		0.7	**0.0219**	0.0152	0.0109	0.0080	0.0062	0.0038	0.0025	0.0017	0.0008
		0.8	0.0205	**0.0153**	**0.0118**	**0.0094**	**0.0078**	**0.0057**	0.0043	0.0032	0.0022
		0.9	0.0145	0.0111	0.0092	0.0078	0.0068	0.0054	**0.0045**	**0.0039**	**0.0029**
		1.0	0	0	0	0	0	0	0	0	0
	α_{v2}	1.0	**0.189**	**0.158**	**0.137**	**0.121**	**0.110**	**0.096**	**0.087**	**0.079**	**0.068**

(1)	(2)	
Triangular load (fixed base)	Triangular load (hinged base)	The hoop forces, vertical moments and radial shears, at depths denoted by z/l_z, are given by the following equations, where l_z is the height of the wall, r the radius to the centre of the wall, z the depth from the top of the wall and γ the unit weight of the liquid.

Hoop force: $n = \alpha_n \gamma l_z r$ (per unit height)

Vertical moment: $m = \alpha_m \gamma l_z^3$ (per unit length)

Radial shear: $v = \alpha_v \gamma l_z^2$ (per unit length)

For α values shown in tables, positive signs indicate for: (α_n) tension, (α_m) tension in outside face, (α_v) force acting inward.

Table C11 Cylindrical Tanks: Elastic Analysis – 2

Load Case	α	z/l_z	\multicolumn{9}{c}{Values of Coefficient α for Values of $l_z^2/2rh$}
			2	3	4	5	6	8	10	12	16
(3) Uniform load (fixed base, free top)	α_{n3}	0	**1.253**	**1.160**	**1.085**	1.037	1.010	0.989	0.989	0.994	1.000
		0.1	1.114	1.112	1.073	1.044	1.024	1.005	0.998	0.997	0.999
		0.2	1.041	1.061	1.057	**1.047**	1.038	1.022	1.010	1.003	0.999
		0.3	0.929	0.998	1.029	1.042	**1.045**	1.036	1.023	1.014	1.003
		0.4	0.806	0.912	0.977	1.015	1.034	**1.044**	1.039	1.031	1.015
		0.5	0.667	0.796	0.887	0.949	0.986	1.026	**1.040**	**1.043**	1.032
		0.6	0.514	0.646	0.746	0.825	0.879	0.953	0.996	1.022	**1.040**
		0.7	0.345	0.459	0.553	0.629	0.694	0.788	0.859	0.911	0.975
		0.8	0.186	0.258	0.332	0.379	0.430	0.519	0.592	0.652	0.750
		0.9	0.055	0.081	0.105	0.128	0.149	0.189	0.226	0.262	0.321
	α_{m3}	0.1	0.0010	0.0007	0.0004	0.0002	0.0001	0.0000	0.0000	0.0000	0.0000
		0.2	0.0036	0.0026	0.0015	0.0008	0.0004	0.0001	0.0000	0.0000	0.0000
		0.3	0.0066	0.0051	0.0033	0.0019	0.0011	0.0003	0.0000	0.0000	0.0000
		0.4	0.0088	0.0074	0.0052	0.0035	0.0022	0.0008	0.0002	0.0000	−0.0001
		0.5	**0.0089**	**0.0091**	0.0068	0.0051	0.0036	0.0018	0.0009	0.0004	0.0001
		0.6	0.0059	0.0083	**0.0075**	**0.0061**	**0.0049**	0.0031	0.0021	0.0014	0.0006
		0.7	−0.0019	0.0042	0.0053	0.0052	0.0048	**0.0038**	**0.0030**	0.0024	0.0012
		0.8	−0.0167	−0.0053	−0.0013	0.0007	0.0017	0.0024	0.0026	0.0022	**0.0020**
		0.9	−0.0389	−0.0223	−0.0145	−0.0101	−0.0073	−0.0040	−0.0022	−0.0012	−0.0005
		1.0	**−0.0719**	**−0.0483**	**−0.0365**	**−0.0293**	**−0.0242**	**−0.0184**	**−0.0147**	**−0.0123**	**−0.0091**
	α_{v3}	1.0	**0.370**	**0.310**	**0.271**	**0.243**	**0.222**	**0.193**	**0.172**	**0.158**	**0.137**
(4) Uniform load (hinged base, free top)	α_{n4}	0	**1.205**	1.074	1.017	0.992	0.989	0.985	0.992	0.998	1.002
		0.1	1.160	1.079	1.037	1.014	1.033	0.996	0.995	0.997	1.000
		0.2	1.121	**1.081**	1.053	1.035	1.023	1.008	1.000	0.997	0.998
		0.3	1.073	1.075	1.067	1.056	1.043	1.024	1.011	1.002	0.999
		0.4	1.011	1.049	**1.069**	**1.069**	1.063	1.043	1.028	1.017	1.003
		0.5	0.934	1.006	1.045	1.062	**1.066**	**1.064**	1.052	1.041	1.021
		0.6	0.819	0.919	0.979	1.017	1.039	1.061	**1.066**	**1.064**	1.050
		0.7	0.669	0.779	0.853	0.906	0.943	0.997	1.030	1.050	**1.064**
		0.8	0.480	0.575	0.647	0.703	0.747	0.821	0.878	0.920	0.976
		0.9	0.251	0.310	0.356	0.394	0.427	0.486	0.533	0.577	0.636
	α_{m4}	0.1	0.0009	0.0004	0.0001	0.0000	0.0000	0.0000	0.0000	0.0000	0.0000
		0.2	0.0033	0.0018	0.0007	0.0001	0.0000	0.0000	0.0000	0.0000	0.0000
		0.3	0.0073	0.0040	0.0016	0.0006	0.0002	−0.0002	−0.0002	−0.0001	0.0000
		0.4	0.0114	0.0063	0.0033	0.0016	0.0008	0.0000	−0.0001	−0.0002	−0.0001
		0.5	0.0158	0.0092	0.0057	0.0034	0.0019	0.0007	0.0002	0.0000	−0.0002
		0.6	0.0199	0.0127	0.0083	0.0057	0.0039	0.0020	0.0011	0.0005	−0.0004
		0.7	**0.0219**	0.0152	0.0109	0.0080	0.0062	0.0038	0.0025	0.0017	0.0008
		0.8	0.0205	**0.0153**	**0.0118**	0.0094	0.0078	0.0057	0.0043	0.0032	0.0022
		0.9	0.0145	0.0111	0.0092	**0.0094**	**0.0078**	**0.0057**	**0.0045**	**0.0039**	**0.0029**
		1.0	0	0	0	0	0	0	0	0	0
	α_{v4}	1.0	**0.189**	**0.158**	**0.137**	**0.121**	**0.110**	**0.096**	**0.087**	**0.079**	**0.068**

(3)
Uniform load
(fixed base)

(4)
Uniform load
(hinged base)

The hoop forces, vertical moments and radial shears, at depths denoted by z/l_z, are given by the following equations, where l_z is the height of the wall, r the radius to the centre of the wall and p the uniform pressure.

Hoop force: $n = \alpha_n p r$ (per unit height)

Vertical moment: $m = \alpha_m p l_z^2$ (per unit length)

Radial shear: $v = \alpha_v p l_z$ (per unit length)

For α values shown in tables, positive signs indicate for: (α_n) tension, (α_m) tension in outside face, (α_v) force acting inward.

Table C12 Cylindrical Tanks: Elastic Analysis – 3

Coefficients for Circumferential Forces and Vertical Moments in Wall of Constant Thickness

Load Case	α	z/l_z	Values of Coefficient α for Values of $l_z^2/2rh$								
			2	3	4	5	6	8	10	12	16
(5) Moment at base (hinged base, free top)	α_{n5}	0	−0.68	−1.78	−1.87	−1.54	−1.04	−0.24	0.21	0.32	0.22
		0.1	0.22	−0.71	−1.00	−1.03	−0.86	−0.53	−0.23	−0.05	0.07
		0.2	1.10	0.43	−0.08	−0.42	−0.59	−0.73	−0.64	−0.46	−0.08
		0.3	2.02	1.60	1.04	0.45	−0.05	−0.67	−0.94	−0.96	−0.64
		0.4	2.90	2.95	2.47	1.86	1.21	−0.02	−0.73	−1.15	−1.28
		0.5	3.69	4.29	4.31	3.93	3.34	2.05	0.82	−0.18	−1.30
		0.6	4.30	5.66	6.34	6.60	6.54	5.87	4.79	3.52	1.12
		0.7	**4.54**	**6.58**	8.19	9.41	10.3	11.3	11.6	11.3	9.67
		0.8	4.08	6.55	**8.82**	**11.0**	**13.1**	**16.5**	19.5	21.8	24.5
		0.9	2.75	4.73	6.81	9.02	11.4	16.1	**20.9**	**25.7**	**34.7**
	α_{m5}	0.1	−0.002	−0.007	−0.008	−0.007	−0.005	−0.001	0.000	0.000	0.000
		0.2	−0.002	−0.022	−0.026	−0.024	−0.018	−0.009	−0.002	0.000	0.000
		0.3	0.012	−0.030	−0.044	−0.045	−0.040	−0.022	−0.009	−0.003	0.002
		0.4	0.034	−0.029	−0.051	−0.061	−0.058	−0.044	−0.028	−0.016	−0.003
		0.5	0.096	0.010	−0.034	−0.057	−0.065	−0.068	−0.053	−0.040	−0.021
		0.6	0.193	0.087	0.023	−0.015	−0.037	−0.062	−0.067	−0.064	−0.051
		0.7	0.340	0.227	0.150	0.095	0.057	0.002	−0.031	−0.049	−0.066
		0.8	0.519	0.426	0.354	0.296	0.252	0.178	0.123	0.081	0.025
		0.9	0.748	0.692	0.645	0.606	0.572	0.515	0.467	0.424	0.354
		1.0	**1.0**	**1.0**	**1.0**	**1.0**	**1.0**	**1.0**	**1.0**	**1.0**	**1.0**
	α_{v5}	1.0	−2.57	−3.18	−3.68	−4.10	−4.49	−5.18	−5.81	−6.38	−7.36
(6) Moment at top (any base, free top)	α_{n6}	0	−13.73	−20.37	−27.20	−33.99	−40.77	− 54.32	−67.88	−81.46	−108.6
		0.1	−7.52	−9.36	−10.74	−11.60	12.04	−11.92	−10.80	−8.93	−3.56
		0.2	−3.04	−2.14	−0.80	0.86	2.76	6.88	11.10	15.16	**22.35**
		0.3	−0.04	2.03	4.15	6.18	8.07	**11.31**	**13.72**	**15.35**	16.66
		0.4	1.79	3.97	**5.79**	**7.18**	**8.18**	9.23	9.34	8.80	6.69
		0.5	2.73	**4.42**	5.52	6.05	6.14	5.55	4.43	3.18	−0.89
		0.6	**3.06**	4.00	4.35	4.20	3.75	2.47	1.22	0.23	−0.89
		0.7	3.02	3.13	2.88	2.37	1.76	0.58	−0.26	−0.71	−0.86
		0.8	2.77	2.07	1.40	0.80	0.30	−0.36	−0.65	−0.67	−0.41
		0.9	2.44	0.96	−0.03	−0.56	−0.79	−0.78	−0.56	−0.33	−0.04
		1.0	2.08	−0.17	−1.41	−1.83	−1.75	−1.04	−0.35	−0.06	0.24
	α_{m6}	0	**1.0**	**1.0**	**1.0**	**1.0**	**1.0**	**1.0**	**1.0**	**1.0**	**1.0**
		0.1	0.943	0.918	0.895	0.872	0.851	0.811	0.774	0.739	0.676
		0.2	0.809	0.740	0.676	0.620	0.570	0.483	0.410	0.348	0.250
		0.3	0.643	0.537	0.446	0.370	0.308	0.211	0.141	0.089	0.023
		0.4	0.476	0.353	0.254	0.179	0.122	0.047	0.004	−0.022	−0.042
		0.5	0.325	0.208	0.118	0.057	0.017	−0.026	−0.041	−0.043	−0.035
		0.6	0.202	0.106	0.037	−0.005	−0.028	−0.042	−0.039	−0.031	−0.015
		0.7	0.109	0.043	−0.001	−0.024	−0.034	−0.033	−0.024	−0.014	−0.003
		0.8	0.046	0.012	−0.010	−0.020	−0.022	−0.017	−0.010	−0.004	0.001
		0.9	0.011	0.001	−0.005	−0.007	−0.007	−0.005	−0.002	−0.001	0.001
		1.0	0	0	0	0	0	0	0	0	0

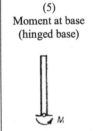

(5) Moment at base (hinged base)

(6) Moment at top (any base)

The hoop forces, vertical moments and radial shears, at depths denoted by z/l_z, are given by the following equations, where l_z is the height of the wall, r the radius to the centre of the wall and M the edge moment per unit length.

Hoop force: $n = \alpha_n M r/l_z^2$ (per unit height)

Vertical moment: $m = \alpha_m M$ (per unit length)

Radial shear: $v = \alpha_v M/l_z$ (per unit length)

Values given for load case (6) are for a semi-infinite cylinder, but may be used for any base with errors that are small for $l_z^2/2rh > 2$, and negligible for $l_z^2/2rh > 8$.

For α values shown in tables, positive signs indicate for: (α_n) tension, (α_m) tension in outside face, (α_v) force acting inward.

Table C13 Cylindrical Tanks: Elastic Analysis – 4

Load Case	α	z/l_z	Values of Coefficient α for Values of $l_z^2/2rh$								
			2	3	4	5	6	8	10	12	16
(7) Shear at top (fixed base, free top)	α_{n7}	0	**5.12**	**6.32**	**7.34**	**8.22**	**9.02**	**10.42**	**11.67**	**12.76**	**14.74**
		0.1	3.83	4.37	4.73	4.99	5.17	5.36	5.43	5.41	5.22
		0.2	2.68	2.70	2.60	2.45	2.27	1.85	1.43	1.03	0.33
		0.3	1.74	1.43	1.10	0.79	0.50	0.02	−0.36	−0.63	−0.96
		0.4	1.02	0.58	0.19	−0.11	−0.34	−0.63	−0.78	−0.83	−0.76
		0.5	0.52	0.02	−0.26	−0.47	−0.59	−0.66	−0.62	−0.52	−0.32
		0.6	0.21	−0.15	−0.38	−0.50	−0.53	−0.46	−0.33	−0.21	−0.05
		0.7	0.05	−0.19	−0.33	−0.37	−0.35	−0.24	−0.12	−0.04	0.04
		0.8	−0.01	−0.13	−0.19	−0.20	−0.17	−0.09	−0.02	0.02	0.05
		0.9	−0.01	−0.04	−0.06	−0.06	−0.01	−0.01	0.00	0.00	0.02
	α_{m7}	0.1	−0.077	−0.072	−0.068	−0.064	−0.062	−0.057	**−0.053**	**−0.049**	**−0.044**
		0.2	−0.115	−0.100	**−0.088**	**−0.078**	**−0.070**	**−0.058**	−0.049	−0.042	−0.031
		0.3	**−0.126**	**−0.100**	−0.081	−0.067	−0.056	−0.041	−0.029	−0.022	−0.012
		0.4	−0.119	−0.086	−0.063	−0.047	−0.036	−0.021	−0.012	−0.007	−0.001
		0.5	−0.103	−0.066	−0.043	−0.028	−0.018	−0.007	−0.002	0.000	0.002
		0.6	−0.080	−0.044	−0.025	−0.013	−0.006	0.000	0.002	0.002	0.002
		0.7	−0.056	−0.025	−0.010	−0.003	0.000	0.002	0.002	0.002	0.001
		0.8	−0.031	−0.006	0.001	0.003	0.003	0.003	0.002	0.001	0.000
		0.9	−0.006	0.010	0.010	0.007	0.005	0.002	0.001	0.000	0.000
		1.0	0.019	0.024	0.019	0.011	0.006	0.001	0.000	0.000	0.000

Coefficients for Circumferential Tensions and Vertical Moments in Wall of Constant Thickness

(7) Shear at top (fixed base)

The hoop forces and vertical moments, at depths denoted by z/l_z, are given by the following equations, where l_z is the height of the wall, r the radius to the centre of the wall and V the outward edge force per unit length.

Hoop force: $n = \alpha_n Vr/l_z$ (per unit height)

Vertical moment: $m = \alpha_m Vl_z$ (per unit length)

For α values shown in tables, positive signs indicate for: (α_n) tension, (α_m) tension in the outside face.

Coefficients for rotational stiffness of wall and FEM for load cases (1), (3) and (7)										
	$l_z^2/2rh$	2	3	4	5	6	8	10	12	16
Stiffness	α_w	0.445	0.548	0.635	0.713	0.783	0.903	1.010	1.108	1.281
FEM	α_{w1}	0.0436	0.0333	0.0268	0.0222	0.0187	0.0146	0.0122	0.0104	0.0079
	α_{w3}	0.0719	0.0483	0.0365	0.0293	0.0242	0.0184	0.0147	0.0123	0.0091
	α_{w7}	−0.019	−0.024	−0.019	−0.011	−0.006	−0.001	0	0	0

Rotational stiffness and FEMs are given by the following equations:

Rotational stiffness of wall (hinged base and free top): $K_w = \alpha_w E_c h^3/l_z$ where E_c is the modulus of elasticity of concrete.

FEMs: load case (1) $M_w = \alpha_{w1}\gamma l_z^3$, load case (3) $M_w = \alpha_{w3}\gamma l_z^3$, load case (7) $M_w = \alpha_{w7}Vl_z$.

Table C14 Cylindrical Tanks: Elastic Analysis – 5

		Coefficients for Bending Moments in a Uniform Circular Slab on an Elastic Foundation											
Load Case	r/r_k	Radial Coefficient α_r for Values of r_x/r						Tangential Coefficient α_t for Values of r_x/r					
		1.0	0.8	0.6	0.4	0.2	0	1.0	0.8	0.6	0.4	0.2	0
(1) Edge moment (no shear restraint)	0	1.0	1.0	1.0	1.0	1.0	1.0	1.0	1.0	1.0	1.0	1.0	1.0
	1	1.0	0.998	0.993	0.988	0.984	0.982	0.993	0.991	0.988	0.984	0.983	0.982
	2	1.0	0.977	0.906	0.828	0.770	0.749	0.908	0.876	0.832	0.789	0.759	0.749
	3	1.0	0.923	0.707	0.481	0.323	0.266	0.723	0.627	0.495	0.375	0.295	0.266
	4	1.0	0.855	0.513	0.207	0.017	−0.046	0.576	0.426	0.236	0.081	−0.015	−0.046
	5	1.0	0.773	0.353	0.059	−0.081	−0.118	0.492	0.308	0.102	−0.034	−0.100	−0.118
	6	1.0	0.680	0.215	−0.018	−0.078	−0.083	0.441	0.232	0.035	−0.057	−0.081	−0.083
	7	1.0	0.583	0.105	−0.051	−0.049	−0.034	0.405	0.175	0.001	−0.049	−0.042	−0.034
	8	1.0	0.488	0.028	−0.055	−0.023	−0.005	0.378	0.132	−0.020	−0.034	−0.014	−0.005
	9	1.0	0.400	−0.020	−0.045	−0.007	0.006	0.358	0.098	−0.025	−0.020	0	0.006
	10	1.0	0.319	−0.044	−0.029	0.001	0.006	0.341	0.072	−0.025	−0.010	0.004	0.006
(2) Edge shear (no edge rotation)	0	−0.250	−0.106	0.006	0.086	0.134	0.150	−0.050	0.022	0.078	0.118	0.142	0.150
	1	−0.249	−0.105	0.006	0.086	0.133	0.149	−0.050	0.022	0.078	0.117	0.141	0.149
	2	−0.240	−0.098	0.009	0.081	0.123	0.137	−0.048	0.021	0.073	0.109	0.130	0.137
	3	−0.211	−0.072	0.017	0.066	0.090	0.097	−0.042	0.019	0.059	0.082	0.094	0.097
	4	−0.172	−0.039	0.025	0.046	0.048	0.047	−0.034	0.017	0.040	0.047	0.047	0.047
	5	−0.139	−0.013	0.029	0.029	0.019	0.014	−0.028	0.014	0.025	0.022	0.016	0.014
	6	−0.116	0.002	0.027	0.017	0.005	0	−0.023	0.012	0.016	0.009	0.003	0
	7	−0.100	0.010	0.021	0.009	0	−0.003	−0.020	0.010	0.010	0.003	0.002	−0.003
	8	−0.088	0.013	0.016	0.003	−0.001	−0.002	−0.018	0.009	0.006	0	−0.002	−0.002
	9	−0.078	0.015	0.011	0.001	−0.001	−0.001	−0.016	0.007	0.003	0	−0.001	−0.001
	10	−0.071	0.015	0.007	0	−0.001	0	−0.014	0.006	0.002	0	0	0

Note: Radial and tangential moments per unit width, at positions denoted by r_x/r, are given by the following equations, where r is the radius of the slab and r_x the distance from the centre of the slab. For α values shown above, positive signs indicate tension at the top, compression at the bottom.

	Radial moment	Tangential moment
Load case (1), where M is edge moment per unit length (rotation inward)	$m_r = \alpha_r M$	$m_t = \alpha_t M$
Load case (2), where Q is edge load per unit length (deflection downward)	$m_r = \alpha_r Q r$	$m_t = \alpha_t Q r$

The radius of relative stiffness r_k is given by the following equation, where E_c is the modulus of elasticity of concrete, h the slab thickness and k_s the modulus of subgrade reaction (see Table B1 for further information):

$r_k = [E_c h^3/12(1-\upsilon^2)k_s]^{0.25}$ where υ is Poisson's ratio.   For $\upsilon = 0.2$, $r_k = [E_c h^3/11.52 k_s]^{0.25}$

Coefficients for Rotational Stiffness of Slab and FEM for Load Case (2)												
	r/r_k	0	1	2	3	4	5	6	7	8	9	10
Stiffness	α_s	0.104	0.105	0.118	0.159	0.222	0.285	0.346	0.407	0.468	0.529	0.590
FEM	α_{s2}	0.250	0.249	0.240	0.211	0.172	0.139	0.116	0.100	0.088	0.078	0.071

Rotational stiffness and FEMs are given by the following equations:

Rotational stiffness of the slab: $K_s = \alpha_s E_c h^3/r$ where E_c is the modulus of elasticity of concrete

FEM for load case (2): $M_s = \alpha_{s2} Q r$ where Q is the edge load per unit length acting downward

Index